Airborne Maritime Surveillance in the North Atlantic

This book offers a comprehensive examination of historical Norwegian and allied airborne maritime surveillance operations in the North Atlantic, from World War II to the present.

Russia's resurgent naval and submarine activity in the High North needs to be closely observed. Norway, together with her closest allies, has monitored such activity in the High North since the end of World War II. One of the key instruments in that effort has been maritime patrol aircraft (MPA), with their anti-submarine warfare (ASW) capabilities. However, after the Cold War, in many nations an atrophy took place within the MPA communities of knowledge, resources, and experience with airborne ASW. This occurred just as the concept of intelligence, surveillance, and reconnaissance (ISR) came to play an increasingly important role in modern warfare, but without sufficient attention being paid to the specific challenges of maritime airborne surveillance and reconnaissance. This book addresses that gap and examines the entwined evolution of airborne maritime surveillance by MPA and ASW in the High North, both during and after the Cold War, with the main focus being on Norway. The discussion identifies the fundamental "building blocks" of these historical surveillance operations through a portrayal of the development of airborne surveillance and reconnaissance operations. These building blocks are then applied to make recommendations for the future for maritime surveillance in the High North.

This book will be of much interest to students of naval history, strategic studies, intelligence studies, Nordic studies, and international relations.

John O. Birkeland is a faculty advisor at the NATO Defence College, Rome, and has a PhD in military science from the University of Glasgow, UK.

Cass Series: Naval Policy and History
Series Editor: Geoffrey Till
ISSN 1366–9478

This series consists primarily of original manuscripts by research scholars in the general area of naval policy and history, without national or chronological limitations. It will from time to time also include collections of important articles as well as reprints of classic works.

68. **Maritime Gray Zone Operations**
Challenges and Countermeasures in the Indo-Pacific
Edited by Andrew S. Erickson

69. **African Navies**
Historical and Contemporary Perspectives
Edited by Timothy Stapleton

70. **Recovering Naval Power**
Henry Maydman and the Revival of the Royal Navy
Edited by John B. Hattendorf and Geoffrey Till

81. **The U.S. Navy and the Rise of Great Power Competition**
Looking Beyond the Western Pacific
James J. Wirtz, Jeffrey E. Kline and James A. Russell

82. **Naval Constabulary Operations and Fisheries Governance**
An Integrated Approach for the Australian Maritime Domain
Sean A. Andrews

83. **Maritime Insurance, Risk and Seafarer Security**
Navigating Piracy in the Indian Ocean
Jessica Kate Simonds

84. **Airborne Maritime Surveillance in the North Atlantic**
Norwegian and Allied Patrols, 1945 to the Present-day
John O. Birkeland

For more information about this series, please visit: www.routledge.com/Cass-Series-Naval-Policy-and-History/book-series/CSNPH

Airborne Maritime Surveillance in the North Atlantic

Norwegian and Allied Patrols, 1945 to the Present-day

John O. Birkeland

Routledge
Taylor & Francis Group
LONDON AND NEW YORK

First published 2026
by Routledge
4 Park Square, Milton Park, Abingdon, Oxon OX14 4RN

and by Routledge
605 Third Avenue, New York, NY 10158

Routledge is an imprint of the Taylor & Francis Group, an informa business

British Library Cataloguing-in-Publication Data
A catalogue record for this book is available from the British Library

ISBN: 978-1-041-03921-1 (hbk)
ISBN: 978-1-041-03932-7 (pbk)
ISBN: 978-1-003-62601-5 (ebk)

DOI: 10.4324/9781003626015

Typeset in Times New Roman
by Apex CoVanatage, LLC

Contents

Tables and figures

Tables

Figures

Abbreviations

A2/AD	anti-access/area denial
AAM	air-to-air missiles
AAR	air-to-air refuelling
AAW	anti-air warfare
ACINT	acoustic intelligence
AGS	alliance ground surveillance
AIMS	advanced imaging multispectral sensor
AIP	air-independent propulsion
AIS	automatic identification system
AOR	area of responsibility
ASuW	anti-surface warfare
ASW	anti-submarine warfare
ATO	air tasking order
BAMS	broad area maritime surveillance
BAMS-D	bams-demonstrator
BDA	battle damage assessment
BLOS	beyond-line-of-sight
C2	command and control
C4 for ASW	command, control, computers, and communications for ASW
CAS	close air support
CHOD	chief of defence
CHR	command history report
CINCEASTLANT	Commander-in-Chief Eastern Atlantic Area
CMAN	Commander Maritime Air Naples
CNA	Centre for Naval Analyses
CNO	Chief of Naval Operations
COIN	counter-insurgency
COMAFNORTH	Commander Allied Forces North
COMINT	communications intelligence
COMMARAIRNATO	Commander Maritime Air NATO
CONMAROPS	concept of maritime operations
COTS	commercial-off-the-shelf

CSIS	Centre for Strategic and International Studies
CTF	Commander Task Force
CTG	Commander Task Group
CV	general type of aircraft carrier
CVA	attack aircraft carriers
CVS	ASW designated aircraft carriers
DCI	director of central intelligence
DELTIC	delayed time compression
DOTMLPFI	doctrine, organization, training, materiel, leadership & education, personnel, facilities, information/interoperability
DSP	Defence Support Program
ELINT	electronic intelligence
EO	electro-optical
EO/IR	electro-optic/infrared
ESM	electronic *support* measures
FDS	fixed distributed system
FISINT	foreign instrumentation and signals intelligence
FMR	Fagmilitært Råd (Norwegian Chief of Defense White Paper)
FOIA	Freedom of Information Act
GIN	Greenland–Iceland–Norway
GIUK	Greenland–Iceland–UK
GPS	Global Positioning System
GRAB	galactic radiation and background
HALE	high altitude/long endurance
HF	high frequency
HUMINT	human intelligence
ICBM	intercontinental ballistic missile
IFF	identification friend or foe
IMINT	imagery intelligence
INF	intermediate nuclear forces
ISAR	inverse synthetic aperture radar
ISR	intelligence, surveillance and reconnaissance
ISTAR	intelligence, surveillance, target acquisition, reconnaissance
IUSS	integrated undersea surveillance system
JAPCC	Joint Air Power Competence Centre
JFC	joint force command
JP	joint publication
KEFTACEX	Keflavik Tactical Exercise
km	kilometre
LACM	land-attack cruise missiles
LAWS	lethal autonomous weapons systems
lbs	pounds (weight)

LTP	long-term plan
MACA	Maritime Air Control Authority
MALE	medium altitude/long endurance
MARCOM	Allied Maritime Command
MC	Military Committee
MHQ	maritime headquarters
MIRV	multiple, independently targeted re-entry vehicles
MAC	multi-static active coherent
MAD	magnetic anomaly detection
MAOC	Maritime Air Operations Centre
MDAP	Mutual Defence Assistance Program
MoD	Ministry of Defence
MoU	memorandum of understanding
MPA	maritime patrol aircraft
NAC	North Atlantic Council
NIS	Norwegian Intelligence Service
NM	nautical miles
NSR	north-eastern sea route
NATO	North Atlantic Treaty Organization
NIE	National Intelligence Estimate
NOSS	Naval Ocean Surveillance System
OAE	Operation Active Endeavour
ONI	Office of Naval Intelligence
ONR	Office of Naval Research
OPINTEL	operational intelligence
PED	processing, exploitation, and dissemination
PfP	partnership for peace
PNI	presidential nuclear initiatives
RAF	Royal Air Force
R&D	research and development
RF	radio frequency (rf)
RPG	rocket-propelled grenades
SACEUR	Supreme Allied Commander Europe
SACLANT	Supreme Allied Commander Atlantic
SAM	surface-to-air missiles
SALT	Strategic Arms Limitation Treaty
SAR	Search-And-Rescue
SBIRS	Space-Based IR System
SDSR	Strategic Defence and Security Review
SATCOM	satellite communication
SIGINT	signals intelligence
SLAM	stand-off land attack missile
SLBM	submarine-launched ballistic missile
SLCM	submarine-launched cruise missile
SLOC	sea lines of communication

SoI	statement of intent
SOSUS	sound surveillance system
SSBN	strategic submarines carrying nuclear ballistic missiles
SSGN	multirole cruise missile submarine
SSK	conventional attack submarine
SSN	nuclear attack submarine
STANAVFORLANT	standing naval force Atlantic
SURTASS	surface-towed array sonar systems
TACCO	tactical coordinator
TCDL	tactical common data link
T-LAM	Tomahawk land attack missile
UHF	ultra-high frequency
UIP	upgrade improvement programme
VHF	very high frequency

1 Introduction

When the Norwegian chief of defence (CHOD) publicized his strategic long-term advice to the Ministry of Defence (MoD) in 2015 and proposed to phase out the fleet of P-3 *Orion* maritime patrol aircraft (MPA) and replace them with drones, his advice indicated a lack of understanding for both the missions being executed and the maturity of drone technology required to execute them. The CHOD's advice on this particular issue was not adhered to by the strategic security policy level at the MoD, and the ensuing long-term plan issued by the ministry the following year laid forward a plan to replace the *Orion* fleet with a fleet of the state-of-the-art MPA, which is the P-8 *Poseidon*. The lack of debate on the CHOD's original advice and the lack of public insight into the decision to discard the CHOD's advice both inspired the writing of this book. As important as maritime intelligence services in the High North have been for Norwegian and NATO security policy during and after the Cold War, the lack of debate was deafening. Perhaps a consequence of a desire to realize the peace dividend after the Cold War and the ensuing lack of public focus on the North Atlantic in the 2000s? Perhaps due to the secrecy that has followed all matters related to intelligence gathering in the Barents Sea and a natural lack of understanding in the public on the matter? Or perhaps, even more mundane, due to the parochialism within the Armed Forces, where maritime air operations have fallen through the cracks, so to speak, being neither owned nor cherished properly by either the Navy or the Air Force? Regardless, the aim of this book is to highlight the debate and provide an unclassified framework to discuss the requirements for airborne intelligence collection in the maritime domain.

This book discusses the airborne surveillance that took place in the North Atlantic during the Cold War, and thus it underlines the place of submarines in Soviet and later Russian naval strategy and naval operations. Based on this, the role of airborne ASW capability can be outlined.

Much of the discussion on airborne ISR is not properly explained, or it remains classified and hidden from public discourse. However, the principles are not classified and should constitute the basic premises for the ongoing discussion on the future of airborne surveillance in the High North. It is an ambition of this book to inform the public debate on these issues. A fundamental question to be answered by this book is, therefore, "What role ASW capabilities play in airborne ISR in the High North?" Through studying the developments of airborne surveillance and

DOI: 10.4324/9781003626015-1

reconnaissance over the past century or so, elements stand out that can be considered the basis for framing the discussion. Underlying questions on maritime airborne ISR that become relevant are how airborne ASW can be explained within a theoretical framework for airborne ISR, whether manned MPA can be replaced by unmanned aircraft and technology, what role force integration and interaction with other units play for modern maritime airborne ISR units, and what role sensor processing and ISR product dissemination play for maritime airborne ISR units. A theoretical framework that brings these different elements together will facilitate, structure, and benefit any future discussion on airborne ISR and ASW.

Where is the "High North"?

There is no internationally recognized definition of the *High North*. The term "north" can in itself be somewhat misleading in general geographical naming conventions, as it can be applied in relative terms: An Italian travels north to Denmark, and a Norwegian travels south to Denmark. The North Sea, which is the basin between the United Kingdom, the European mainland, and Scandinavia, is situated south of the Norwegian Sea. The High North is, for the sake of this book, defined as the *ocean and land areas north of the Arctic circle in the North Atlantic*.[1] This refers to the ocean areas east of Greenland, east and north of Iceland and the United Kingdom, and west and north of Norway and Russia, stretching all the way to the North Pole. This excludes the American, Canadian, Danish, and Pacific Northern ocean and land areas, which the US Department of Defense (DoD) in their latest Arctic strategy terms "the North American Arctic."[2] The High North in this book is therefore synonymous with what many term the "the European Arctic."

The text in this book will also at times refer to the "Arctic." This is done when discussing geopolitical dynamics, and the term refers to any land and ocean areas north of the Arctic circle, writ large. At times, the text will also refer to the "North Atlantic." This is where most of the operations during the *Battles of the Atlantic* took place, specifically along the transit routes of the Soviet strategic submarines between the US East Coast and the Russian Kola Peninsula. In literal terms, the North Atlantic refers to the parts of the Atlantic north of the equator, but in practical terms we are discussing the ocean areas north of a straight line between the US East Coast and the main ports of Western Europe.

Security policy framework

Norwegian foreign and security politics have, since the latter half of the twentieth century, rested on a balancing act between *integration/deterrence* and *screening/reassurance*:[3] Integration/deterrence with alliance partners, both bilaterally with the United States and multilaterally through NATO, and screening/reassurance through the denial of a permanent presence of foreign troops and bases on Norwegian soil. Surveillance missions flown on a case-by-case basis by allied partners from the Norwegian bases have been restricted to stay west of 24°E, justified by the requirement to maintain low tensions with Russia. This in turn means that in essence it has been Norwegian aircrafts that have monitored Russian activity on and outside the Kola Peninsula. The Kola Peninsula is the home of the Russian Northern Fleet, the most powerful fleet in the Russian Navy, which enjoys year-round ice-free passage to the Atlantic. Among the Russian naval capabilities, submarines played a key role in naval strategy throughout the Cold War and increasingly do so again today, in not only assuring a nuclear second-strike capability but also posing a threat towards Western surface forces and carrier strike groups in the Atlantic. Moreover, the Russian Navy has historically been, and still is, carrying out subsurface, surface, and air testing and development of weapons systems and platforms just outside the Russian coastline in the North. By imposing the mentioned restrictions on foreign military activity from Norwegian soil, Norway has enjoyed somewhat of a monopoly on conducting airborne surveillance and reconnaissance missions against her much more powerful neighbour to the East in the High North.

This book is not so much about the development of Soviet and Russian naval strategy or NATO naval strategy, as it is a story about the place of airborne surveillance capabilities in between those spheres as they relate to both. The discussion pivots around the strategic and security policy of one small nation, Norway, who is manoeuvring between much more powerful and capable Allies and neighbouring foes. At the core, the book is a story about a niche military capability and how it interacts with security policy in an international sense, and to some degree how international cooperation surrounding that capability has been a manifestation of security policy in the High North for Norway. This book shares hitherto untold details on how Allied cooperation, particularly between the United States, the United Kingdom, and Norway took place, within the context of airborne surveillance and ASW in the North Atlantic.

The requirement for military surveillance in the High North has become something of a dogma for Norwegian authorities – it is regularly taken as self-evident and self-explanatory. The main reasons emphasized are unexploited natural resources, fishery surveillance, increased maritime traffic along the coast, and perhaps most importantly: Russian military activities in the High North stemming from the bases on the Kola Peninsula. The surveillance missions have been the responsibility of the Norwegian Armed Forces. Historically, the ability to monitor the operational movements of the Soviet and Russian submarines has been given a high priority by NATO nations. The battle between the hunting ASW forces and the hunted submarines was, for several decades, conducted in favour of the hunter: NATO nations were in the lead technologically and had the geographical advantage of staging areas along the Norwegian, Scottish, and the Icelandic coasts for their MPA. In the 1980s, however, the quieting measures in Russian submarine manufacturing took hold, which made hunting for Russian submarines significantly more complex. After a decade of downturn in activity and force levels on both sides following the collapse of the Soviet Union, Russian submarines are once again patrolling the Atlantic, this time more capable than ever. The new Russian submarines and the renewed aggressive Russian posture in the Atlantic are questioning NATO's ability to hold them at risk.

The Norwegian CHOD's strategic military advice to the MoD in 2015 in preparation for the new long-term plan to be issued the following year highlighted strategic intelligence as the first tier of defence for Norway. Situational awareness in the High North is additionally brought to the front by military as well as political leaders when discussing military capabilities and requirements for Norway. The CHOD's advice proposed changes to how the Norwegian Armed Forces would approach surveillance in the High North through phasing out the P-3 *Orion* MPA, and acquiring new unmanned systems to take over much of the intelligence, surveillance, and reconnaissance (ISR) portfolio of the P-3 fleet. There is, however, a crucial gap between the intelligence collection and types of warfare that a modern MPA can perform and those that an unmanned aircraft can perform. This gap is constituted by acoustic intelligence and ASW. In practical terms, the CHOD proposed to phase out the Norwegian long-range, airborne ASW capability in the North.

The proposal of the Norwegian CHOD to phase out the P-3 *Orion* raised questions regarding the importance of the ability to monitor subsurface activity in the High North. What role have the Russian submarines of the Northern Fleet played in the High North? What role has the MPA fleet played historically in monitoring Russian submarines operating in the Atlantic? What role does the modern, emerging Russian submarine missile threat play in the High North today? What are the elements required for conducting efficient and credible maritime airborne surveillance and reconnaissance in the High North today and in the future?

In order to properly understand the development of MPA technology and tactics that took place during the Cold War and beyond, we need to have an understanding of the basic building blocks of airborne surveillance and reconnaissance. This book lays out these building blocks in a structured manner, in order to facilitate an informed discussion on the current and future requirements for airborne intelligence operations. This understanding will also facilitate an informed discussion on the questions presented above. The ambition of this book is therefore to inform the public debate on these issues. Much of the discussion is either taken for granted and not properly explained, or it remains classified and hidden from public discourse. However, most of the history is not classified, nor is the principled discussion on requirements for airborne ISR. Rather, these should constitute the basic premises for the ongoing discussion on the future of airborne surveillance in the High North.

Security politics are also changing in parallel to increased Russian activity. China, too, is eyeing infrastructure investments in the Arctic, in support of their own ambitions for the High North.[4] Sweden and Finland are now full members of NATO, increasing significantly the military power of what has traditionally been termed the Alliance's Northern Flank. Factors such as climate change, new security policy actors in the North, and resource management in the Arctic are all important to the overall international dynamics of the region.[5] These factors, however, are not the focus of this book. The role of the Russian surface fleet and airborne platforms based on the Kola Peninsula is also worthy of thorough discussions but would take away too much focus from our discussion of airborne

ISR. The submarines in the Atlantic have posed unique challenges with regard to surveillance technology and the organizational framework set up to monitor their movements. In an overarching context, it is impossible to detach the most important element of the Russian defensive bastion in the High North from other supporting and supported elements without losing important understanding of how the Russian military operations in the North Atlantic are organized. The focus here, however, is on the unique challenge posed by airborne assets conducting ISR in the subsurface domain. The discussion thus focuses on the submarines of the Northern Fleet and their interaction with airborne surveillance and reconnaissance historically, today, and in the future.

Anti-submarine warfare

Any discussion on maritime airborne surveillance, reconnaissance, and patrol will (or should) include a discussion on the requirement for ASW capabilities, as the surveillance of and warfare against adversary submarines have historically been a significant, though not the only, part of maritime airborne surveillance. NATO defines ASW as "operations conducted with the intention of denying the enemy the effective use of their submarines."[6] This definition has deliberately been set up so that the focus of it is not necessarily the *destruction* of enemy submarines, but the active denial or restriction of their use.

The latter can be achieved in many ways, not only by *sinking* enemy submarines. At the height of the Cold War and the Western debate on how to approach the Soviet bastion strategy and employment of strategic submarines, the naval strategist and historian Jan Breemer presented three classical *strategies* for achieving this denial of efficient use of the enemy's submarines: A strategy of *destruction*, a strategy of *containment*, and a strategy of *limiting the warfighting efficiency* of the enemy submarines. Furthermore, the operational level of ASW is concerned with where and how to execute these strategies. Typically, the three categories of geography relevant for this discussion are those that enable the execution of the strategy at the "source" of the enemy submarines, i.e. their home ports and bases, in the transit areas of the submarines, or at the patrol areas of the submarines. Finally, the execution of ASW at the tactical level consists of the different phases of surveillance and reconnaissance, detection, tracking, and attack.[7] These phases correlate in many ways to the NATO targeting sequence: Find, fix, track, target, engage, exploit, assess – as we shall see in later chapters.

Destruction at the source means destroying or sinking the submarines in their home port, at the shipyard or destroying the industry that delivers technology that is critical for their construction. The main benefit of this approach is that one bypasses the single most challenging factor in ASW, which is finding the submarine in the first place. But even with modern precision weapons, this type of attack is difficult to execute with success, especially since the submarines themselves are often hidden in hardened shelters. Most of these shelters are so durable that they will require a nuclear attack for them to burst. *Destruction in the transit areas* depends on the local geography and the presence of choke points where the hunter can, to some degree, be certain that the enemy submarines will pass through. Choke points can be used for intensified patrol with weapon-carrying vessels or aircraft, or they can be extensively mined. These tripwires are often termed "ASW barriers." *Destruction in the patrol areas* is highly dependent on either tracking the submarine all the way to the patrol areas and during the submarine patrol, or dependent on strategic intelligence cueing of some sort. Most often, this type of cueing is based on static passive arrays on the seabed that will relay sound frequencies that have been picked up back to shore acoustic analysis centres, which in turn will cue other platforms to search for and eventually destroy the submarines. But the cueing can also stem from electronic intelligence (ELINT), as was often the case during World War II, with interception of radio traffic among enemy submarines and between the submarines and their high command. Destruction in the patrol areas can be defensive, in the sense that the enemy submarines are detected and destroyed as a part of defending either line of communication or a convoy, or a port to which marine traffic is funnelled. It can also be offensive, in the sense that the enemy submarine is actively tracked down and neutralized in their patrol areas, irrespective of their relative whereabouts to friendly units or key infrastructure. This offensive task will be executed by attack submarines, previously known as "hunter/killer" (HUK) submarines, perhaps with the support of airborne surveillance assets.

Containment at the source has traditionally relied on mining the waters close to the home port of the enemy submarine. There are also historical examples of placing physical barriers in place in order to prevent the submarines from setting to sea from their home base, for example, through sinking ships that have obstructed their exit. *Containment in the transit areas* can be to lay a minefield in an ocean channel (e.g. as was done in the English Channel in World War I and II) and thus force the enemy submarine to divert her path to a given transit area. The tracking of Russian submarines in transit to their patrol areas during the Cold War can also be termed containment, or in this case: Holding-at-risk. *Containment in the patrol areas* can be executed through laying minefields or a field of acoustic buoys that the enemy submarine will aim to avoid. These fields can surround the patrol area, both creating a no-go zone for the submarine, as well as communicating to her that there are hostile ASW forces working in the area.

Limiting the submarine's war-fighting efficiency at the source can be achieved by mining enemy crew training areas, or aiming bombing efforts at elements of submarine shore facilities and production areas. *Limiting the efficiency in the transit and patrol areas* can involve setting up an active acoustic buoy barrier or field, which might have the enemy submarine pick a more time-consuming route to her patrol area. Airborne reconnaissance units have traditionally been considered efficient assets in degrading submarines' operational efficiency: The presence of patrol aircraft has forced enemy submarines to operate closer to shore and coastal traffic and slowed them down in order to blend in more with other maritime traffic.

None of these tasks have been, are, or will be carried out by single assets alone. They have been, and are, more or less without exception conducted either in cooperation in real time between, for example, an attack submarine and a surface ship or an aircraft, or the submarine or aircraft have been cued by strategic intelligence, for example, by static acoustic systems on the seabed. ASW execution at the tactical level, with the phases of surveillance and reconnaissance, detection, tracking, and attack, can thus be executed by a chain of assets working together: Static arrays cueing patrol aircraft, the aircraft conducting surveillance and detecting the enemy submarine, the aircraft tracking the submarine and handing it over to another aircraft, the friendly attack submarine receiving targeting information for the airborne asset, and the hunter-killer submarine finally going in for the kill.

A structured approach to ASW took form particularly during convoy escorts across the Atlantic during World War II. In these cases, surface ships escorted merchant and passenger ships and defended against the German wolf packs of submarines aiming to interrupt Allied logistics. These escorts were supported by long-range aircraft flying from ashore, providing over-the-horizon coverage through airborne surveillance. This escort and defence function has morphed into two main types of defensive ASW operations: The *protection of sea-lines of communication* (SLOC), which, in practical terms, is a modern form of convoy escort, and into what we now term *force protection* of, for example, a carrier strike group. Both these defensive constellations consist of surface ships with powerful hull-mounted sonars and towed array sonars, surface ships with ASW helicopters on board, attack submarines that are dedicated to forward presence of the force, fixed-wing ASW aircraft taking off from the carrier, and dedicated surface-towed array sonar system (SURTASS) ships that operate at some distance from the main force in order not be affected by the sounds emitted by the group of friendly ships. Many modern frigates have ASW helicopters on board as an integrated part of their ASW sensor system. They may also carry a passive towed array system that will complement their active sensors. The use of SURTASS ships saw somewhat of a decline post-Cold War, but has been an integral part of NATO ASW exercises since the late 2010s.

The other main ASW operation that has taken place is one outside of war and exercises: The peacetime operation of holding-at-risk enemy submarines as they leave their homeports, as they are in transit, and as they operate in their patrol areas. This task has traditionally been the responsibility of the triad of hunter-killer submarines, passive array systems (mostly static systems on the seabed, but also mobile SURTASS ships), and long-range maritime patrol aircraft (MPA). The task of Allied forces dedicated to this operation has in essence been to make sure that decision makers at all times are aware of the whereabouts of Russian submarines, be they attack, guided missile, or strategic submarines. This, in turn, would have enabled Allied forces to neutralize them if the requirement arose in a crisis or in the lead-up to war. As the story presented in this book will make clear, the efforts of long-range aircraft were indeed supported by, and very often dependent on, cueing from the other two elements of the ASW intelligence triad. However, this book will not focus on the collective efforts that go into force protection or specific SLOC operations, nor on the hunter-killer submarines or the passive array systems that were so important to the operation as a whole. The main focus is on the strategic, operational, and tactical elements of the airborne ASW function, with sporadic nods to the mentioned collaborating elements.

Intelligence, surveillance, and reconnaissance

The original role of military aircraft stemmed from the desire to observe the enemy from an elevated position. Chapter 2 of this book lays out the historical development of airborne surveillance and reconnaissance, from early manned kites and balloons, to modern high-tech satellites and unmanned aircraft supporting strategic and tactical decision makers in all domains of warfare. This type of airborne operations is usually termed "ISR missions." The term "ISR" was not in use until the 1990s. Before this, the common terms were "intelligence collection," "surveillance," and "reconnaissance," often prefixed by the word "airborne." In the maritime domain, the terms "maritime surveillance and reconnaissance" and "maritime patrol" have been used interchangeably. The concept of ISR evolved at a time where maritime surveillance and reconnaissance were overshadowed by modern expeditionary warfare in the Balkans and the Middle East in the 1990s. As warfare has evolved since the Cold War, so has our understanding of intelligence support to operations and its integration into operations. In the beginning of the 1990s, the main foe for the West seemingly disappeared, and the decisive naval battles, the great land campaigns, and the fight for air superiority seemed to turn into anachronisms. Warfare became limited in time, scope, and space, and armed forces were forced to think *joint*. Expeditionary warfare in the 1990s (Gulf War I, Somalia, the Balkans) became the contemporary way of fighting wars, and intelligence stepped up as more of a core tenet of modern warfare as opposed to the more classic interpretation of intelligence as having an auxiliary, supporting role.[8] In addition, there was an explosion in the use of unmanned aircraft in support of ground operations in the 1990s and early 2000s. The availability of ISR products from the tactical to the strategic level has been a giant leap forward for military organizations, although the immense amounts of information accessible also pose significant challenges. It was in this context that "ISR" grew into its own, and became an integrated part of the "network-centric warfare" movement. "ISR" as a term came into daily usage in the mid-1990s, after first being coined by the US Vice Chairman of the Joint Chiefs of Staff, Admiral Owens, pointing to ISR as a vital component of modern warfare, implemented through the concept of network-centric warfare.[9] A network-centric approach to warfare is essentially the military embodiment of information age concepts. It is the linkage of computers, communications, sensors, and military units in order to improve efficiency and effectiveness of military operations.[10] Or as John Ferris puts it, it is "the idea that armed forces will adopt flat structures, working in nets on the internet, with soldiers at the sharp end able to turn data processing systems at home into staffs through 'reachback', real time, immediate and thorough inter-communication."[11] In short, it is about gathering, interpreting, disseminating, and acting upon information in the battlefield faster than your opponent. The tenets of network-centric warfare are thus a networked force in order to improve information sharing, which in turn will enhance the quality of information and shared situational awareness, which will enable collaboration, self-synchronization, sustainability, and speed of command, which finally will increase mission effectiveness.[12] Critical voices in the early and mid-2000s pointed to important elements of command and warfare in general being eroded in favour of an emphasis on quantitative measurements (number of targets hit, numbers of areas covered through surveillance, etc.) and what some commentators have termed the "fetishisation of speed and tacticisation of strategy."[13] Network-centric warfare is not the topic of discussion for this book, however, any discussion of modern ISR must take into account developments of modern armed conflict as a whole, in order to make the gathered intelligence products actionable to the rest of a joint force.

The development of ISR in the 1990s occurred with maritime airborne ISR standing on the sidelines. As warfare became more expeditionary and focused on counterinsurgency, and the level of foreign submarine activity was in free-fall, the open ocean MPA community found itself in search of a mission. This sudden identity crisis led to more and more MPA executing surface and over-land surveillance and reconnaissance, in place of what historically had been their primary mission – ASW.[14] This ambiguity is one of the factors that makes the study of modern maritime ISR an interesting topic for discussion, all the more so as Russian submarine activities are once more on the rise.

To the extent possible, the discussion in this book will rest on definitions endorsed by NATO. NATO defines *surveillance* as "the systematic observation of aerospace, surface, or subsurface areas, places, persons, or things."[15] This definition covers those activities concerned with observing objects, people, and areas *over time*, in support of building empirical knowledge on an adversary. *Reconnaissance*, meanwhile, is defined as "a mission undertaken to obtain . . . information about . . . a particular area."[16] Reconnaissance is thus information gathering in more specific terms. An asset or a person is tasked to observe a *specific* adversary person, unit, or activity, at a *given time and place*. Finally, *intelligence* is built on these collection activities and is defined as "the product resulting from the processing of information concerning foreign nations, hostile or potentially hostile forces or elements, or areas of actual or potential operations."[17] Collection is carried out through the overarching activities of *surveillance* and *reconnaissance* – the products stemming from those activities will then be analysed and contextualized, eventually facilitating *intelligence*.[18] The three are thus inherently indivisible.[19] This indivisibility has matured into the *concept* of "ISR" over the past two decades, although never maturing and developing into a theory. Doctrine, yes eventually, but not *theory*, as an academic would understand it. Although the term *ISR* has not yet matured into an official NATO definition, the United States Department of Defence defines *ISR* as "an activity that synchronizes and integrates the planning and operation of sensors, assets, and processing, exploitation, and dissemination systems in direct support of current and future operations."[20] This definition underlines the integrated relationship between ISR and ASW. ISR is the foundation for ASW, and it is futile to discuss ASW without a thorough understanding of ISR.

There is no internationally acknowledged definition of *airborne ISR*. However, as the definition of ISR has been given above, the question then arises as to what type of asset this activity can be executed with if it is *airborne*. Without going into the technical distinctions between classifications of aircraft, as commonly established by aviation authorities, this book simply refers to aircraft that depends on lift from the air. For the sake of this discussion, it is important to note that satellites do not fly through the air. They are in orbit outside the Earth's atmosphere, and adhere to a different set of laws of physics to maintain their line of movement than aircraft do. Satellites have, however, become very significant contributors to overall intelligence gathering from an elevated position, as we shall see.

As part of an introduction to maritime airborne ISR, it is also necessary to have a common understanding of the term *patrol*, as it is used extensively to describe the role of many maritime airborne ISR missions. NATO defines the term *patrol* as "a detachment of ground, sea, or air forces sent out for the purpose of gathering information or carrying out a destructive . . . mission."[21] The term *maritime patrol* is ingrained in military terminology, but is not properly defined in military doctrine. The understanding of maritime patrol is one of active observation of a given area, conducted by a unit that is capable of *engaging* hostile units within that area. The more passive acts of surveillance, reconnaissance, and ISR in general are not considered *patrol* in the full sense of the term, however, the surveillance and reconnaissance parts comprise significant parts of the act of patrolling an area. Mainly, the difference between a maritime *surveillance* aircraft and a MPA is just this: The ASW capability. As a continuation of this, if the aircraft is capable of searching for, finding, tracking, and subsequently *engaging* enemy submarines, the aircraft can be considered an MPA. The reason that this matters is that NATO members poured significant money and resources into procuring, maintaining, and operating MPA during the Cold War, but the inventory of MPA took a significant downturn after the Cold War. An important part of discussions related to future maritime airborne surveillance is the requirement for an ASW capability. With all the costs associated with operating an MPA fleet, this is an important political and military strategic decision in either direction. This question goes to the core of the discussion on future ISR in the High North.

Airborne ASW is complicated business. The flying crew must understand the environment they are working in, and they need to have a working knowledge of the particular equipment they use and how well it might function in the given circumstances. Speed of reaction, flexibility to change operating areas quickly and efficiently, and the ability to deploy sophisticated buoys are all an advantage to the aerial ASW platform. The aircrew has an array of additional sensors on board that will help them accomplish their mission, be it covertly following the enemy submarine or hunting it down in order to neutralize her. As submarines have become quieter, in addition to operating in unpredictable patterns and large operating areas, the task of searching for, locating, and tracking a modern submarine has grown more challenging, even with state-of-the-art sensors and computers. It is important to note that ASW is not necessarily about neutralizing the enemy submarine by sinking it with a torpedo. It also involves putting on an extensive search that the submarine commander is aware of, and thus has to evade or move around or through in order to accomplish his or her mission. The use of active sonar sensors in the water is one way of informing the submarine of the ASW force's presence. And when a naval force must traverse a given part of the ocean to reach an operational area, the water space ahead of the force must be searched for any threats. This is often termed *sanitization*. ASW forces will operate ahead of the naval force to ensure that there are no enemy submarines operating en route. These are but a few examples of how airborne ASW assets can hinder a submarine from operating effectively.

This book is about the airborne operations carried out on a daily basis by MPA in the North Atlantic after World War II until the present. Together with an historic portrayal of the development of airborne ISR in general, this discussion aims to inform the discussion on future requirements for airborne ISR in the High North.

Structure of the book

The first three chapters of the book discuss the historical development of airborne surveillance and reconnaissance writ large, upon which a novel framework for discussing this capability in principle is presented. Chapter 2 provides a chronological depiction of the development of airborne surveillance and reconnaissance. These historical developments then provide the foundation for a novel theoretical framework for airborne ISR presented in Chapter 3. The benefit of this framework is that it provides tools to properly discuss the requirements for surveillance and reconnaissance, and ASW in the North Atlantic, and how those requirements can be met.

Chapters four through eight of the book is a story of the development of the Russian submarine threat in the High North and the Atlantic, how naval strategy developed on both sides of the Iron Curtain, how Norwegian security policy developed in tandem with these, and how the airborne maritime surveillance capabilities developed both technologically and operationally. Chapter 4 establishes the historical backdrop to the maritime surveillance effort in the North Atlantic during the early Cold War. Chapter 5 is a case study of the NATO procurement programme for a common MPA, namely the *Atlantic*, which started in 1957. Chapter 6 analyses the 1960s, when Norway was given new maritime aircraft by the Americans to monitor Russian activity in the High North, but was not sufficiently capable of delivering intelligence from the subsurface domain. Chapter 7 then examines the evolution of airborne ASW in the North Atlantic during the final decades of the Cold War. The United States, the United Kingdom, and Norway all flew capable aircraft and cooperated to hold the Russian submarines at risk. Chapter 8, meanwhile, focuses on the post-Cold War era, which saw a sharp downturn in Russian naval activity following the collapse of the Soviet Union. This led to fundamental changes in how Western MPA were utilized in operations and for peacetime surveillance. The chapter ends with a changed geopolitical situation and a renewed focus on the maritime domain and the subsurface threat.

Finally, the last two chapters of the book discuss contemporary issues, challenges, and future recommendations in all matters just presented. Chapter 9 goes through the Russian submarine threat of today and attempts to see trends in operating patterns and weapons that will shape the future. The book then culminates in

Chapter 10, with the main discussion of the role of ASW in airborne ISR in the High North in the future. Here, the historical elements together with technological characteristics, in the context of the framework for discussion, provide the building blocks for the main arguments of the way forward for airborne ISR in the High North.

Notes

1 The Arctic Circle is situated at approximately 65 degrees North.
2 US DoD, "2024 Arctic Strategy," 2, June 21, 2024, https://media.defense.gov/2024/Jul/22/2003507411/-1/-1/0/DOD-ARCTIC-STRATEGY-2024.PDF#:~:text=This%20new%20DoD%20strategy%20directs%20the%20Department%20to,to%20build%20readiness%20for%20operations%20at%20high%20latitudes.
3 Johan J. Holst, "Norsk Sikkerhetspolitkk i Strategisk Perspektiv," *Internasjonal Politikk* 24, no. 5 (1966): 463–90; Rolf Tamnes, "Integration and Screening – The Two Faces of Norwegian Alliance Policy," *Forsvarsstudier* 6 (1987).
4 John Simpson, "How Greenland Could Become China's Arctic Base," *BBC News Online*, December 18, 2018, www.bbc.com/news/world-europe-46386867.
5 See for example Pan Min, "Fisheries Issue in the Central Arctic Ocean and Its Future Governance," *The Polar Journal* 7, no. 2 (2017); Mariia Kobzeva, "China's Arctic Policy: Present and Future," *The Polar Journal* 9, no. 1 (2019); Arild Moe, "The Northern Sea Route: Smooth Sailing Ahead?" *Strategic Analysis* 38, no. 6 (2014); Alexander Sergunin and Valery Konyshev, "Russia in Search of Its Arctic Strategy: Between Hard and Soft Power?" *The Polar Journal* 4, no. 1 (2014).
6 AAP-6, *NATO Glossary of Terms and Definitions* (NATO Standardization Agency, April 3, 2013), 2-A-16.
7 Jan S. Breemer, "Anti-Submarine Warfare: A Strategy Primer," Naval Postgraduate School, July 1988, Dudley Knox Archive, NPS, Monterey, CA.
8 Gregory Elder, "Intelligence in War: It Can Be Decisive," *Studies in Intelligence* 50, no. 2 (2006); Brian P. Tice, "Unmanned Aerial Vehicles – The Force Multiplier of the 1990s," *Air & Space Power Journal* 5, no. 1 (1991).
9 David A. Deptula and R. Greg Brown, "A House Divided – The Indivisibility of Intelligence, Surveillance and Reconnaissance," *Air & Space Power Journal* 22, no. 2 (Summer 2008).
10 Clay Wilson, *Network Centric Warfare: Background and Oversight Issues for Congress* (Washington, DC: Congressional Research Service, June 2, 2004), CRS-2.
11 John Ferris, "Netcentric Warfare, C4ISR, and Information Operations: Towards a Revolution in Military Intelligence?" *Intelligence and National Security* 19, no. 2 (2004): 199–225.
12 John Luddy, *The Challenge and Promise of Network-Centric Warfare* (Arlington, VA: Lexington Institute, February 2005), 3.
13 See for example Ferris, "Netcentric Warfare, C4ISR, and Information Operations," 201.
14 Scott Jasper, "Does Maritime Patrol Have a Future?" *United States Naval Institute Proceedings* 123, no. 4 (April 1997); John O. Birkeland, "The Potential of LIDAR as an Anti-Submarine Warfare Sensor" (M. Phil (R), War Studies, Glasgow, University of Glasgow, 2009), 54–55, http://theses.gla.ac.uk/1252/1/2009birkelandmphil.pdf; William Perkins, *Alliance Airborne Anti-Submarine Warfare – A Forecast for Maritime Air ASW in the Future Operational Environment* (Joint Air Power Competence Centre (JAPCC), June 2016), 45–46.
15 AAP-6, *NATO Glossary of Terms and Definitions*, 2-S-15.
16 AAP-6, 2-R-4.
17 AAP-6, 2-I-6.

18 The collection of intelligence is carried out in the framework of the respective intelligence disciplines (HUMINT, GEOINT, SIGINT, OSINT, MASINT, TECHINT, ACINT), which will not be elaborated on at this stage. For details, see for example US DoD (2012) Joint Pub. 2-01 "Joint and National Support to Operations".
19 Deptula and Brown, "A House Divided."
20 Joint Publication (JP) 1-02, *Department of Defense Dictionary of Military and Associated Terms* (United States Department of Defense, February 15, 2016), 116, www.dtic.mil/doctrine/new_pubs/jp1_02.pdf.
21 AAP-6, *NATO Glossary of Terms and Definitions*, 2-P-2.

Bibliography

AAP-6. *NATO Glossary of Terms and Definitions*. NATO Standardization Agency, April 3, 2013.

Birkeland, John O. "The Potential of LIDAR as an Anti-Submarine Warfare Sensor." M. Phil (R), War Studies, University of Glasgow, 2009. http://theses.gla.ac.uk/1252/1/2009birkelandmphil.pdf.

Breemer, Jan S. "Anti-Submarine Warfare: A Strategy Primer." Naval Postgraduate School, July 1988. Dudley Know Archive, NPS, Monterey, CA.

Deptula, David A., and R. Greg Brown. "A House Divided – The Indivisibility of Intelligence, Surveillance and Reconnaissance." *Air & Space Power Journal* 22, no. 2 (Summer 2008).

Elder, Gregory. "Intelligence in War: It Can Be Decisive." *Studies in Intelligence* 50, no. 2 (2006).

Ferris, John. "Netcentric Warfare, C4ISR, and Information Operations: Towards a Revolution in Military Intelligence?" *Intelligence and National Security* 19, no. 2 (2004): 199–225.

Holst, Johan J. "Norsk Sikkerhetspolitkk i Strategisk Perspektiv." *Internasjonal Politikk* 24, no. 5 (1966): 463–90.

Jasper, Scott. "Does Maritime Patrol Have a Future?" *United States Naval Institute Proceedings* 123, no. 4 (April 1997).

Joint Publication (JP) 1-02. *Department of Defense Dictionary of Military and Associated Terms*. United States Department of Defense, February 15, 2016. www.dtic.mil/doctrine/new_pubs/jp1_02.pdf.

Kobzeva, Mariia. "China's Arctic Policy: Present and Future." *The Polar Journal* 9, no. 1 (2019).

Luddy, John. *The Challenge and Promise of Network-Centric Warfare*. Arlington, VA: Lexington Institute, February 2005.

Min, Pan. "Fisheries Issue in the Central Arctic Ocean and Its Future Governance." *The Polar Journal* 7, no. 2 (2017).

Moe, Arild. "The Northern Sea Route: Smooth Sailing Ahead?" *Strategic Analysis* 38, no. 6 (2014).

Perkins, William. *Alliance Airborne Anti-Submarine Warfare – A Forecast for Maritime Air ASW in the Future Operational Environment*. Joint Air Power Competence Centre (JAPCC), June 2016.

Porter, Air Commodore (r) Garfield "Garry". *British Maritime Patrol Aircraft Operations During and After the Cold War*. London: Royal Air Force Club, May 31, 2017.

Sergunin, Alexander, and Valery Konyshev. "Russia in Search of Its Arctic Strategy: Between Hard and Soft Power?" *The Polar Journal* 4, no. 1 (2014).

Simpson, John. "How Greenland Could Become China's Arctic Base." *BBC News Online*, December 18, 2018. www.bbc.com/news/world-europe-46386867.

Tamnes, Rolf. "Integration and Screening – The Two Faces of Norwegian Alliance Policy." *Forsvarsstudier* 6 (1987).

Tice, Brian P. "Unmanned Aerial Vehicles – The Force Multiplier of the 1990s." *Air & Space Power Journal* 5, no. 1 (1991).
USDoD. "2024 Arctic Strategy." June 21, 2024. https://media.defense.gov/2024/Jul/22/2003507411/-1/-1/0/DOD-ARCTIC-STRATEGY-2024.PDF#:~:text=This%20new%20DoD%20strategy%20directs%20the%20Department%20to,to%20build%20readiness%20for%20operations%20at%20high%20latitudes.
Wilson, Clay. *Network Centric Warfare: Background and Oversight Issues for Congress*. Washington, DC: Congressional Research Service, June 2, 2004.

2 A short history of airborne ISR

The history of aerial surveillance and reconnaissance is an exciting one, filled with daring experiments and avantgarde technology. But as we navigate through the following overarching historical portrayal of this development, it is essential to identify the enduring characteristics that are fundamental to an aircraft's ability to support planning and the conduct of operations with information. This historical overview provides the necessary background for the discussion of the principles of aerial surveillance and reconnaissance that follows.

Early airborne reconnaissance

Among the first descriptions of airborne reconnaissance are those of Marco Polo from the fourteenth century, where he wrote about Chinese man-lifting kites that probably were used for military reconnaissance operations.[1] But it was not until the late 1700s that manned free-flight with balloons would catch on, of which the first was recorded in Paris in October 1783. The first manned balloon was really the beginning of manned flight, or as air power historian Tyler Morton puts it, "the watershed event from which all future military aviation grew."[2] The French experimentation spurred interest internationally, and both the British and Americans initiated experiments of their own, the latter led by Benjamin Franklin. The military benefits became increasingly prominent, and the French established their first balloon corps a decade later with the objectives of reconnaissance and long-range signalling. The first aerial reconnaissance mission utilizing a balloon took place on 26 June 1794, and the balloon observers communicated to the ground by dropping notes in small sand bags to the troops below, and using hand signals.[3] The purpose of the balloons was, first and foremost, reconnaissance, and the Army of the French Republic used them for observing Austrian troop movements and dispositions in operations throughout that campaign.[4] After the Napoleonic Wars, the development of manned flight was scarce. A plausible explanation for this was the 1815 Congress of Vienna and the following relatively peaceful decades, with war-weary nations reluctant to spend time and treasure to explore new war fighting technologies.[5] Significant developments did, however, take place in the 1800s that would forever shape the development of aerial information gathering. In the latter half of the century, the first aerial photography was captured, and the first wired telegraph

DOI: 10.4324/9781003626015-2

communications from a balloon took place. These two foundational, technological developments would shape the beginning of what could be perceived as near real-time reconnaissance and dissemination – a concept that will be discussed further in the next chapter.

The concept of aerial reconnaissance proliferated, and further records of utilizing balloons for observation emerged during the American Civil War in 1861–1865, the siege of Paris in 1870, British operations in Africa in the 1880s, and the Boer War at the turn of the century.[6] The Union Army of the American Civil War used tethered balloons, and the observers would bring a telegraph key into the basket under the balloon, enabling them to transmit reports directly back to headquarters through the wire linking them to the ground.[7] The Union Army established a Balloon Corps during the Civil War, but the initiative lost momentum and the corps was disbanded soon after the war. But it was a combination of positive experiences of utilizing balloons for observations during the Civil War and the lack of a coherent effort of balloon observation in the war with Spain in 1898 that led the Americans to pursue early doctrine and a framework for military aerial observation in war.[8] Like the French, the British had established a permanent balloon section within the British Corps of Royal Engineers a decade before the Boer War started. The Boers, however, considered the technology unfair, and thought the British were not playing by the rules.[9]

Increased artillery effectiveness and the static position of the balloons gave momentum to ideas of the *dirigible* balloon, the mobile observation platform that would be harder to take down and would provide better intelligence due to flexibility in positioning. Much effort was given to developing dirigibles, eventually leading to likely the most famous model, the *Zeppeliner*.[10] However, with the airplane making its entrance in the early 1900s, the dirigibles never stood a chance in the long run.

Before the arrival of the military's first aircraft, significant amount of thought had gone into aerial observation. In 1910, the US Army *Field Service Regulations* described an aerial company that would support every corps-size unit, but left the specification of mission to be described after more experience had been accumulated from the use of aircraft.[11]

When armed forces started pouring resources into airplane development, doctrine and ideas on the utilization of observation aircraft were already in place. In what can be considered a rare occurrence of doctrine racing ahead of actual practice and tactics, the aircraft capabilities of the armed forces could not live up to the missions stated in the Field Regulations of 1910. Aircraft flying procedures and techniques were in their infancy, and aircraft crashed at an alarmingly high rate. But experimentation with aircraft continued nonetheless. The move from balloons to aircraft ushered the requirement for wireless communications, which, for the first time, took place from an aircraft in 1910 – this would cement the aircraft as the platform of choice for elevated, flexible, and mobile military observation for the future.[12] The combination of aerial photography and wireless communication was to shape the impact that aircraft would make on modern warfare.

The Italians paved the way for using aircraft for military reconnaissance in 1911, when fighting against the Turks over the provinces of present-day Libya, with the first heavier-than-air reconnaissance mission being flown on 23 October that year.[13] The French, Mexicans, Bulgarians, Turks, and the Americans followed suit in other campaigns in 1912 and 1913.[14] Before the outbreak of World War I, the British experimented with aerial photography through a designated experimentation squadron of the British Royal Flying Corps. The British held regular army exercises using reconnaissance aircraft, prompting an officer in 1912 prophetically to state that "it seems to me impossible for troops to fight while hostile aeroplanes are able to keep up their observations . . . warfare will be impossible unless we have the mastery of the air."[15]

In 1914, the updated US Army Field Regulations promulgated three basic missions for aerial observation: Strategic reconnaissance, tactical reconnaissance, and artillery observation. A fourth mission was implied, namely the protection of friendly observation aircraft through the use of armed aircraft. The latter point became increasingly prominent with the development of enemy fighter aircraft that with ease could take down the highly flammable hydrogen-inflated balloons.[16] Although balloons would be used in the coming war, the era of the airplane had begun.

World War I

Although the development of aircraft had come a long way in the decade leading up to the war, there was extensive use of balloons for observation in the beginning stages of World War I (WWI). A line of balloons paralleled the front lines, connected to the ground by field telephones.[17] The static nature of the balloons gave the observers an oblique angle of vision, a challenge that would be overcome by the increasing use of aircraft flying directly above the object under reconnaissance. Wireless communication was in its infancy, and primitive communication between the air units and the ground units hampered the efficiency of the observation effort. Bursts of static Morse Code sent over a wide range of frequencies would often times be masked by the static produced by the aircraft engines.[18]

British experiments with observation aircraft did, however, pay dividends in early 1915, when aerial photography performed by the Royal Flying Corps observed German trenches and troop dispositions, reconnaissance that contributed to the success of Allied attacks in January that year.[19] The Germans had success with aerial observation of their own, when air units performed surveillance of areas in Eastern Prussia, locating the Russian advance and enabling the cutting off of the main Russian forces.[20] And in an early air power landmark event, Allied aerial reconnaissance provided the basis for planning and execution of the counter-attack against the German advance on Paris, leading to the First Battle of the Marne.[21] The stalemate trench warfare that followed would also shape the use of aerial observation. Although long-range reconnaissance remained important, there was a shift towards tactical support of artillery units as the ground troops dug in. With observation aircraft executing both reconnaissance and artillery spotting, both sides of the trenches began to improve their anti-aircraft weapons. This further led to a requirement to operate at a higher altitude, which in turn increased the observer's field-of-view. With increased altitude rose the requirement for mechanical optics to support the actual observation. Momentum was thus given to improvements in photographic technology.[22]

The reconnaissance aircraft would initially fly with handheld cameras, then with bigger cameras mounted on the rear machine gun ring, eventually leading to the engineers cutting a hole in the rear fuselage of the aeroplane to fit more powerful cameras.[23] The requirement for intelligence processing of the thousands of photographs that were taken led to a new category of military specialists being born, namely that of the photographic interpreter. These specialists would find their place in the intelligence cycle, and, as Colonel Terrance Finnegan states, would "reinvent the way that modern battle was envisioned, planned and executed."[24] Photography had clear advantages over visual observation, and these provided some of the basic facilitating factors of airborne surveillance and reconnaissance. First, photography allowed the observer to give more focus and attention to *collecting intelligence*, rather than interpreting it as the mission progressed.[25] Second, the photographs provided *objective data* that were not influenced by the observer's subjective understanding, or lack thereof.[26] Specifically trained photographic interpreters would analyse the products, contextualize them, notice changes over time, and so on. Third, the photographs were much more detailed and accurate than any notes that could be taken by any observer, and thus provided much more information than any manual observation could facilitate.[27] These factors cemented aerial photography as one of the most trusted sources of intelligence in WWI.

The aerial photographic units of WWI executed their missions mainly in direct support of the ground troops. The characteristics of most of the aircraft involved were simply tactical in nature (short range, light weight, manoeuvrable) and prevented aircraft to be utilized in a strategic sense.[28] Examples of strategic reconnaissance were those missions executed by the German airships in support of the German High Seas Fleet and the Russians flying the Sikorsky-developed *Muromets*, a combined reconnaissance-bomber aircraft, far beyond the German front lines to gather information on German activities.[29] The latter aircraft is often considered to be the world's first strategic reconnaissance aircraft. But the dirigibles that supported the German Navy are an interesting case for anyone studying maritime surveillance and reconnaissance. Because of the extent to which the balloons and dirigibles were vulnerable to enemy fighter aircraft, the notion that they would be an effective reconnaissance asset for tactical ground troops in WWI quickly diminished as the war progressed. The German Navy, however, recognized the potential for the *Zeppelin* dirigible as a long-range, maritime surveillance and reconnaissance asset. The threat level was necessarily less over the high seas, and the Germans used the *Zeppelin* to a great extent for defensive surveillance and reconnaissance over the North Sea.[30] Douglas Robinson argues that the undependability of the *Zeppelin*, together with the lack of proper doctrine for its use, render the German dirigibles in WWI a failure. But Tyler Morton sees beyond these factors, and points to two important basic tenets of airborne surveillance that came out of the *Zeppelin* operations. First, the Germans demonstrated the ability to conduct *long-range* airborne naval reconnaissance with accuracy, given permissive weather. Second, the *Zeppeliners* were able to communicate their observations back to higher headquarters more or less in *real time*.[31] Through this they set high expectations early on for what future airborne surveillance and reconnaissance assets should be able to do. Experimentation with maritime aircraft took place before the outbreak of the war in 1914, and initially the so-called floatplanes were introduced to service in the Royal Navy in 1915.[32] Later, flying boats were developed which were bigger and had greater capacities in all respects, and these would play a significant role in the war against hostile navies, particularly submarines. From barely being a part of the naval inventory at the beginning of the war, the Allies commanded more than 500 aircraft for maritime surveillance and the search and destruction of enemy submarines at the end of the war, in what has been termed the *Battle of the Atlantic*.[33]

In the early stages of WWI, aircraft were, in essence, used exclusively for observation and reconnaissance. And as the understanding of the importance of aerial observation gained momentum, significant efforts were put into denying the enemy opportunities to exploit the air for reconnaissance and surveillance. The development of weaponized aircraft became an increasing threat to balloons, dirigibles, and observation aircraft that lacked the ability to protect themselves. Photographic reconnaissance became so vital to operations that the aircraft carrying out the missions had to be protected, which in turn led to the development of fighter escorts for the reconnaissance aircraft. This is thought to have given a significant momentum for the development of fighter aircraft as a whole.[34]

The inter-war years

After WWI, the nations demobilized and conducted a significant drawdown of forces to include forces for gathering intelligence from the air.[35] However, as manpower dwindled and the armed forces scaled back after the war, doctrinal discussions kept their pace, especially with regard to the usefulness of airpower. In addition, experiences from the war fed raging debate on the role of airpower in modern warfare. The wartime chief of the Air Service in the American Expeditionary Forces, General Patrick, stated in 1919 that "an air service in war ought to consist of 80 percent observation and 20 percent other aircraft." But 7 years after this statement, with further debate on the role of airpower and development of aircraft in general, he reversed these proportions.[36] Brigadier General William Mitchell, regarded by many as the original thinker on air power in the United States, considered observation as "the least important Air Service mission" at the same time as General Patrick's initial statement. Observation, he argued, was simply a phase in the early development of the air arm. The American Air Corps Act of 1926 – establishing the Air Corps under the army – cemented the important role of air power, but as army generals made budget cuts to aviation units in order to prioritize ground troops, development dwindled. On both sides of the Atlantic, observation fell into an intellectual and professional backwater within the air forces.[37]

As theories of strategic bombardment increased their dominance in air power theory as a whole in the 1920s, observation doctrine stagnated as a consequence. As Edgar Raines underlines, the reason for this was that the very existence of flying observation units presumed the continued importance of ground armies in future conflicts.[38] In many ways, observation units were doctrinally tied to ground troops and their tactical battle requirements. Air power theories taken to the extreme, on the other hand, would relegate armies to the role of guarding air bases, as the strategic bombing forces would win the war. Tactics, techniques, and procedures for aerial observation thus remained at the 1918 level, with very little room for innovative thinking. With army officers actively being left out of the development of aircraft that could support ground troops, doctrinal and institutional stagnation continued for the observation units. There was a potential for a proper synthesis between newly evolved doctrine, improved radios, and new light aircraft, had there only been someone there to pull the strings together. But focus was on the supremacy of air power and strategic bombing.

Overall, for aircraft in general, there took place a mechanical revolution in the 1920s and 1930s that paved the way for capable, high-flying aircraft. Engines became more powerful, more reliable, and lighter in weight. The internal fuselage structure became stronger, cockpits were closed through the use of canopies, and landing gears became retractable. This led to faster and higher-flying aircraft with longer endurance.[39] These improvements serve as a bridge between the early *Zeppeliners* for long-range maritime surveillance and reconnaissance during World War I, and the modern aircraft conducting bombing and long-range reconnaissance in the following hot and cold conflicts. Also, the long range and endurance facilitated maritime reconnaissance in a new way, a point that will be elaborated on later in this chapter.

In 1935, further doctrinal delineation between different types of observation was written by the US Army Air Corps. The main emphasis was given to the difference between long-range and short-range observation: *Short-range* observation was to belong to and support the *ground troops* and be their main source of battlefield reconnaissance. *Long-range* observation was to belong to the *Air Corps* and consist of airborne reconnaissance far beyond the enemy lines, supporting targeting efforts for strategic bombing.[40] Consequently, long-range reconnaissance missions, which until then were a part of army observation units' portfolio, were given to light bombardment units in 1939.[41] This move would solidify the ties between airborne reconnaissance and surveillance, and strategic bombing – a tie that became particularly evident after World War II (WWII).

The Germans, although restrained by the Treaty of Versailles with regard to rearmament and weapons development, established the Luftwaffe in the shadows of their civilian aircraft industry.[42] The Germans developed highly capable aircraft constructed for strategic photographic reconnaissance, with dedicated, powerful aircraft carrying state-of-the-art Zeiss photographic lenses.[43] Along with rearmament came doctrinal development, to include observation units – regarded as a significant resource in modern warfare by the Germans. In the late 1930s, the Commander-in-Chief of the German Army stated that "the next war will be won by the military organization with the most effective photographic reconnaissance."[44] And like the Americans and the British, the Germans came to similar conclusions: Short-range observation aircraft would be allocated to army commands, and long-range observation aircraft would solely be commanded by the *Fliegerkorps*.[45]

In Britain, the former MI6 spy turned chief of the Air Intelligence Service, Wing Commander Winterbotham, took initiative to purchase dedicated aircraft to monitor the German developments. In conjunction with the foreign intelligence service MI6, he initiated a project that would be led by Sidney Cotton, who flew a modified Lockheed 12A aircraft at high altitude in order to obtain images of Germany in the late 1930s.[46] Cotton's philosophy was to evade any threat rather than to defeat it, and he recommended that due to modern air defences, modern observation aircraft had to be high-altitude, high-speed, long-range, and have low-observable characteristics.[47] Another associate of Winterbotham and Cotton, Maurice Longbottom, wrote a memorandum in 1939 entitled *Photographic Reconnaissance in Enemy Territory in War*, where the British, for the first time, delineated between the emerging categories of reconnaissance flying: Tactical work in the immediate vicinity of the front line, and strategic reconnaissance of enemy territory behind the zones of conflict.[48] Supporting his colleague, he underlined the importance of avoiding enemy fighters and aerial defences. He advocated the use of "*a single small machine relying solely on its speed, climb, and ceiling to avoid detection.*"[49] This led to the Royal Air Force (RAF) placing modified Spitfires and Mosquitos as their primary reconnaissance and surveillance platforms in the beginning of WWII.

The Japanese were among the first to recognize the importance and utility of long-range aerial reconnaissance for military purposes. They used aircraft extensively for long-range reconnaissance against the Chinese, and the experiences gained provided valuable intelligence that specifically supported the planning of operations in the early 1940s, such as the attack on Pearl Harbour.[50] There was a constant flow of intelligence reaching Tokyo, and aircraft were used to provide verification of that information. The historian Robert Jackson considers the Japanese, although their early efforts in this field tend to be overlooked by historians, to be the "true precursors of the clandestine intelligence-gathering mission that would be flown a decade and more later."[51]

Maritime air reconnaissance after WWI was partly executed by land-based aircraft, but increasingly by flying boats. The mentioned mechanical revolution in the 1930s facilitated very long endurance, and the *Short Sunderland* flying boat that entered service in 1938 had an endurance of 12,5 h and a range of 850 nautical miles.[52] Doctrinally, the maritime air reconnaissance domain belonged to the flying boat.

World War II

When WWII started, most nations engaged in the war commenced utilizing specialized aircraft for surveillance and reconnaissance, although these were usually modified bomber or fighter aircraft. Operations at the outset of the war would be based on a doctrine that had been developed in the 1930s, and the respective nations' understanding of airborne surveillance and reconnaissance naturally matured and further evolved during the early 1940s. Discussions continued into the war on whether ground troops would be better suited procuring and thus controlling their own aircraft in support of ground operations, as opposed to organizationally separated air units supporting ground troops during operations. These discussions, however, mostly took place in very large nations, primarily in the United States.[53]

The Japanese continued their use of aircraft for long-range reconnaissance. As the war in the Pacific intensified, they conducted reconnaissance of American and British task forces that provided information that facilitated the infamous kamikaze attacks.[54] The German Luftwaffe focused on high-altitude, long-range reconnaissance for strategic intelligence gathering, with aircraft flying at what was at that point unprecedented heights – 42,000 feet. Air defences, both ground- and air-based, had come a long way since the end of the last war, and the altitude for observation reflected this. As did the choice of aircraft for tactical reconnaissance: The main aircraft that were used for tactical support to ground troops were modified variants of the top-performing fighter aircraft of the Luftwaffe. This underlines the understanding of the increasingly dangerous environment that fighter aircraft flying close to the ground were operating in. The Soviets had a steep learning curve at the beginning of the war. They had no dedicated aircraft, nor personnel, for executing reconnaissance missions. But as operations progressed, the importance of photographic intelligence became increasingly obvious, and as a consequence more effort was put into establishing a surveillance and reconnaissance force within the air forces. Initially, the task was given to non-specialized units flying improvised reconnaissance missions over German positions, but as time passed, more and more experienced personnel became available, in the form of both pilots and photo interpreters. Towards the end of the war, every single Soviet operation was preceded by photographic reconnaissance missions.[55] The British built on their doctrine and experiences from the late 1930s with regard to aerial observation. Based in Heston, Middlesex, the photographic reconnaissance unit (PRU) was established in 1939.[56] Among other missions, units from the PRU flew dedicated reconnaissance missions to locate and map German advances and manoeuvres following the retreat from Dunkirk. Modified *Spitfires*, termed *photographic reconnaissance (PR) Spitfires*, flew missions into occupied territory, regardless of the weather, to obtain critical information on adversary dispositions in a critical phase of the war.[57] These missions echo the challenges that modern planners see in operating in so-called *denied areas* more than 70 years after WWII.[58] With the establishment of the PRU, the unit was authorized to act as a strategic reconnaissance unit on behalf of the RAF.[59] But WWII also saw the emergence of tactical reconnaissance not only in support of ground forces, but also in support of airborne attacks. One example of this is the so-called "outrider tactics," which built on experiences from World War I. The British Coastal Command struggled early in the war with attack formations arriving too late on target to act on recent changes to target intelligence, particularly moving maritime targets. The outrider tactics remedied this by sending fast-flying reconnaissance aircraft ahead of the attack formation in order to attain detailed updates on target positions, course, type, activity, and other relevant data. They would then convey the updates over radio to the incoming formation of attack aircraft, who would attack based on the updated information.[60]

In the late 1930s, the US Army Air Forces were given increased autonomy from the army intelligence organization and set out to establish their own intelligence structure. For these efforts, the Americans relied heavily on the already established training and analysis elements of the British for airborne reconnaissance and surveillance, support that would assist in solidifying the close relationship between the two nations with regard to exchange of intelligence, training, and doctrine.[61] As the British clashed with the Germans and had put their training and doctrines to the test in live combat, the Americans were still not a part of the war in 1940. American officers visited the PRU in Britain, and learning from their close ally they established their own training structure for analysts and interpreters. British planning and execution of operations underlined the importance of aerial surveillance and reconnaissance in modern warfare. As the war progressed, the pressure to deliver relevant intelligence for operational planning grew, as did the few institutions trusted with photographic intelligence interpretation and analysis. The interpretation unit at Heston airfield was moved and co-located with others at Medmenham as the Central Interpretation Unit (CIU), and later the Allied CIU (ACIU).[62] The ACIU was to become the cradle of institutionalized photographic interpretation and analysis for the Allies, and played a crucial role in supporting the Allied bombing campaign in general, and, among other specific operations, the search for the manufacturing and launch sites of the V-1 and V-2 flying bombs.[63] With ever-developing technology facilitating new *modus operandi* for air operations, WWII was really the coming of age of the intelligence disciplines that we know so well today. Imagery intelligence (IMINT) was already somewhat established at the beginning of the war, with several nations having established a structure for handling the interpretation and analysis of the returning photographs, of which the most prominent was the ACIU. But WWII is the cradle of efforts to counter adversary utilization of the electromagnetic spectrum, and Robert Jackson terms the years between 1936 and 1945 with regard to aerial combat, "The Counter-measures war."[64] Operations in WWII included more and more signals intelligence (SIGINT), which can be subdivided into ELINT and communications intelligence (COMINT).

The Battle of Britain was the first arena for aerial combat where both sides executed ELINT operations. When the British commenced building radars along the coast, with the objective of providing warning of any German aerial attacks, the Germans on their side equipped their airships with antennae and radio equipment that would pick up the signals from the newly invented British radar sets.[65] Although unsuccessful due to German malfunctions, the British were able to follow the airship using their new equipment, and the electronic war had begun. Further enhancing the exploitation of the electromagnetic spectrum, the Germans started using radio beams to guide their bombers flying from the continent towards British cities. Placing transmitters on the French coast, the bomber pilots would be guided by radio beams pointing them in the right direction. Crossing beams would indicate range from the target and the drop point. Night after night, throughout the winter of 1940 and 1941, British cities were bombed by aircraft executing beam-guided attacks.[66] The British were thus forced to come up with counter-measures in the fight against the attacks. The answer was the use of *jamming*, the disturbance and disruption of electromagnetic signals, and dedicated ELINT flights that would determine the location and nature of the transmitters, providing crucially important intelligence for counter-attacks. There were also examples of real-time exploitation of the sensor data in the field of SIGINT, where COMINT operators listened in to German radio traffic, and then jammed the signals and frequencies they were using through the use of radio counter-measures. The pressures of combat and the technology and techniques derived from those experiences and operations formed the basis for all future forms of electronic warfare.[67] The Americans started experimenting with ELINT equipment on aircraft in the Pacific in order to defeat the Japanese advances and manoeuvres. Soon after the Pearl Harbour attack, it became apparent that the US Armed Forces were seriously deficient in imagery reconnaissance capabilities.[68] But the development progressed quickly, from visual observation from flying boats to dedicated, long-range photographic and ELINT aircraft towards the end of the war. There were even experiments with imagery aircraft large enough to house a laboratory so that photographs could be processed in flight.[69] The aircraft never became operational, and post-landing analysis would continue to be the norm. The Americans placed ELINT equipment in bomber aircraft and commenced the collection of signals emitted from Japanese radars in the Pacific. The collected information provided the basis for localizing the radars for attack, and building counter-measures such as jamming equipment. Although not used extensively, the ELINT flights also provided the basis for the evolving art of COMINT in the Pacific. The ranges involved in the Pacific theatre of war made modified bomber aircraft a feasible solution with regard to the choice of airframes, laying the foundation for the post-war fleet of IMINT and ELINT aircraft when it comes to range and size.

As mentioned, a few years into the war, the British placed linguists on board their ELINT aircraft to intercept and analyse hostile radio traffic, establishing the discipline of COMINT. It soon became common practise to bring linguists on board all ELINT flights.[70] The linguists monitored on known frequencies for enemy communications, warned friendly aircraft of adversary movements, and called in aerial bombardment with accuracy based on what they had heard. In addition, Allied analysts could enter the post-mission logs of the linguists to determine German and Japanese order-of-battle (OOB) and force dispositions, significantly enhancing their overall understanding of hostile operational and strategic intent.[71] There were challenges when it came to placing recorders on board the aircraft, and for several years the recordings were executed by handwritten notes by the linguists. And in the Pacific theatre, inter-service rivalry delayed some of the advancement in operational deliveries, when the Navy and the Army Air Forces fought over which organization should control operational intelligence (OPINTEL) services.[72] However, towards the end of the war, COMINT and ELINT aircraft were integrated in bombing raids; they provided crucial information on enemy dispositions and thus facilitated planning of future operations.

Maritime air reconnaissance aircraft supported the war in many capacities, most notably through ASW. The key to defeating the German submarine wolfpacks in the Atlantic was the introduction of radar-equipped ASW MPA in 1942. The essence of the Wolfpack tactics was the submarines' ability to operate on the surface close to the convoys crossing the Atlantic. However, the new MPA were able to operate at a significantly expanded radius around the convoys.[73] Through the signals intelligence and code-breaking efforts in the Pacific theatre (codename: ULTRA) and the fusion of OPINTEL in the maritime domain in general, the long-endurance MPA were to play a crucial part in the defeat of the adversary submarine menace in the so-called *Second Battle of the Atlantic*.

The post-war years

During WWII, the Allies created huge intelligence hubs of analysts, like the one in Medmenham, England, that played a crucial role in servicing the Allied bombing campaign. But as the war ended, an extensive drawdown of the massive wartime organizations took place. Manpower vanished from many elements of the intelligence organization, as the requirements for substantial intelligence analyst support elements in peacetime diminished.[74] Michael Warner terms this a "collapse of intelligence support for air power," and it was not until two years into the Korean War that the US Armed Forces relearned the importance of an all-source imagery intelligence, targeting, and battle-damage assessment capability.[75] The sources that provided the basis for targeting intelligence in the United States on Soviet targets directly following WWII were the Library of Congress, interviews with Russian émigrés, Tsarist era maps, debriefings of German prisoners of war recently released from Russian prison camps, and captured German air force aerial photographs.[76] In total, these sources provided a good basis for intelligence, but they were not accurate enough for targeting. The limitations in and lack of US reconnaissance capabilities immediately following WWII shaped war planning efforts following the war. Emergency war plans post-1945 included plans to take out entire cities through a nuclear attack, with the aim of crippling specific industries of the adversary.[77] The need to annihilate entire cities in order to ensure the destruction of relatively small elements of that city was due to the lack of detailed and specific target intelligence. This again was a direct consequence of the lack of adequate reconnaissance assets, reconnaissance operations, and an associated processing and analysis organization. This lack of accurate, strategic intelligence-gathering assets after WWII was eventually recognized. Based on the increasing Soviet air defence threat to surveillance flights, Western powers initially used higher and faster flying aircraft to avoid more capable air defences. Initially, strategic surveillance and reconnaissance missions were flown along the national borders of the adversary, or along the "periphery." These flights were thus called *peripheral flights* and were initiated as early as the fall of 1945. The requirements for detailed information from targets deep inside adversary territory only grew, which in turn pointed towards the possibility of penetrating sovereign airspace in order to gather information. The intersection between the increasing threat to own aircraft and the requirement for information first manifested itself in the use of balloons again. Balloons fitted with cameras and ELINT equipment were sent into Soviet territory from the European side, with the hope that winds would carry them all the way across to the Pacific side of the Soviet Union.[78] The project was a failure, with failing equipment and unsuccessful recovery. Many balloons landed inside Soviet territory.

The first direct overflight by an American aircraft over Soviet territory was executed in August 1948, when a former bomber aircraft modified for reconnaissance flew a 19-h sortie deep into Siberia.[79] Through the use of modified versions of the most capable aircraft available at the end of the war, these top-secret flight operations were executed in the years following WWII in order to gather as much detailed, up-to-date information as possible on the adversary. These operations were meant to map bombing routes across the Arctic, which is the shortest route between the United States and the former Soviet Union, and to gather information on specific targets far into adversary territory. In parallel, top-secret projects were initiated to meet the increasing challenge of acquiring intelligence from inside hostile territory.[80] These would manifest themselves in the strategic capabilities that emerged in the late 1950s, as we shall see.

The Americans required strategic intelligence to support the development of their nuclear strategy and flew systematically along the Soviet periphery and over Soviet territory to gather this information.[81] Aerial reconnaissance aircraft, their associated intelligence units, and the intelligence community as a whole were channelling their efforts towards strategic thinking and planning.

With regards to naval warfare, long-range land-based MPA had through humble beginnings in WWI, peacetime reconnaissance during the inter-war years, and most crucially through contributing significantly to winning the *Second Battle of the Atlantic* during WWII put themselves centre stage with regard to maritime reconnaissance and surveillance. A Jack-of-all-trades, from ELINT to COMINT to search-and-rescue, and particularly contributing to the art of ASW, Allied MPA would play a crucial role in Cold War reconnaissance in the North Atlantic, as the coming chapters will show.

The Korean War

When war broke out on the Korean peninsula, Western forces were not prepared for conducting airborne intelligence collection in support of operations. There were few tactical reconnaissance aircraft and fewer intelligence professionals in the Air Force, and the individuals that were there were poorly trained and lacking in experience.[82] The outbreak of the war forced strategic air forces to support tactical situations again, although they were neither designed nor equipped for this. Strategic reconnaissance assets that were just getting accustomed to gathering strategic intelligence on Soviet were now utilized for tactical purposes where the tactical and operational requirements of the day shaped the utilization of the asset.[83] The tactical reconnaissance units that were available were summoned and tasked initially to conduct mapping of the Korean peninsula in order to provide up-to-date tactical maps for air force units. By 1951, the communists had installed more sophisticated radar early warning systems, and an important mission for reconnaissance units was to locate and photograph these sites for targeting.[84]

In the beginning of the conflict, imagery that had been gathered had to be flown to Japan in order to be analysed due to the lack of embedded facilities and personnel at the air bases in Korea.[85] There were even fewer SIGINT aircraft and knowledgeable SIGINT personnel involved in the beginning, and significant changes in dispositions had to be made to support the Korean operations. However, a few key persons with experience from WWII established analysis centres and a regime for requesting, validating, prioritizing, and authorizing intelligence targets for collection, what is now called information request management (IRM). After 1 year into the conflict, many more assets and a systematic approach to airborne intelligence collection was more or less in place.[86] It was not until well into the war that the Western forces relearned the importance and need for an all-source imagery intelligence, targeting, and battle-damage assessment capability. Another important development was "Hammer Operations," where reconnaissance aircraft would report over radio in real time enemy troop dispositions to friendly artillery and fighter/bombers, increasing the accuracy of friendly fire. Visual sightings were often as important as imagery.[87] Yet another significant move forward for the intelligence community's support to operations was the use of tactical COMINT, where airborne linguists would monitor enemy communications and warn friendly aircraft of any approaching threats. Although a shift in Chinese *modus operandi* from high-frequency (HF) to very-high-frequency (VHF) communications caused some initial challenges, this tactical COMINT contributed significantly to the lopsided kill ratio of fighter aircraft in favour of Western fighter aircraft (F-86) against aircraft flying for the North (MiG-15).[88] COMINT capabilities were further refined, and some aircraft were tasked to perform deep-penetration flights behind enemy lines to gather COMINT. In support of these operations, there were other aircraft that stayed further behind and relayed the information that the deep-penetrating aircraft were collecting. These relay aircraft became airborne hubs of intelligence that were able to relay information and direct friendly air operations – whilst airborne. They thus became the precursor of the modern-day airborne command and control aircraft such as the E-2 *Hawkeye* and the E-3 *Sentry*, command and control aircraft that also have inherent collection capabilities. A final aspect with airborne intelligence collection during the Korean war was that it was common for collection aircraft to drop their tapes of intelligence collection to the ground in parachutes, due to the lack of an inflight analysis capability. Information that could be seen and heard was reported over the radio in real time, but information that had to be processed was either dropped in parachutes to a waiting analysis centre on the ground or brought by a courier to that analysis centre post-landing.[89]

Surveillance and reconnaissance after the Korean War

After the Korean War, tactical reconnaissance units were often used for overflights of the Soviet Union and Chinese territory to gather strategic intelligence. Following the war, these units trained and experienced in evading surface-to-air threat that might hinder collection of intelligence. After these flights, the films were flown to Japan for processing and analysis. By the 1960s, it was commonplace for tactical reconnaissance squadrons to have several jet aircraft in their inventory specifically dedicated for the delivery of picture prints to analysis facilities and users of the gathered intelligence.[90] What effectively shut down these series of overflights was the emergence of the capable MiG-19 interceptor fighter aircraft, posing a very real threat to all the tactical reconnaissance aircraft in the Western inventory. The RF-100 was eventually able to fly at a very high altitude (50,000 feet) at almost the speed of sound (Mach 0.96) but was only able to do so with the afterburner on. This led to an unacceptably short operational range for the aircraft.[91] US Air Force officers such as Richard Leghorn, by many thought to be one of the original thinkers on strategic aerial intelligence, advocated the use of high-flying aircraft that would be hard, if not impossible, to detect and shoot down, for gathering intelligence on adversary nations.[92] A prominent factor in his thoughts on airborne intelligence was that such systematic information gathering would be one of the primary tools to preserve peace in the new, post-WWII world order.[93] His thoughts, along with those of others, facilitated the push for investigating potential technologies and systems for strategic surveillance and reconnaissance, work that first led to the fielding of the U-2 strategic reconnaissance aircraft and then to the CORONA series of surveillance satellites.[94] The U-2 surveillance and reconnaissance aircraft became operational in 1956 and was intended to escape any threats by flying *above* Soviet air defences, in addition to being very hard to see on any radar.[95] Although there were gaps in radar coverage that the U-2 could slip through, the missions were detected.[96] And although the U-2 flew too high for the air defence ground systems and fighter aircraft in 1956, only a few years into the programme, the Russians caught up. Eventually, the U-2 didn't fly high enough to escape the increasingly capable Soviet air defence systems. Thus, the U-2 missions were executed with a severe degree of risk, both politically and militarily, the latter being manifested through the inherent risk of flying within the range and territory of enemy air defences. The nature of these flights demanded that they be approved by the highest military and political level.[97] The US Air Force acknowledged the possibility of a U-2 being shot down, and started to look for alternative measures to gather intelligence over the Soviet Union. One such idea was to convert a jet-powered target drone into a reconnaissance drone that would fly well above the reach of known air defence systems at several times the speed of sound.[98] Although the initial work stalled and was cancelled due to various reasons, the seed was sown. Finally, a U-2 flown by Gary Powers was shot down on 1 May 1960 over Soviet, which could have killed the pilot, and which led to significant political tensions between the United States and the USSR. The information gathered by aircraft flying close to the border of enemy territory, or even *into* enemy airspace, gave information about strategic enemy capabilities, and thus supported the notion that aerial intelligence gathering was inherently strategic. The famous strategic reconnaissance aircraft U-2 had a relatively short career in the overflight business. The aircraft became operational in 1956, and missions penetrating Soviet sovereign airspace were effectively shut down following the Powers shoot-down. With the increasing capacity of air defence systems, overflight operations became too risky both in human and political terms. After 1960, strategic manned aerial intelligence was essentially conducted along the border of other nations, as opposed to penetrating other nations' sovereign airspace.

Almost as a planned handover between assets, the *CORONA* surveillance satellites became operational in the fall of 1960, and the era of satellite surveillance had begun.[99] The *CORONA* system was able to cover the Soviet Union land mass and primarily search for intercontinental ballistic missile (ICBM) installations and updates to these. The resolution in 1961 ranged between 10 and 15 feet.[100] In the absence of technology for transmitting the raw sensor data from the satellites to the ground by electronic means, the *CORONA* satellites dropped capsules of film from their orbit onto the earth. A parachute would deploy as the capsule reached 50,000 feet in order to slow its descent, and a specially designated C-119 aircraft took off from Hawaii to snatch the capsule in mid-air. The aircraft towed a long nylon loop with which the recovery crew would catch the parachute and reel it into the aircraft. If a capsule was missed and landed in the water, it would float for a few days until ships could recover it. After 3 days, the capsule was designed to fill with water and sink, in order for it not to be recovered by anyone else.[101] Early satellites showed the great potential they had compared to manned aircraft. The swath width of both the U-2 aircraft and the CORONA satellite was about 190 km, but the satellite orbited the earth every 90 min, enabling it to cover the entire Soviet Union in one day. In fact, the first CORONA mission took more imagery of the Soviet Union than all the U-2 flights combined.[102] The CORONA satellites would, by August 1960, provide photographic evidence that no Soviet ICBMs were yet emplaced, removing fears of the so-called "missile gap" that was thought to have emerged.[103] But US photo interpreters and military planners required higher resolution in order to carry out more detailed analysis of Soviet capabilities, and the development of the *GAMBIT* system was ordered. This was a high-resolution, spotting satellite system that could focus on specific targets for imagery, as opposed to the *CORONA* system, which performed a wide-area search with lower resolution. The first *GAMBIT* satellite flew in July 1963 and provided resolution that generally varied between 3.5 and 10 feet, and at times much better. People could be made out on the pictures, and at one point the place kicker meant to hold the ball in place could be seen on an American football field.[104] The *CORONA* and *GAMBIT* systems worked closely in tandem in order to provide diverse imaging perspectives for intelligence collection.

The operationalization of satellite reconnaissance in the early 1960s would change reconnaissance and surveillance forever. Satellites more or less assured one of the basic tenets of aerial ISR: Survivability. Although discussions and programmes have been in progress with regard to neutralizing adversary satellites since the early satellites of the 1950s, surveillance satellites have essentially been performing their missions without interference. And, crucially, satellites were able to fly uninterrupted over any land mass to include hostile territory. To an increasing degree, this would, alongside constantly improving sensor technology, eliminate the requirement for overflight missions using aircraft to penetrate hostile airspace in peacetime.

Development started in the late 1950s of the first dedicated SIGINT aircraft that was not a modified aircraft meant for another purpose. The C-130A-II had 10 dedicated COMINT operator stations and flew along the Soviet periphery in the late 1950s and early 1960s.[105] The C-130 platform was considered to be too slow and have too short a range, and the SIGINT community required a transition into the jet age. Conversions of Boeing 707s began in 1961, turning them into the jet-powered and capable RC-135, arguably the most significant strategic surveillance and reconnaissance assets in the American inventory of the past century. First operational in 1962, the RC-135 SIGINT platform has flown in peacetime and operations ever since, including the Vietnam War.[106] With improvements to communications technology over the years, the RC-135 has become increasingly capable of communicating with ground forces. Initially, this was done via voice communications, and eventually methods evolved to include the use of data links and downlinks of sensor data similar to the one used on board U-2 aircraft over Vietnam, which is explained below.

Just a few years into the 1960s, momentum for reconnaissance drones was regained after it had been dropped during U-2 development, when yet another U-2 was shot down over Cuba in 1962. Previous ideas of converting jet-powered target drones into high-speed reconnaissance drones now materialized. In 1959, Ryan Aeronautical investigated how they could extend the range of its *Firebee* target drone and turn it into a reconnaissance drone. Several variants of ISR drones were developed over the coming years, most famous of which were the *Fire Fly* and the *Lightning Bug* drones.[107] The high-speed, high-altitude *Lightning Bug* drone almost saw action over Cuba in 1962, but flew its first operational mission over Vietnam in 1964 on a reconnaissance mission to monitor the Chinese-Vietnamese border.[108] Although based on the U-2 cameras, the sensor products were rudimentary. But they gave insight into adversary ground movements, so the drone missions continued. Two of the *Lightning Bugs* were shot down in the fall of 1964, but their downing did not have the same political impact as that of the U-2s.[109] This points to an important factor for deploying unmanned systems as opposed to manned systems: The lowered risk assumed is not only in human terms, but it is also manifested in political terms. Following the shoot-down of an American ELINT aircraft flying along the border of North Korea in 1969, killing all 31 crew members on board, special ELINT equipment was installed on several *Lightning Bug* drones. They would pick up the dangerous missions along the Korean coastline and transmit their recorded signals to analysts in ground stations via data links. These were invented especially for these aircraft, but were later transferred to be used by aircraft such as the U-2 in other missions.

Tactical reconnaissance naturally fell somewhat into the background of strategic reconnaissance in peacetime. Day-to-day operations included participation in exercises in order to keep the skills of the crew at an acceptable level. There was some frustration in the tactical reconnaissance community in the late 1950s and early 1960s because their units were often utilized as enemy attack formations, as opposed to exercising their primary reconnaissance missions. However, training and exercises were crucial elements of being able to provide a credible tactical reconnaissance force should there arise a requirement for "hot" missions.[110] In Europe, for example, the combination of evading enemy surface-to-air Missiles (SAMs) through the element of surprise and staying below the weather in order not to get any clouds between the camera and the target, demanded training at high speeds and at low altitude. When the next war came, tactical reconnaissance units stationed in Europe were prepared for action.[111]

The Vietnam War

Although the overarching focus of the military airborne ISR community returned to a strategic focus post-Korea, it was forced back into the tactical realm again during the Vietnam War. But this time, the airborne intelligence units were prepared. Tactical reconnaissance units that had trained and exercised in Europe were used to lack of airspace restrictions, bad weather, and constantly being intercepted by own friendly forces – in addition to sitting on alert status with live nuclear weapons.[112] This is also evident in the presence of traditionally strategic intelligence aircraft collecting IMINT and ELINT close to Vietnam before there were ground troops committed to the hostilities.[113] Tactical reconnaissance units, for example, took part in Project Field Goal, which was reconnaissance support to the Government of Laos in its fight against North Vietnamese aggression before the United States became fully involved in the conflict.[114] By the time ground units arrived, the airborne units had further refined the system from the Korean War of processing, validating, prioritizing and tasking intelligence requirements in a systematic manner.[115] This led to timely delivery of intelligence to the war fighters on the ground, the quintessential purpose of airborne tactical reconnaissance. There were imagery analysis centres at all main air bases, and intelligence products were disseminated by courier to the appropriate requesting unit.

Early in the conflict, the US Air Force began deploying airborne early warning aircraft as command and control platforms, and they placed linguists on board to monitor enemy communications. The combination of the airborne radar and COMINT provided valuable support to other aircraft. Further building on this concept, the first data link system, called *Combat Lightning*, was developed.[116] This was a system that aimed a disseminating near-real-time tactical information between aircraft on other units' whereabouts, incoming enemy aircraft, potential threats, and other data that would enhance the situational awareness of the flying aircrew. The system struggled to get online due to classification issues and thus paradoxically challenges with sharing information, but the notion of a tactical data link set the precedent for systems to come in the future.

Another precedent was *Project Teaball*, which integrated the COMINT information that was downlinked from U-2 surveillance aircraft with command and control units for fighter aircraft. Later, the command and control units also started receiving ELINT data from RC-135s and radar data from Navy ships, leading to an unprecedented level of data fusion in support of tactical operations.[117] New radars and communication capabilities made it possible to relay acquired tactical information to ground command and control stations, which in turn could relay this information to other aircraft and troops. With this information and situational awareness, the fighter pilots were able to avoid enemy ambushes, and *Project Teaball* contributed to a decrease in friendly fighters lost to the very capable North Vietnamese air defences.[118] Tactical reconnaissance aircraft that flew in Vietnam flew missions along and behind enemy lines to gather intelligence. Crucial information included battle damage assessment (BDA), where the task was gathering detailed pictures of bombing and artillery raids. After the flights, the aircraft would return to base or the carrier ship and offload the camera films, and these would be developed for analysis. After this analysis, the intelligence would be disseminated to the respective units in need of the sensor products. One new improvement to this process was the utilization of new and small inertial navigation systems (INSs), which would print a small data matrix on the picture just taken with the position and altitude of the aircraft at the time of capturing the image, as well as the time. This improved the analysis significantly.[119]

The previously mentioned *Lightning Bug* drones were extensively used for reconnaissance missions during the Vietnam War. From 1965 to 1972, Lightning Bug drones flew 3,450 sorties carrying out IMINT and ELINT collection, in addition to decoy and leaflet-drop missions.[120] The *Lightning Bugs* had limited range and were vulnerable to enemy air defences, but the drones' overall success gave significant momentum to find unmanned systems that could fly higher, faster, and farther.

During the Vietnam War, tactical reconnaissance units perfected their *modus operandi* not only by flying fast and low, but also in the manner they executed their missions in concert with other aircraft. Tactical reconnaissance aircraft were integrated parts of strike packages, both performing pre-strike reconnaissance and providing updates on targets to the strike aircraft, and then performing reconnaissance for BDA immediately after the strike. One common tactic was to fly fast at low level towards the target, then at 5 nautical miles prior to the target ascend quickly, with afterburner, to 15,000 feet for optimal pictures of the target. Then quickly descend to low level and return to home base at high speed. It would also be increasingly common for tactical reconnaissance aircraft to carry electronic countermeasures (ECM) equipment, which greatly increased their survivability. But operating the jamming equipment often rendered the missile warning system ineffective, necessitating operating with several other aircraft at a time, each functioning in separate roles.[121]

Intelligence gathered through the use of satellites was considered strategic in nature and does not seem to have been released to tactical forces during the Vietnam War. The fear was that dissemination of intelligence stemming from satellites would be compromised and thus reveal strategic intelligence capabilities.[122] In the years following the Vietnam War, some of these restrictions would be removed.

The 1960s, 1970s, and 1980s

Apart from and after the Vietnam War, an overarching strategic focus shaped the daily routines of the aerial surveillance and reconnaissance community, with periodic anomalies of tactical support during short crisis, episodes, and wars. One example of such an incident is the paramount importance of intelligence on Egyptian armed forces ahead of the deadly effective strikes by Israeli air force fighters in the Six-Day War in 1967. The success of the Israeli air strikes rested on a mixture of strategic and tactical intelligence, resulting in the Israeli Air Force neutralizing three-quarters of the Egyptian Air Force within the first couple of hours of the conflict. The Israeli use of unmanned ISR assets for fingerprinting hostile SAM systems was crucial in this context.[123]

American aircraft developers had no illusions about the fast-evolving Soviet air defence technology. This led the developers of the U-2 to initiate a project that would improve the flight characteristics of their revered strategic reconnaissance aircraft, this time to the extreme. The so-called "flight envelope" of what was to become the A-12 was an initial flight altitude of 84,500 feet, climbing to 97,600 feet at the end of the sortie. The aircraft would be capable of flying at Mach 3.2 – more than three times the speed of sound.[124] The A-12 had a powerful camera, ELINT systems, and a jamming system for incoming missiles. The CIA intended to fly the A-12 over mainland China, but the very capable SA-2 air defence system in North Vietnam was deemed to pose too much of a threat to several of the U-2 missions and *Lightning Bug* and *Fire Fly* drone missions, and the A-12 was brought in to fly in their place. The first operational mission was flown in May 1967, providing crucial intelligence on North Vietnamese air defence systems, the main threat to American aircraft. The wet film from the aircraft was first flown by a courier aircraft to the camera factory in Rochester, New York, for processing and then on to Washington for photo interpretation and analysis.[125] The CIA-operated A-12 programme was shut down only a year later due to costs. The SR-71 *Blackbird* was a slightly larger aircraft of similar shape and capabilities and took over from where the A-12 left off. The first operational SR-71 mission took place in March 1968, and the sortie provided a wide array of intelligence products from the aircraft's multi-mode radar and several cameras. A year later, the aircraft was equipped with automatic ELINT receivers. The US Air Force operated the SR-71s and deployed larger intelligence units together with their aircraft. Initially, the sensor products gathered in Asia were flown from the *Blackbirds'* home-base in Japan to Hawaii for processing. This process would normally take approximately 40 h. With more automated systems on base, the intelligence units were able to produce reports within 11 h, and provide Washington with the most pressing, oral reports within a few hours of landing.[126]

With authorization for manned, deep-penetrating overflights removed, efforts began in the early 1960s to develop capable drones for strategic reconnaissance, in parallel to the manned aircraft such as the A-12 and the SR-71. The D-21 *Tagboard* was one such drone. It would be carried on the back of a launch aircraft, and after launch the drone would accelerate to Mach 3.5–4 and cruise at 95,000–100,000 feet and perform its mission. The D-21 would then return to a planned recovery area and, similar to the recently deployed satellites, the drone would release a capsule containing the photographic film in a parachute to be retrieved by other units in mid-air. The drone would then continue its descent and self-destruct when given a signal by the barometric system.[127] The first flight over mainland China occurred in late 1969, but due to malfunctions and operational failures on all four operational flights executed (none of the films were recovered), the D-21 missions were cancelled in 1972, right before President Nixon's visit to China.[128] An improved version of the *Lightning Bug* was made operational for overflight missions in the early 1970s, but the requirement for its services evaporated with Nixon's diplomatic missions to China. Also, the constant development and improvements in satellite technology now allowed non-provoking reconnaissance over other nations and thus removed the requirement for overflights by aircraft. The programme for the improved *Lightning Bugs* was cancelled due to the lack of operational requirements, and the aircraft were destroyed in order to assure the denial of the technology to other nations in 1973.[129] The experiences with the *Lightning Bug* and the *Tagboard* would, however, serve as a foundation for the efforts that would lead to the modern unmanned aircraft that we see today. The early 1970s also saw improvements being made to the U-2 fleet, putting new sensors and the data link equipment used on the *Lightning Bugs* onto the already capable manned reconnaissance aircraft. This put a significant brake on the development of unmanned reconnaissance aircraft in the 1970s. It did not help for the advocates of unmanned systems that capability planners knew that drones would not be allowed to fly in controlled airspace together with civilian airliners the way that manned reconnaissance aircraft were able to. The very fast reconnaissance aircraft SR-71 was originally meant to take over from the U-2, but as the U-2 proved cost-efficient and scalable to support more and new sensor suites, they were both kept in the inventory for some time. Due to its speed, the SR-71 could cover much ground in a single mission and fly through or past a high-threat environment in a very short amount of time. But in spite of these extreme performance parameters and extreme cost of operating, the SR-71 did not have data links or a sensor downlink installed. This meant that the aircraft had to land and have her photographic images developed and processed at an appropriate analysis facility.[130] Declassified documents have revealed that the SR-71 flew missions all around the world, to include the High North and the Barents Sea, gathering intelligence on the Soviet Northern Fleet on the Kola Peninsula.[131] By the middle of the 1960s, in terms of strategic surveillance and reconnaissance, there were satellites monitoring restricted areas deep inside sovereign countries; there were SR-71s performing quick passes of disputed areas, and there were U-2s conducting stand-off intelligence collection. There was little need for exotic drones for strategic reconnaissance, and few, if any, reconnaissance drones were developed through the late 1970s and early 1980s that became fully operational.[132] In the early 1980s, concerns arose with respect to new and mobile Russian launchers of ICBMs. The *QUARTZ* drone programme was meant to be able to fly behind enemy lines into a high-threat environment in order to locate and keep track of these mobile launchers. Few details about the drone programme are known, except that it ran until it was cancelled in 1989, without producing any operational aircraft.[133] According to available unclassified documentation, the SR-71 never overflew foreign, sovereign territories on reconnaissance missions without approval of the country being overflown. The SR-71 was eventually cancelled in the fall of 1990, much to the dismay of the operational planners of the Gulf War that came only months later.[134]

In the late 1970s, both US Army and US Air Force initiated projects to meet their requirements for airborne sensors to locate movements on the ground, so-called moving target indicators (MTIs). In 1982, the programmes were consolidated, and throughout the 1980s avantgarde radar technology was further developed, and culminated in the joint surveillance target attack radar system (J-STARS). Built on a Boeing 707 airframe, the J-STARS first saw operational action in Operation Desert Storm in Iraq in 1991. J-STARS utilizes MTI technology to detect moving targets and synthetic aperture radar technology to provide image maps and location of static targets. In addition to encrypted voice communication with ground units, the J-STARS is able to transmit complete MTI and synthetic aperture radar data over the surveillance and control data link (SCDL) and can transmit situational awareness data that have been processed on board in support of the tactical picture over the joint tactical information distribution system (JTIDS).

These systems that materialized in the late 1970s must be seen in the context of the American *second Offset Strategy*. A strategic push for high-tech, avantgarde systems existed for an unprecedented real-time integration of information on ground, air, and naval units, which, among other things, manifested itself through the merging of data links, with ISR units providing these systems with targeting information and updates on enemy positions. An operational "reconnaissance-strike complex" would enable the Americans to "see deep" and "strike deep."[135] A crucial piece of technology that was to significantly shape future air combat was thus the further development of a tactical data link. Building on important lessons from the Vietnam War with the *Combat Lightning* system, the JTIDS project was initiated in the mid-1970s. Further developed in the 1980s, the system provides jamming-resistant communications and secure sharing of tactical information, built on known enemy information as well as real-time updates.[136] It provides a "God's eye view" of the battlefield and portrays to the operator a combination of data derived from on-board and off-board the aircraft.[137] The technology used would provide a backbone architecture for tactical data links that have been used well into the 2000s and are still in use today. The success of the air units in *Operation Desert Storm* in Iraq in 1991 was largely due to this new technology, enabling more or less real-time sharing of battlefield information among friendly units.

The *CORONA* imagery satellite programme was originally only intended to last for 2 years, but it ran for a decade. The wide-area CORONA search satellites were replaced by one of the largest satellites ever built, the *HEXAGON* (later known to the American public as "*Big Bird*"), which was 60-feet long, had a diameter of 10-feet, and weighed 30,000 pounds. The HEXAGON could cover several thousand square miles and capture high-resolution mapping imagery. The first launch took place on 15 June 1971, with the first film capsule returning for analysis 5 days later.[138] The HEXAGON satellites were able to conduct surveillance of 80%–90% of the Sino-Soviet landmass in a single mission.[139] Due to both the immense costs involved and the time required to get the films back to earth and have them processed and analysed, both *GAMBIT* and *HEXAGON* programmes were cancelled in the 1980s in favour of near-real-time imaging satellites.[140] Imaging satellites had an impact on military planning and operations since the day they were launched and delivered clear advantages by the late 1970s. Surveillance satellites covered strategically important missile sites and naval bases that were hard to reach by surface and air units, such as shipyards on the Kola Peninsula.[141] Particularly, satellites were able to follow the construction of large vessels, and whether ships were in their designated places in military ports. Significant interest was shown by both superpowers in ASW conducted from space. Necessarily, this would entail non-acoustic sensors, such as radars, imaging, and ELINT. There were no real breakthroughs in sensor experimentation, and satellites have been mostly associated with some imaging for ASW but mostly functioned as crucial nodes in a network-centric approach to warfare. Continuous presence of a communications satellite would facilitate quick – and later real-time – dissemination of tactical information, including for ASW units at sea communicating with shore-based units.[142] An important development within the sphere of satellite surveillance, particularly in the United States, is the making of satellite intelligence increasingly available to tactical units starting in the late 1970s under a series of Tactical Exploitation of National Capabilities (TENCAP) programmes. This has clearly been to the benefit of tactical units, but satellite pictures were utilized during the failed hostage rescue attempt in Iran in 1980, and pictures were left in the desert after the evacuation of the forces. This revealed the capabilities of the KH-11 satellite to the Iranians.[143]

As will be discussed in the coming chapters, Western maritime air units experienced much success in tracking Soviet submarines and other naval units during most of the Cold War. MPA flew peripheral ISR missions along the Soviet border, and conducted ASW missions in the open ocean in concert with static acoustic systems on the seabed, ASW submarines, SURTASS ships, surface forces, and other ASW units. By the 1980s, the Soviet Union had initiated programmes to quieten their submarines, and the Western acoustic advantage over Russian submarines was diminishing. Still, throughout the Cold War, the MPA force was the preferred tool for airborne intelligence gathering in the maritime domain and keeping pressure on Soviet submarines on deployment.

The 1990s to the mid-2010s

The Gulf War of 1991 was such a landmark in modern military warfare that even the Chinese describe admiringly in their doctrinal writings the way the American-led coalition fought.[144] After decades of technological development with regard to satellite communications, satellite navigation, satellite observation, data link to connect the forces, and improved sensors for surveillance and reconnaissance in general, it was a formidable force that expelled the Iraqis from Kuwait. In addition to supporting other types of technology and data fusion, satellites also played an important role in intelligence gathering for tactical units during the Gulf War. Operations in 1991 were supported by satellites with side-looking imaging radars, by capable military photographic satellites, as well as by civilian imaging satellites with lesser resolution.[145] The JTIDS data link was in many ways able to bring the legacy of the *Teaball* project from the Vietnam War into a well-functioning integration of tactical information. Airborne COMINT platforms sanitized their information and translated it into language readable by the data link; airborne radar information was integrated in the same way, as were the ground radar tracks of enemy movements. All this information was integrated with information on friendly aircraft, which gave the airborne units an unprecedented situational awareness for their respective missions.[146] The 1991 Gulf War was a milestone in near-real-time intelligence fusion for tactical air units. The data links that were utilized cemented an enduring tenet of airborne intelligence: Integrating airborne intelligence assets into the joint force, thereby facilitating increased speed of dissemination of the sensor products. Unmanned aircraft was also extensively used, such as the *Pioneer* drone flown from US Navy ships in support of artillery spotting and BDA.[147] The speed of operations during the Gulf War also facilitated fundamental discussions and doctrinal change as to the level of control of surveillance and reconnaissance assets that could and should be given from the strategic level of command and control. "Real-time intelligence" was now more or less a reality, which led to discussions as to the necessity of channelling sensor products through the chain-of-command before it could be fed to the tactical war-fighter.[148] The technological possibilities that came with the modern weapons of the early 1990s seem to have initiated a demand for speed of delivery of intelligence products that have continued to this day. The military commander demands as much information as possible, in real time. This facet of modern operations will be discussed further in the theoretical framework to follow.

Technological development continued unabated after the Gulf War, with sensors and data links improving alongside improvements in data technology in general. Many aircraft were upgraded with modern sensors, modern data links, and modern communication suites, as opposed to being replaced by an entirely new aircraft. The workhorses in Western aerial surveillance and reconnaissance, the RC-135, the J-STARS, the U-2, the EP-3, the P-3, and of course an increasing number of intelligence satellites, were upgraded and not necessarily replaced. The arena that really experienced a revolution was the field of unmanned aircraft.

There seems to be a clear link between actual, tangible operational requirements, and actual investments into servicing those requirements. In most of the Cold War, when the requirements for airborne intelligence were mostly strategic in nature, it was mostly strategic intelligence aircraft that were built. Then, as the Cold War ended and warfare, in a Western sense, turned more expeditionary in nature, different requirements came to the fore: An increased demand for immediate support to tactical ground units, demands for less casualties for both civilians and friendly forces, and a focus on limitations in time, space, and scope for any conflict. With the Gulf War in 1991 setting precedence in low numbers in casualties in modern warfare, and more and more conflicts involving increasing aspects of asymmetric warfare, these new operational aspects led to new operational requirements. Western powers were less inclined to take any casualties, and tactical persistence in surveillance was needed for airborne ISR integration into counter-insurgency (COIN) operations.[149] Together with improved satellite communications technology, the stage was thus set for an explosion in the development of unmanned airborne ISR systems.

The 1990s saw heavy utilization of airborne surveillance and reconnaissance assets, particularly during the conflicts in the Balkans. Airborne surveillance assets played a key role in the support of planning, they were integrated to a high degree in the execution of operations, and they played a key role in combat assessment phases of the air campaign over Croatia, Bosnia and Herzegovina, and Kosovo.[150] The potential of unmanned aerial vehicles was recognized long before the end of the Cold War but started to gain proper momentum with maturing technologies in the early 1990s.

Ever since the first shoot-down, the U-2 programme had continually faced the threat of cancellation. However, the aircraft was able to support both strategic intelligence gathering as well as support to tactical operations, and this flexibility kept the U-2 programme alive. As mentioned, there was a significant push towards development of drones in the 1960s and early 1970s, as can be seen by the resources spent on the drones in the Vietnam War, for instance. However, with other capable, manned aircraft available, the momentum withered away. With the SR-71 retired, and new demands for intelligence support to the battlefield being manifested through Gulf War I, studies were undertaken in the early 1990s that eventually led to the development of the *Global Hawk* drone. Prior experience with programmes such as the *Firebee*, *Lightning Bug*, *Compass Arrow*, and *Compass Cope* unmanned systems informed the development process.[151] Development continued throughout the 1990s, and it was *prototypes* that were eventually sent to Afghanistan and Iraq in the early 2000s to support the Middle Eastern campaigns of the new millennium. Nearly being cancelled several times, the *Global Hawk* finally evolved into an operational capability, albeit being under continuous pressure due to the immense costs associated with the programme.[152] The development of the *Predator* drone had been going on for several decades, but it was in the mid-1990s that the technology was mature enough to come together and materialize into an operational capability. The *Predator* drone was first used for surveillance over Kosovo in the late 1990s, and the operational experiences over the Balkans served as a foundation for its more (in)famous use in the aftermath of 9/11.[153]

As will be discussed later in this book, the demise of the Soviet Union led to a steep decline in Russian naval deployments outside coastal waters. This led to much less operational activity for Western maritime ISR unites and consequentially to the questioning of the *raison d'etre* of many maritime units, particularly MPA forces. At first, operational focus moved from blue-water, open-ocean missions to near-shore, littoral operations. Over time, the traditionally *maritime* units were also being utilized *over land* during the Balkan conflicts in the late 1990s and the Middle Eastern campaigns of the early 2000s, particularly for SIGINT and IMINT.[154] As Russia again increased her naval posture both through more outward looking and proactive strategies as well as through operational deployments, the traditional MPA mission of ASW and surface surveillance was revitalized in the 2010s.

The utilization of manned MPA for overland IMINT missions in Afghanistan and Iraq was hardly cost-efficient. The momentum for developing unmanned aircraft for ISR missions only increased with Western rediscovery of COIN operations and associated requirements for up-to-date tactical intelligence. In addition to unmanned assets, the use of very small, manned aircraft with capable sensors was permissible due to the general low surface-to-air threat during COIN operations. One example is the *King Air* aircraft, which is a small, two-propeller aircraft with two pilots, two sensor operators, and carries IMINT and SIGINT sensors similar to many MPA.[155] The philosophy behind the *King Air* is clearly to utilize existing aircraft with new and improved sensors, creating a system that is more than likely to be capable of servicing most of the specific OPINTEL requirements, if not all of them. Compared to unmanned aircraft of similar size, it is significantly lacking in endurance capacities, but it is still a very capable aircraft.[156] The alternative is to develop a new design tailored for this specific mission, a process associated with significant costs.

Yet another development that was facilitated by the low surface-to-air threat environment and the type of warfare being conducted in the early 2000s was the re-emergence of tethered balloons in Afghanistan and Iraq. These static, aerial ISR assets have been named *aerostats*. Western forces are mounting capable EO/IR sensors on tethered balloons that function as sentinels around bases for the armed forces, providing persistent surveillance in support of force protection.[157] Yet another resurrection is an advanced balloon that is not tethered, but is able to navigate using winds and changes in altitude. Modern computers can achieve relatively accurate navigation by performing altitude changes based on modelled and observed winds. If stationed at altitudes between 60,000 and 120,000 feet, the balloons would generally be out of reach of most SAM systems and adversary aircraft.[158] With sufficient volume for lift, the balloons will be able to carry significant amount of ISR equipment, enabling them to conduct very long-endurance ISR missions. As modern technology evolves, we should expect to see more and more of the old ideas fielded a century ago in the name of cost-efficiency.

In the mid-1990s there was a fundamental shift in dynamics in the space domain. This shift is by many termed the shift from "old space" to "new space." Old space refers to the government-led, large, expensive space and satellite programmes, concerning capabilities that could only be afforded by very rich nations (such as the United States), or by nations that were willing to pay the immense costs associated with operating in space, even if it meant financial ruin in the long run (such as the Soviet Union). There took place an increase in collaboration between governmental agencies on both sides of the former Iron Curtain, in addition to a significant increase in commercial interest in an emerging market. Space-specific, military technologies from among others the recently disestablished Soviet Union were made available on the world market, facilitating the dawn of the new, fundamentally *commercially* driven space age. With miniaturization of technology, higher processing capacities, more cost-efficient, faster and better production methods, and much more affordable launch services, a new sphere of commercial space capabilities has taken off in full. The development over the past two decades has led to the emergence of a wide range of commercial companies that provide space-based services. The company *Planet*, for example, has approximately 200 microsatellites in a constellation that can offer images twice daily of a given position on earth, with a resolution of 72 cm.[159] This has further led to militaries around the world looking to the commercial sector for potentially satisfying a large part of the national intelligence requirement. Additionally, the new technology has opened the door for smaller nations, either alone or in partnership with other nations, to build or acquire small satellites that are launched in order to meet intelligence requirements. Suffice it to say, the *New Space* environment has facilitated, is facilitating, and will in the future facilitate intelligence services that, 50 years ago, were only accessible to immensely costly space programmes belonging to space faring nations. We should expect to see space being utilized even more in the future with regard to surveillance and reconnaissance.

A study of the development of aerial ISR highlights fundamental aspects of ISR integration into operations, such as communication challenges between supported and supporting units, command and control architecture for unmanned systems, and flexible use of scarce satellite bandwidth. The Balkan conflicts of the 1990s are also regarded as the starting point of efforts for proper ISR asset integration and coordination, initial use of modern link systems for airborne assets, and a general increase in investments in ISR technology, linked to the new operational requirements, as mentioned above.[160] Thus, the term "ISR" became synonymous with *intelligence support for operations* in the late 1990s and early 2000s. Healthy discussions on efficient tasking of airborne ISR assets, increased flexibility in their utilization, and the optimal use of these assets are ongoing. Also, there are significant changes in materiel inventories afoot in several Western nations that will shape the future of airborne surveillance and reconnaissance, to a significant degree, to include the maritime domain. The introduction of the F-35 *Lightning II* fifth-generation fighter aircraft, the fielding of the P-8 *Poseidon* MPA, and the growing procurement of the unmanned maritime ISR aircraft MQ-4 *Triton*, the maritime derivative of the *Global Hawk*, stand out as some of the most prominent.

Enduring tenets of airborne ISR

The evolution of airborne ISR is filled with avantgarde technology, flight envelopes of aircraft pushed to the limit, and ingenious methods of disseminating sensor products to the force. The details are many, so are the anecdotes, but there are some – a few – timeless elements that stand out as enduring factors since the first ISR mission until today. These fundamental aspects will likely never cease to be crucial for conducting ISR missions by airborne units.

The first is the aspect sought by man for improved observation in the first place: *Altitude*. In times of war, it is a crucial aspect for not only getting a better view but also avoiding surface threats. In peacetime, it is necessary to get the best perspective from afar, whilst remaining outside adversary sovereign airspace.

Second: *Speed*. When flying low, this is essential for the element of surprise. When flying high, it is an important aspect of evading air threats. It is also an important factor for responsiveness to events that must be monitored. But it is also an aspect that can be downplayed when operating in low-threat environments.

Third: *Range*. This has played a crucial role in reaching areas beyond enemy lines, as well as for conducting ISR in areas not necessarily supported by friendly support bases, or when flying in particularly sensitive areas, such as the High North. The elements of height, speed, and range are also considered to be the basic tenets of airpower. And then come the more ISR-specific fundamentals.

Fourth: *Survivability*. In the 1930s doctrinal writings, particularly in the West, altitude and speed were viewed as the requirements for evading threats. This was also the rationale behind the SR-71. But as bigger aircraft are bearing the brunt of the day-to-day ISR missions (aircraft such as the P-3, EP-3. J-STARS, RC-135, U-2), it is crucial that these are able to either defend themselves or that they be used in lower-threat environments. Operational planners can also place fighter aircraft in defence positions in order for the ISR aircraft to perform their missions with top cover.

Fifth: *Communications*. Quite obviously, this is crucial to the relevance of the sensor data and the purpose that these can be used for. Must the camera roll be developed and analysed by strategic specialists, or can the live footage be transmitted in real time to tactical ground forces?

Sixth: *Sensors*. Sensors are on a span of complexity and scope, from high-resolution cameras to low-resolution radars, from in-flight linguists interpreting tactical communication to post-flight analysis of telemetry between the firing units and the cruise missile. The sensor suite is fundamental to the type of intelligence that can be delivered.

This has provided but a short portrayal of an incredible evolution in technology and a specific type of military operations. The proliferation and development of miniaturized elements of surveillance technology has led to unmanned aircraft and satellite surveillance and reconnaissance being integrated elements of modern military operations. These two types of surveillance and reconnaissance stand out as likely the most crucial elements of future ISR constellations. The following chapter will break down the specific elements of aerial ISR so as to provide a framework for the discussion on future requirements. The historical portrayal indirectly distinguishes the building blocks of airborne ISR and facilitates a discussion on the effects that airborne ISR can and should support.

Notes

1 Robert Jackson, *High Cold War – Strategic Air Reconnaissance and the Electronic Intelligence War* (Sparkford: Patrick Stephens Ltd., 1998), 11; Norman Polmar and John F. Bessette, *Spyplanes – The Illustrated Guide to Manned Reconnaissance and Surveillance Aircraft from World War I to Today* (Minneapolis: Quarto Publishing Group Inc, 2016), 9.

2 Tyler Morton, "Ears and Eyes in the Sky: The Evolution of Manned Airborne ISR" (Master's thesis, School of Advanced Air and Space Studies, 2012), 12, www.dtic.mil/get-tr-doc/pdf?AD=AD1019401.

3 Paul Lashmar, *Spy Flights of the Cold War* (Phoenix Mill: Sutton Publishing Ltd., 1996), 8; Polmar and Bessette, *Spyplanes*, 9; Morton, "Ears and Eyes in the Sky," 18.

4 Edgar F. Jr. Raines, *Eyes of Artillery – The Origins of Modern U.S. Army Aviation in World War II* (Honolulu: University Press of the Pacific, 2002), 5; Jackson, *High Cold War*, 11.

5 Morton, "Ears and Eyes in the Sky," 20.
6 Lashmar, *Spy Flights of the Cold War*, 8–9; Jackson, *High Cold War*, 11.
7 Polmar and Bessette, *Spyplanes*, 9; Raines, *Eyes of Artillery*, 5.
8 Raines, *Eyes of Artillery*, 7–8.
9 Lashmar, *Spy Flights of the Cold War*, 9.
10 Morton, "Ears and Eyes in the Sky," 29–31.
11 Raines, *Eyes of Artillery*, 8.
12 Morton, "Ears and Eyes in the Sky," 12.
13 Polmar and Bessette, *Spyplanes*, 10.
14 Lashmar, *Spy Flights of the Cold War*, 9.
15 Polmar and Bessette, *Spyplanes*, 10.
16 Polmar and Bessette, 9.
17 Raines, *Eyes of Artillery*, 10.
18 Raines, 10–11; Morton, "Ears and Eyes in the Sky," 48.
19 Lashmar, *Spy Flights of the Cold War*, 9; Jackson, *High Cold War*, 11.
20 Eric Lawson and Jane Lawson, *The First Air Campaign* (Conshoshocken, PA: Combined Books, Inc., 1996), 39.
21 Basil Collier, *A History of Air Power* (Oxford: Macmillan Publishing Co., 1974), 50.
22 Morton, "Ears and Eyes in the Sky," 46–48.
23 Polmar and Bessette, *Spyplanes*, 10.
24 Terrence J. Finnegan, *Shooting the Front* (Stroud: History Press, 2011), 13.
25 Morton, "Ears and Eyes in the Sky," 49.
26 Collier, *A History of Air Power*, 52.
27 Morton, "Ears and Eyes in the Sky," 49.
28 Lashmar, *Spy Flights of the Cold War*, 10.
29 Jackson, *High Cold War*, 16–17; Polmar and Bessette, *Spyplanes*, 15.
30 Douglas H. Robinson, *Giants in the Sky* (Seattle, WA: University of Washington Press, 1973), 91.
31 Morton, "Ears and Eyes in the Sky," 51–52.
32 Chris Ashworth, *Avro's Maritime Heavyweight: The Shackleton* (Bourne End: Aston Publications Ltd., 1990), 8.
33 David Owen, *Anti-Submarine Warfare – An Illustrated History* (Annapolis, MD: Naval Institute Press, 2007), 41. As this book will show, more battles of the Atlantic were to follow, making this the "first". The second came in Word War II, and historians of ASW have termed the Cold War search for Soviet submarines on patrol in the Atlantic as the "Third Battle of the Atlantic". Now, some are discussing the re-emergence of Russian submarines in the 2000s as the run-up to the "Fourth Battle of the Atlantic", see for example James Foggo and Alanik Fritz, "The Fourth Battle of the Atlantic," *United States Naval Institute Proceedings* 142, no. 6 (June 2016).
34 Jackson, *High Cold War*, 11; Polmar and Bessette, *Spyplanes*, 15.
35 Doug Gordon, *Tactical Reconnaissance in the Cold War* (Barnsley: Pen & Sword Aviation, 2006), 9.
36 Raines, *Eyes of Artillery*, 14.
37 Raines, 18–20; Allan Williams, *Operation Crossbow – The Untold Story of the Search for Hitler's Secret Weapons* (London: Arrow Books/Random House, 2014), 27; Jackson, *High Cold War*, 12.
38 Raines, *Eyes of Artillery*, 17.
39 Raines, 18.
40 Morton, "Ears and Eyes in the Sky," 63.
41 Raines, *Eyes of Artillery*, 20; Morton, "Ears and Eyes in the Sky," 63.
42 Williams, *Operation Crossbow*, 14–16; Lashmar, *Spy Flights of the Cold War*, 10.
43 Polmar and Bessette, *Spyplanes*, 16–17; Jackson, *High Cold War*, 12, 16–17.
44 Williams, *Operation Crossbow*, 10; Jackson, *High Cold War*, 12.

45 Morton, "Ears and Eyes in the Sky," 65–66.
46 Polmar and Bessette, *Spyplanes*, 17; Williams, *Operation Crossbow*, 16–17.
47 Lashmar, *Spy Flights of the Cold War*, 11–12.
48 Jackson, *High Cold War*, 14–15.
49 Jackson, *High Cold War*, 15; Williams, *Operation Crossbow*, 30–31.
50 Jackson, *High Cold War*, 17; Polmar and Bessette, *Spyplanes*, 19–21.
51 Jackson, *High Cold War*, 19.
52 Ashworth, *Avro's Maritime Heavyweight*, 8.
53 Raines, *Eyes of Artillery*, 19, 28.
54 Polmar and Bessette, *Spyplanes*, 20–21.
55 Polmar and Bessette, 19–22.
56 Jackson, *High Cold War*, 15–16.
57 Lashmar, *Spy Flights of the Cold War*, 12–13.
58 Denied areas in terms of airpower are element of airspace that pose a significant threat to aircraft of being shot down, or where the electromagnetic spectrum is being manipulated and thus denying normal operations for entering aircraft. This will be discussed further in later chapters.
59 Jackson, *High Cold War*, 16.
60 Håvard Klevberg, "Maritime Surveillance in the North: 333 Squadron in Norwegian Security Politics" (PhD, History, Oslo, University of Oslo, 2011), 129–30.
61 Morton, "Ears and Eyes in the Sky," 68–69.
62 Williams, *Operation Crossbow*, 55, 81, 84.
63 For an extensive account of the Allied search for Hitler's V-1 and V-2 weapons, see Williams, *Operation Crossbow*.
64 Jackson, *High Cold War*, 20–30.
65 Polmar and Bessette, *Spyplanes*, 26.
66 Jackson, *High Cold War*, 20.
67 Lashmar, *Spy Flights of the Cold War*, 17–19; Jackson, *High Cold War*, 30.
68 Jackson, *High Cold War*, 32.
69 Jackson, 35.
70 Morton, "Ears and Eyes in the Sky," 80; Jackson, *High Cold War*, 28.
71 Larry Tart and Robert Keefe, *The Price of Vigilance – Attacks on American Surveillance Flights* (New York: Ballantine Books, 2001), 164–76; Morton, "Ears and Eyes in the Sky," 82.
72 Morton, "Ears and Eyes in the Sky," 82–86.
73 Owen R. Jr. Cote, "The Third Battle – Innovation in the U.S. Navy's Silent War Struggle with Soviet Submarines," *Naval War College Newport Papers* 16 (2003): 10–11.
74 Michael Warner, "The Collapse of Intelligence Support for Air Power, 1944–52," *Studies in Intelligence* 49, no. 3 (2005): 3; Lashmar, *Spy Flights of the Cold War*, 20–21; Gordon, *Tactical Reconnaissance in the Cold War*, 10.
75 See also John T. Furquhar, *A Need to Know: The Role of Air Force Reconnaissance in War Planning, 1945–1953* (Maxwell AFB, Montgomery, Alabama: Air University Press, 2005), 133–36.
76 Robert S. III Hopkins, *Spyflights and Overflights – US Strategic Aerial Reconnaissance – Volume I, 1945–1960* (Manchester: Hikoki Publications, 2016), 28.
77 Furquhar, *A Need to Know*, 32–37.
78 Norman Polmar and Thomas B. Allen, *Spy Book: The Encyclopedia of Espionage* (New York: Random House, 1998), 12.
79 Morton, "Ears and Eyes in the Sky," 98.
80 A prominent example is the project that led to the development of the CORONA imaging satellite. For a detailed account, see Albert D. Wheelon, "Corona – A Triumph of American Technology," in *Eye in the Sky – The Story of the CORONA Spy Satellites* (Washington, DC: Smithsonian Institution Press, 1998).

81 Lashmar, *Spy Flights of the Cold War*, 34–42.
82 Furquhar, *A Need to Know*, 133–36; Morton, "Ears and Eyes in the Sky," 123–24.
83 Polmar and Allen, *Spy Book*, 12; Furquhar, *A Need to Know*, 133.
84 Gordon, *Tactical Reconnaissance in the Cold War*, 25–27.
85 Morton, "Ears and Eyes in the Sky," 124–25.
86 Furquhar, *A Need to Know*, 134–36; Morton, "Ears and Eyes in the Sky," 126–27.
87 Morton, "Ears and Eyes in the Sky," 128–29; Gordon, *Tactical Reconnaissance in the Cold War*, 21.
88 Tart and Keefe, *The Price of Vigilance*, 195–96; Morton, "Ears and Eyes in the Sky," 130–33. For a discussion on the kill ratio of American fighter aircraft in Korea see Robert Coram, *Boyd – The Fighter Pilot Who Changed the Art of War*, reprint ed. (New York: Back Bay Books, 2004).
89 Morton, "Ears and Eyes in the Sky," 132–36.
90 Gordon, *Tactical Reconnaissance in the Cold War*, 184.
91 Gordon, 44–73.
92 Philip Taubman, *Secret Empire – Eisenhower, the CIA, and the Hidden Story of America's Space Espionage* (New York: Simon and Schuster, 2003), 53–56; Jackson, *High Cold War*, 97; Williams, *Operation Crossbow*, 349.
93 Dwayne A. Day, John M. Logsdon, and Brian Latell, eds., *Eye in the Sky – The Story of the CORONA Spy Satellites* (Washington, DC: Smithsonian Institution Press, 1998), 90–92; Lashmar, *Spy Flights of the Cold War*, 40.
94 For an account of this development, see Taubman, *Secret Empire*; Wheelon, "Corona – A Triumph of American Technology," 29–32.
95 Western powers continued to use aircraft that were fielded during World War II in the post-war decade. But as air defences developed, higher requirements for survival were put against existing aircraft, and a spiral of weapons development had the development of aircraft and air defences attempting to outpace each other. Several aircraft deserve mentioning in this context, but are not included in the scope of this discussion. For further details see for example Hopkins, *Spyflights and Overflights*.
96 Tart and Keefe, *The Price of Vigilance*, 139–41.
97 Hopkins, *Spyflights and Overflights*, 37; Mason Redfearn and Richard J. Aldrich, "The Perfect Cover: British Intelligence, the Soviet Fleet, and Distant Water Trawler Operations, 1963–1974," *Intelligence and National Security* 12, no. 3 (1997): 166–77; Polmar and Bessette, *Spyplanes*, 34–45.
98 Rick Thomas, *Global Hawk – The Story and Shadow of America's Controversial Drone* (Great Britain: Amazon, 2015), 33–34.
99 Day, Logsdon, and Latell, *Eye in the Sky*, 31–39.
100 Gerald Haines, *Critical to US Security: The Gambit and Hexagon Satellite Reconnaissance Systems Compendium* (Washington, DC: Center for the Study of National Reconnaissance, January 2012), 2.
101 Day, Logsdon, and Latell, *Eye in the Sky*, 36, 51.
102 Pat Norris, *Spies in the Sky – Surveillance Satellites in War and Peace* (Chichester, UK: Praxis Publishing Ltd, 2008), 68.
103 Robert Perry, *A History of Satellite Reconnaissance*, vol. 1 (National Reconnaissance Office (NRO), October 1973), xx, ee.
104 Haines, *Critical to US Security*, 2–14.
105 Morton, "Ears and Eyes in the Sky," 112–13. Early versions actually were fitted with imagery equipment (IMINT), but as satellite technology matured the mission for photographic intelligence were increasingly taken over by satellites.
106 James P. Marshall, *Near-Real Time Intelligence on the Tactical Battlefield*, Research Report (Maxwell AFB, Montgomery, AL: Air University, January 1994), 16; Walter J. Boyne, "The Teaball Tactic," *Air Force Magazine*, July 2008, 69.
107 Dave Sloggett, *Drone Warfare – The Development of Unmanned Aerial Conflict* (New York: Skyhorse Publishing, 2015), 76–82.

108 Thomas, *Global Hawk*, 36–37.
109 Thomas, 37.
110 Gordon, *Tactical Reconnaissance in the Cold War*, 92, 203.
111 Gordon, 121.
112 Gordon, 132.
113 Tart and Keefe, *The Price of Vigilance*, 162–63.
114 Gordon, *Tactical Reconnaissance in the Cold War*, 214.
115 Morton, "Ears and Eyes in the Sky," 136–37.
116 Morton, 138–39.
117 William J. Nichols, *From Teaball to Fastball: The Value of Real-Time Intelligence Support to the Cockpit* (School of Advanced Air and Space Studies, May 12, 1992), 11; Doyle E. Larson, "Direct Intelligence Combat Support in Vietnam: Project Teaball," *American Intelligence Journal* 15, no. 1 (Spring–Summer 1994): 56; Boyne, "The Teaball Tactic," 67–70.
118 Michael Warner questions the causality of this, pointing to the fact that the North Vietnamese concurrently prioritized the use of missiles in-stead of sending fighter aircraft to fight the air war. This, Warner implies, led to a safer environment for friendly aircraft that could stay safe by staying out of reach of the missiles. See Michael Warner, "US Intelligence in Vietnam: The Official Version(s)," *Intelligence and National Security* 25, no. 5 (2010): 630.
119 Robert R. Powell, *RA-5C Vigilante Units in Combat* (Oxford: Osprey Publishing, 2004), 11–12.
120 Thomas, *Global Hawk*, 37–38.
121 Gordon, *Tactical Reconnaissance in the Cold War*, 230–37.
122 Norman Friedman, *Seapower and Space* (Annapolis, MD: Naval Institute Press, 2000), 110–11.
123 Elder, "Intelligence in War," 6; Tice, "Unmanned Aerial Vehicles," 43.
124 Paul F. Crickmore, *Lockheed Blackbird – Beyond the Secret Missions*, revised ed. (Oxford: Osprey Publishing, 2016), 30–47.
125 Crickmore, 104–24.
126 Crickmore, 268–69.
127 Crickmore, 130–31.
128 Polmar and Allen, *Spy Book*, 153–54; Thomas, *Global Hawk*, 37–39; Crickmore, *Lockheed Blackbird*, 132–39.
129 Thomas, *Global Hawk*, 42–43.
130 Thomas, 50.
131 Crickmore, *Lockheed Blackbird*, 305–7.
132 Thomas, *Global Hawk*, 45–46.
133 Thomas, 45–46.
134 Thomas, 50.
135 Luis Simon, "The 'Third' US Offset Strategy and Europe's 'Antiaccess' Challenge," *Journal of Strategic Studies* 39, no. 3 (2016): 425.
136 Morton, "Ears and Eyes in the Sky," 142–44.
137 Marshall, *Near-Real Time Intelligence on the Tactical Battlefield*, 25–26.
138 Haines, *Critical to US Security*, 22–31.
139 Haines, 36.
140 Haines, xv.
141 Friedman, *Seapower and Space*, 99–100.
142 Friedman, 205.
143 Friedman, 111, 333.
144 See for example Qiao Liang and Wang Xiangsui, *Unrestricted Warfare* (Beijing: PLA Literature and Arts Publishing House, 1999), 4–11.
145 Friedman, *Seapower and Space*, 98.
146 Marshall, *Near-Real Time Intelligence on the Tactical Battlefield*, 25–26.

147 Thomas, *Global Hawk*, 52–53.
148 Thomas, 58–59.
149 Gjert Lage Dyndal and John O. Birkeland, "Autonome Droner Og Våpensystemer: Endres Krigen Og Måten vi Fører Krig På?" in *Når Dronene Våkner*, ed. Tor Arne Berntsen, Gjert Lage Dyndal, and Sigrid Redse Johansen (Oslo: Cappelen Damm Akademiske, 2016), 39.
150 Kevin M. Coyne, "Developing US European Command's Intelligence, Surveillance, and Reconnaissance Strategy for Fiscal Year 2010 Through 2015," *Air & Space Power Journal* 24, no. 4 (Winter 2010): 82; David A. Deptula and Michael Fransisco, "Air Force ISR Operations – Hunting Versus Gathering," *Air & Space Power Journal* 24, no. 4 (Winter 2010): 14; Frank Strickland, "The Early Evolution of the Predator Drone," *Studies in Intelligence* 57, no. 1 (2013): 2; Robert P. Jr. Haffa and Anand Datla, "Joint Intelligence, Surveillance, and Reconnaissance in Contested Airspace," *Air & Space Power Journal* 30, no. 3 (Fall 2016): 31.
151 Thomas, *Global Hawk*, 80–81.
152 For a detailed account from the inside of the development process of the Global Hawk drone, see Thomas, *Global Hawk*.
153 For an extensive account of the development of the Predator drone, see Richard Whittle, *Predator – The Secret Origins of the Drone Revolution* (New York: Henry Holt and Company, 2014).
154 Birkeland, "The Potential of LIDAR as an Anti-Submarine Warfare Sensor," 54–55; Perkins, *Alliance Airborne Anti-Submarine Warfare*, 45.
155 Deptula and Fransisco, "Air Force ISR Operations," 14; Air Force Technology, "MC-12W Liberty Intelligence, Surveillance and Reconnaissance (ISR) Aircraft," *Airforce-Technology.Com*, accessed July 23, 2018, www.airforce-technology.com/projects/mc-liberty/.
156 The King Air can operate with an endurance of approximately 8 h, modern MPA can operate with between 10–12 h of endurance, and the Predator and Global Hawk unmanned aircraft can operate with between 24 and 30 h.
157 Graham Bowley, "Spy Balloons Become Part of the Afghanistan Landscape, Stirring Unease," *New York Times*, May 12, 2012, www.nytimes.com/2012/05/13/world/asia/in-afghanistan-spy-balloons-now-part-of-landscape.html; Lockheed Martin Corporation, "Always There: Always on," *Lockheedmartin.Com*, accessed July 23, 2018, www.lockheedmartin.com/en-us/products/unmanned-aerostats-airships-and-lighter-than-air-technology.html.
158 Collin Fox, "If It Floats, It Fights," *United States Naval Institute Proceedings* (June 2018): 46; Donna McKinney, "NRL Scientists Update Horizontal Wind Model with Eye Toward Next-Generation Space Weather Prediction Systems," July 10, 2015, www.nrl.navy.mil/news/releases/nrl-scientists-update-horizontal-wind-model-eye-toward-next-generation-space-weather.
159 Planet.com, "Planet Monitoring – Daily Cadence, Global Coverage," accessed August 1, 2018, www.planet.com/products/monitoring/.
160 Coyne, "Developing US European Command's Intelligence, Surveillance, and Reconnaissance Strategy for Fiscal Year 2010 Through 2015," 82–83.

Bibliography

Air Force Technology. "MC-12W Liberty Intelligence, Surveillance and Reconnaissance (ISR) Aircraft." *Airforce-Technology.Com*. Accessed July 23, 2018. www.airforce-technology.com/projects/mc-liberty/.

Ashworth, Chris. *Avro's Maritime Heavyweight: The Shackleton*. Bourne End: Aston Publications Ltd., 1990.

Birkeland, John O. "The Potential of LIDAR as an Anti-Submarine Warfare Sensor." M. Phil (R), War Studies, University of Glasgow, 2009. http://theses.gla.ac.uk/1252/1/2009birkelandmphil.pdf.
Bowley, Graham. "Spy Balloons Become Part of the Afghanistan Landscape, Stirring Unease." *New York Times*, May 12, 2012. www.nytimes.com/2012/05/13/world/asia/in-afghanistan-spy-balloons-now-part-of-landscape.html.
Boyne, Walter J. "The Teaball Tactic." *Air Force Magazine*, July 2008.
Collier, Basil. *A History of Air Power*. Oxford: Macmillan Publishing Co., 1974.
Coram, Robert. *Boyd – The Fighter Pilot Who Changed the Art of War*, reprint ed. New York: Back Bay Books, 2004.
Cote, Owen R. Jr. "The Third Battle – Innovation in the U.S. Navy's Silent War Struggle with Soviet Submarines." *Naval War College Newport Papers* 16 (2003).
Coyne, Kevin M. "Developing US European Command's Intelligence, Surveillance, and Reconnaissance Strategy for Fiscal Year 2010 Through 2015." *Air & Space Power Journal* 24, no. 4 (Winter 2010).
Crickmore, Paul F. *Lockheed Blackbird – Beyond the Secret Missions*, revised ed. Oxford: Osprey Publishing, 2016.
Day, Dwayne A., John M. Logsdon, and Brian Latell, eds. *Eye in the Sky – The Story of the CORONA Spy Satellites*. Washington, DC: Smithsonian Institution Press, 1998.
Deptula, David A., and Michael Fransisco. "Air Force ISR Operations – Hunting Versus Gathering." *Air & Space Power Journal* 24, no. 4 (Winter 2010).
Dyndal, Gjert Lage, and John O. Birkeland. "Autonome Droner Og Våpensystemer: Endres Krigen Og Måten vi Fører Krig På?" In *Når Dronene Våkner*, edited by Tor Arne Berntsen, Gjert Lage Dyndal, and Sigrid Redse Johansen. Oslo: Cappelen Damm Akademiske, 2016.
Elder, Gregory. "Intelligence in War: It Can Be Decisive." *Studies in Intelligence* 50, no. 2 (2006).
Finnegan, Terrence J. *Shooting the Front*. Stroud: History Press, 2011.
Foggo, James, and Alanik Fritz. "The Fourth Battle of the Atlantic." *United States Naval Institute Proceedings* 142, no. 6 (June 2016).
Fox, Collin. "If It Floats, It Fights." *United States Naval Institute Proceedings*, June 2018. https://www.usni.org/magazines/proceedings/2018/june/if-it-floats-it-fights.
Friedman, Norman. *Seapower and Space*. Annapolis, MD: Naval Institute Press, 2000.
Furquhar, John T. *A Need to Know: The Role of Air Force Reconnaissance in War Planning, 1945–1953*. Maxwell AFB, Montgomery, AL: Air University Press, 2005.
Gordon, Doug. *Tactical Reconnaissance in the Cold War*. Barnsley: Pen & Sword Aviation, 2006.
Haffa, Robert P. Jr., and Anand Datla. "Joint Intelligence, Surveillance, and Reconnaissance in Contested Airspace." *Air & Space Power Journal* 30, no. 3 (Fall 2016).
Haines, Gerald. *Critical to US Security: The Gambit and Hexagon Satellite Reconnaissance Systems Compendium*. Washington, DC: Center for the Study of National Reconnaissance, January 2012.
Hopkins, Robert S. III. *Spyflights and Overflights – US Strategic Aerial Reconnaissance – Volume I, 1945–1960*. Manchester: Hikoki Publications, 2016.
Jackson, Robert. *High Cold War – Strategic Air Reconnaissance and the Electronic Intelligence War*. Sparkford: Patrick Stephens Ltd., 1998.
Klevberg, Håvard. "Maritime Surveillance in the North: 333 Squadron in Norwegian Security Politics." PhD, History, University of Oslo, 2011.
Larson, Doyle E. "Direct Intelligence Combat Support in Vietnam: Project Teaball." *American Intelligence Journal* 15, no. 1 (Spring–Summer 1994).
Lashmar, Paul. *Spy Flights of the Cold War*. Phoenix Mill: Sutton Publishing Ltd., 1996.
Lawson, Eric, and Jane Lawson. *The First Air Campaign*. Conshoshocken, PA: Combined Books Inc., 1996.

Liang, Qiao, and Wang Xiangsui. *Unrestricted Warfare*. Beijing: PLA Literature and Arts Publishing House, 1999.

Lockheed Martin Corporation. "Always There: Always on." *Lockheedmartin.Com*. Accessed July 23, 2018. www.lockheedmartin.com/en-us/products/unmanned-aerostats-airships-and-lighter-than-air-technology.html.

Marshall, James P. *Near-Real Time Intelligence on the Tactical Battlefield*. Research Report. Maxwell AFB, Montgomery, AL: Air University, January 1994.

McKinney, Donna. "NRL Scientists Update Horizontal Wind Model with Eye Toward Next-Generation Space Weather Prediction Systems." July 10, 2015. www.nrl.navy.mil/news/releases/nrl-scientists-update-horizontal-wind-model-eye-toward-next-generation-space-weather.

Morton, Tyler. "Ears and Eyes in the Sky: The Evolution of Manned Airborne ISR." Master's thesis, School of Advanced Air and Space Studies, 2012. www.dtic.mil/get-tr-doc/pdf?AD=AD1019401.

Nichols, William J. *From Teaball to Fastball: The Value of Real-Time Intelligence Support to the Cockpit*. School of Advanced Air and Space Studies, May 12, 1992.

Norris, Pat. *Spies in the Sky – Surveillance Satellites in War and Peace*. Chichester, UK: Praxis Publishing Ltd., 2008.

Owen, David. *Anti-Submarine Warfare – An Illustrated History*. Annapolis, MD: Naval Institute Press, 2007.

Perkins, William. *Alliance Airborne Anti-Submarine Warfare – A Forecast for Maritime Air ASW in the Future Operational Environment*. Joint Air Power Competence Centre (JAPCC), June 2016.

Perry, Robert. *A History of Satellite Reconnaissance*, vol. 1. National Reconnaissance Office (NRO), October 1973.

Planet.com. "Planet Monitoring – Daily Cadence, Global Coverage." Accessed August 1, 2018. www.planet.com/products/monitoring/.

Polmar, Norman, and Thomas B. Allen. *Spy Book: The Encyclopedia of Espionage*. New York: Random House, 1998.

Polmar, Norman, and John F. Bessette. *Spyplanes – The Illustrated Guide to Manned Reconnaissance and Surveillance Aircraft from World War I to Today*. Minneapolis: Quarto Publishing Group Inc., 2016.

Powell, Robert R. *RA-5C Vigilante Units in Combat*. Oxford: Osprey Publishing, 2004.

Raines, Edgar F. Jr. *Eyes of Artillery – The Origins of Modern U.S. Army Aviation in World War II*. Honolulu: University Press of the Pacific, 2002.

Redfearn, Mason, and Richard J. Aldrich. "The Perfect Cover: British Intelligence, the Soviet Fleet, and Distant Water Trawler Operations, 1963–1974." *Intelligence and National Security* 12, no. 3 (1997): 166–77.

Robinson, Douglas H. *Giants in the Sky*. Seattle, WA: University of Washington Press, 1973.

Simon, Luis. "The 'Third' US Offset Strategy and Europe's 'Antiaccess' Challenge." *Journal of Strategic Studies* 39, no. 3 (2016): 417–45.

Sloggett, Dave. *Drone Warfare – The Development of Unmanned Aerial Conflict*. New York: Skyhorse Publishing, 2015.

Strickland, Frank. "The Early Evolution of the Predator Drone." *Studies in Intelligence* 57, no. 1 (2013).

Tart, Larry, and Robert Keefe. *The Price of Vigilance – Attacks on American Surveillance Flights*. New York: Ballantine Books, 2001.

Taubman, Philip. *Secret Empire – Eisenhower, the CIA, and the Hidden Story of America's Space Espionage*. New York: Simon and Schuster, 2003.

Thomas, Rick. *Global Hawk – The Story and Shadow of America's Controversial Drone*. Great Britain: Amazon, 2015.

Tice, Brian P. "Unmanned Aerial Vehicles – The Force Multiplier of the 1990s." *Air & Space Power Journal* 5, no. 1 (1991).

Warner, Michael. "The Collapse of Intelligence Support for Air Power, 1944–52." *Studies in Intelligence* 49, no. 3 (2005).
Warner, Michael. "US Intelligence in Vietnam: The Official Version(s)." *Intelligence and National Security* 25, no. 5 (2010).
Wheelon, Albert D. "Corona – A Triumph of American Technology." In *Eye in the Sky – The Story of the CORONA Spy Satellites*. Washington, DC: Smithsonian Institution Press, 1998.
Whittle, Richard. *Predator – The Secret Origins of the Drone Revolution*. New York: Henry Holt and Company, 2014.
Williams, Allan. *Operation Crossbow – The Untold Story of the Search for Hitler's Secret Weapons*. London: Arrow Books/Random House, 2014.

3 Airborne intelligence, surveillance, and reconnaissance

Framing the discussion

After the Cold War, and as we discussed in the previous chapter, warfare became limited in time, scope and space, and armed forces were forced to collaborate and think *joint*. Expeditionary warfare in the 1990s (Gulf War I, Somalia, Balkans) was an indication of a new type of warfare, and with the shift to counter-insurgency warfare, intelligence stepped up as more of a core tenet of modern warfare as opposed to the more classic interpretation of intelligence as having an auxiliary, supporting role. This is where "ISR" grew into its own and became an integrated part of the "network-centric warfare" movement. The term "ISR" entered daily usage in the mid-1990s, after first being coined by the US Vice Chairman of the Joint Chiefs of Staff Admiral Owens, who highlighted ISR as a vital component of modern warfare, implemented through the concept of network-centric warfare.[1] A network-centric approach to warfare is essentially the military embodiment of information age concepts. It is the linkage of computers, communications, sensors, and military units in order to improve efficiency and effectiveness of military operations.[2] Or, as John Ferris puts it, it is "the idea that armed forces will adopt flat structures, working in nets on the internet, with soldiers at the sharp end able to turn data processing systems at home into staffs through 'reachback', real time, immediate and thorough inter-communication."[3] In short, it is about gathering, interpreting, disseminating, and acting upon information in the battlefield – faster than your opponent. The tenets of network-centric warfare thus facilitate a networked force in order to improve information sharing, which in turn will enhance the quality of information and shared situational awareness. This will enable collaboration, self-synchronization, sustainability, and speed of command, thereby enhancing mission effectiveness.[4] The most prominent critique of an exaggerated emphasis on network-centric warfare is the focus on technology that usually encompasses such an emphasis. Critical voices in the early and mid-2000s pointed to important elements of command and warfare in general being eroded in favour of an emphasis on quantitative measurements (number of targets hit, numbers of areas covered through surveillance, etc.), and what some commentators have termed the "fetishisation of speed and tacticisasion of strategy."[5] Network-centric warfare is not the topic of discussion for this book; however, any discussion of modern ISR should take into account developments of modern armed forces as a whole, in order to make the gathered intelligence products actionable to the rest of the joint force.

DOI: 10.4324/9781003626015-3

Military developments in the new millennium have been characterized and shaped by Western military engagements in the Middle East. The focus has largely been on COIN operations, with a significant decline in focus on other areas of warfare, such as ASW, especially when compared to the Cold War.[6] Discussions over the past decade highlight a concern with *how ISR fits into new modes of warfare* and have been the subject of many articles. The arguments emphasized that ISR must be approached as a "synergistic whole" and how ISR provides battlespace awareness and further decision superiority when utilized and integrated properly.[7] There were warnings a few years into the Middle Eastern campaigns that the ISR community was heading in the wrong direction.[8] The focus seemed to be exclusively on real-time, tactical processing of sensor data, which, the argument went, was eroding the analysts' ability to conduct long-range predictive analysis.[9] John Ferris commented that " 'computers' have eaten qualities once assigned to 'command' while 'intelligence' has diminished, as an idea connoting 'to think' slips into one meaning 'to sense.' "[10] One important element from the discussions is that ISR is all about delivering *effect*: The decision maker, at the various levels, must be informed with actionable intelligence in order to facilitate effect in operations. The emphasis of ISR efforts must be on *qualitative effects*, and not a *quantitative focus* on the number of areas and targets surveyed.[11] The balance between focus on actions and focus on effects, an overarching, strategic, joint approach to the development of ISR, intelligence sharing between services and nations, and the role of leaders' sound direction and guidance for ISR operations are all key factors to proper integration of ISR into warfare.[12]

Another prominent factor for discussion in the 2000s with regard to ISR was the *efficient tasking and utilization* of ISR assets. The insatiable desire for more information and expectations of ISR support by any ground commander executing any mission, combined with a constant shortage of ISR assets to service all requests, meant that the question of efficient tasking took centre stage in discussions within the intelligence community. A central challenge pertains to integrating ISR assets into the air tasking order (ATO), a 72-h tasking cycle revolving around attack aircraft and targeting.[13] Some claim that the ATO process is outdated, static, and built around assumptions and dynamics from the Cold War; it is based on symmetry in warfare, pre-planned targeting, and peer-to-peer warfighting.[14] That one must move away from the quantitative placement of assets towards targets, to a qualitative focus on the commander's intent – the underlying *purpose* of the operation must be at the forefront of the execution of any operation.[15] This is based on the relatively high level of authorization required for the use of lethal force in aerial targeting. One proposed solution is to embed "ISR tactical controllers" with ground units for optimal utilization to the supported entity.[16] This would provide an important "translation" of terminology between the ground unit and the air unit, and likely improve the use of specific assets in support of specific units on specific missions, much like the forward air controller vis-a-vis close air support (CAS).[17]

Never before have intelligence products been as available as they are now to military units, whether they belong to the tactical, operational, or strategic level. This immense amount of information does pose significant challenges. The need to process and analyse significant caches of data ("big data") has prompted further discussions pertaining to the requirement for on-board computers, to autonomously process sensor information into actionable intelligence, through so-called autonomous *processing, exploitation, and dissemination* (PED). Autonomous PED will eventually provide the foundation for the in-flight analysis and transmission of processed intelligence. Other challenges, such as control of unmanned systems beyond-line-of-sight (BLOS), survivability in contested airspace, common command and control infrastructure, open architecture, and standardized data link systems, are all lessons from the past 20 years that are shaping discussions for the future development of airborne ISR.[18]

The chapter aims to structure the characteristics of airborne ISR under what really matters: The effect delivered to the force. By analysing the history of airborne ISR in the previous chapter, manned or unmanned, aircraft or satellite, it has become possible to identify three categories or tiers that should structure the discussion of the characteristics and capabilities involved. The lowest and most basic tier includes the factors that facilitate airborne ISR in the first place. They have been narrowed down to altitude, speed, range, communications, and survivability. The second tier consists of the sensors that the airborne asset is carrying, facilitating the respective intelligence products that can be delivered. The top tier involves the processing and eventual dissemination of these products, in a given *time* and with a given *product maturity*. The actual sensor data processing is an important factor for airborne ISR, as it is for all intelligence collection and dissemination. In cases where there is no on-board processing capability, neither human nor automated, the sensor data are transmitted raw. If there is a processing capability on board, whether it be humans or computers, the airborne asset becomes capable of disseminating processed intelligence to some degree of maturity. Real-time dissemination of sensor data more or less exclusively consists of raw sensor data. Strategic intelligence, on the other hand, has been processed, validated, and contextualized, but it takes time to reach that maturity level. Together, these three tiers of airborne ISR frame the discussion on airborne ISR.

Tier 1: Characteristics of airborne ISR

The basic characteristics of airborne ISR are the building blocks that facilitate the missions that aircraft are tasked to execute. The exploitation of the air rests on the three enduring tenets of airpower: Altitude, speed, and range.[19] In addition, there are two basic facilitators for battlefield integration, i.e. communications and survivability.

Height, or *altitude*, was the original attribute required by surveillance aircraft in order to gain the proper perspective. Altitude is a matter of giving the sensor the appropriate elevation for the right angle and placement against a target. The higher the altitude, the further the sensor is able to see and detect signals. The higher the altitude the less impact elevated features such as tall buildings and mountains have on the ability to gather intelligence on the target. As we have also seen, extremely high altitudes also provided some degree of safety from both detection and air defence systems. However, higher altitudes demand more capacity from the sensors – for optical sensors, this means a requirement for a larger aperture. Higher altitudes also mean more interference for the RF antennae attempting to distinguish between many more signals in the air than is the case at lower altitudes. Very high altitudes also demand special engines for the aircraft flying in ultra-thin air. And, of course, the higher the altitude the greater the chance of weather and cloud coverage obscuring the target for optical sensors. Traditional MPA have been operating at very low altitudes, and have been equipped with turbo-propeller engines, which are more efficient at low altitudes. Traditional MPA tactics, as the ones utilized during the tracking of Soviet submarines in the 1970s and 1980s, demand low altitudes in order to plot the sonobuoys drifting in the water onto the tactical overview. Low altitude has also facilitated a higher responsiveness with regard to attacking the target as well as getting close enough for a visual confirmation of any target close to or on the surface. Jet aircraft have traditionally flown at much higher altitudes because they operate more efficiently in thinner air. But at very high altitudes, this changes again, and the thin air requires specialized jet engines. Lower altitudes give a more recognizable picture to the interpreter of the imagery, both for technical analysis and by untrained operators. The low altitude of small, tactical unmanned systems used extensively in the past two decades has lowered the threshold for "layman" interpretation of imagery, simply because the aspect of the imagery is familiar to the operator.[20]

Speed is another original tenet of airpower in general, and of airborne ISR specifically. Speed allows you to cover a larger area, respond rapidly to an incident, and increase survivability. An extreme example of speed is the SR-71 *Blackbird* ISR aircraft, which flies at more than three times the speed of sound. Likewise, modern jet-powered MPA fly faster and are thus more responsive compared to older, propeller-driven models. The ability to respond to a situation or incident somewhere can be crucial for intelligence collection. The last known position of a submarine, or datum, is a good example of this. In operations where ISR aircraft have flown fast and low in order to avoid detection and enemy fire, it was at times a challenge to get images with clarity and proper resolution.[21] Lower speeds and higher altitudes have mitigated this issue. Finally, speed facilitates the coverage of a larger area for surveillance in a mission. MPA have evolved from relatively slow-flying propeller aircraft to much faster jet-powered MPA. This will facilitate both shorter reaction times and coverage of larger areas.

Range – the ability to fly beyond enemy lines, to observe remote locations, to monitor the vast oceans – has always played a key role in airborne intelligence collection. Range is also one of the characteristics of modern ISR aircraft that influences how ISR is integrated into over-land operations. Some unmanned aircraft can stay aloft for more than 24 h, which facilitates more or less continuous monitoring of targets, which again leads to greatly improved battle-space awareness.[22] Aircraft endurance is not necessarily connected to the level of warfare, although most short-range aircraft are utilized tactically, and most strategic aircraft have a long endurance. In the middle, we find aircraft servicing the operational level, that boast of immensely long endurance but can provide surveillance over one specific area for the tactical or operational level. Maritime surveillance and reconnaissance usually require long endurance due to the vast surveillance areas or the remote target location. Maritime airborne ISR also entails peripheral intelligence flights along the border of foreign countries, normally requiring long endurance as well. As we have seen, range and lack of range have also been instrumental in the geographical delineation of responsibility between different MPA nations in the North Atlantic. Furthermore, increased range has the potential of changing international dynamics in the High North with regard to airborne surveillance, where new technology can bypass any Norwegian national guidelines that might apply to operating in the Arctic.

The most central aspect with *communications* for airborne ISR assets is the ability to transmit sensor data to a receiving unit for action. The time it takes for sensor products to be transmitted to a receiving unit on the battlefield is correlated to the level of battlefield integration. Following this, the requirement for urgent transmission of intelligence has led to ever-increasing demands for "real-time intelligence." This means that the ISR aircraft has its communications suite connected to it sensor(s) or sensor processing systems, which enables the airborne asset to transmit either real time, or parts of the gathered sensor data after processing it to a certain degree in near-real-time. As the previous chapters have shown, communications for MPA in the High North have traditionally not been a major issue, due to the nature of the operations and the lack of any requirement for detailed, high-bandwidth, instant communication. However, new adversary technology and *modus operandi* are posing new requirements for communication solutions in the Arctic. In an increasingly complex battlespace, there is an imperative to be connected with the rest of the force through the use of data link systems. Link systems provide awareness of friendly and enemy positions, awareness of threats to own unit and the force as a whole, and updates to battlespace developments. They provide an automated ability to communicate and coordinate with the rest of the force.[23]

The reciprocal element of transmitting data from the aircraft is receiving updated information from the ground. This usually involves directions and guidance for employment of the ISR asset and can involve updated orders in case of a manned aircraft, or electronic steering and detailed waypoints for an unmanned aircraft. This can be summarized as the ability to execute command and control (C2) of an ISR asset. In the context of airborne ISR, C2 is about the level of the ability to communicate intentions and requirements, and receive orders and requirements from other units.[24] Dynamic C2 rests on the ability to communicate, and the ability to change areas for surveillance and reconnaissance and the coordination and de-confliction with other aircraft such changes entail. The way any ISR aircraft is tasked for its mission also falls under C2 and shapes the flexibility in the utilization of that aircraft. Most historical strategic reconnaissance and surveillance were executed with minimal input from any higher entity: They flew on orders given before the flight and received few amendments *in situ*. Most modern airborne ISR aircraft are in continuous contact with their commanding authority, which either makes them flexible to strategic or operational demands during the mission, or has them answer to orders from the supported tactical ground unit at any given time. It must be stressed that in the context of this discussion, the focus is the ability to *integrate* C2, rather than on C2 *per se*. It is about the ability to receive orders and changes *in situ*, not about how orders have been developed or from what level in the military hierarchy those orders were transmitted. This aspect is particularly important when we discuss autonomy in military units, manned or unmanned.

Survivability is another core characteristic of any aircraft operating close to, or within adversary or contested airspace. Although survivability was of great importance in ASW missions during the Cold War, a generation of warfighters have, in the new millennium, operated in war zones in the Middle East with little to no air threat, nor threats from the ground or surface. Most planners and analysts agree that future warfare will not necessarily entail such freedom of manoeuvre in the air, and threats from enemy air, ground and maritime units will demand a much higher degree of survivability through self-protection than has been and is the case for many military ISR aircraft. However, the mission being executed shapes the necessity for self-protection: Tactical maritime reconnaissance aircraft far out to sea usually have fewer concerns about being shot down, but aircraft approaching the firing-range of hostile SAMs will have to function with an increased level of risk, or be able to protect themselves against the threat.

This is an important aspect of any warfighting aircraft, but the dangers associated with even peacetime reconnaissance during times of tension are shown through all the ISR aircraft that have been shot down outside of war.[25] Survivability can be improved through a host of measures. Speed and altitude were historically the main features of survivability, but now – in response to the advances made in the field of missile technology – counter-measures within the airframe of an aircraft designed to neutralize or defend against incoming attacks have come to play a more important role. Classic examples are physical measures such as deploying chaff, which are streams of metal that disturb the radar picture of incoming missiles, and flares, which are chemical elements deployed burning at an extremely high temperature, in order to disturb heat-seeking missiles. Further, there are jamming devices that will disturb the radio frequency spectrum for enemy aircraft and missiles and lasers that dazzle hostile sensors attempting to track an aircraft. The discussion around the *Atlantic* procurement programme showed that weapons, too, play a role in an aircraft's self-defence.

The core characteristics of airborne ISR can thus be listed as the following:

Table 3.1 Core characteristics of airborne ISR (Tier 1).

Altitude of the asset	Communication capabilities
Range of the asset	Survivability & self-protection
Speed of the asset	

With every airborne ISR mission and aircraft through history we see these characteristics to differing degrees. They do vary by airframe, which shapes aircraft capability in total. However, every core characteristic has implications for the others. Long endurance potentially denotes a significant amount of data that must be processed and analysed. The level of survivability affects areas the asset can cover and fly into, with implications for C2 and the ability to communicate with friendly units. These are a few examples, but they show an interconnectedness among the basic building blocks of airborne ISR, as highlighted by the negotiations to procure a common MPA for NATO during the late 1950s.

Tier 2: Sensors

The second tier covers the type of intelligence *products* an airborne asset is capable of developing. These products depend on the *sensors* an aircraft is able to carry. Sensors, then, are the key to any type of remote sensing, and the utilization of sensors essentially comes down to efficient exploitation of the electromagnetic spectrum. The improvement in sensor technology over the past century has been immense, from the early handheld cameras to today's hyperspectral sensors that are able to distinguish between different types of materials through frequency analysis.

In the history of airborne ISR, the first sensor utilized for elevated observation was, of course, the human eye. But as soon as technology evolved, cameras were utilized, providing the basis for IMINT.[26] The most common airborne ISR sensors are electro-optical (EO) cameras that provide still and moving imagery. Today, IMINT includes advanced mapping and imagery radars, such as synthetic aperture radar, and a host of other imaging sensors, such as infrared (IR) sensors, spectral imagery, and LiDARs. Further, as interception and exploitation of electronic signals transmitted through the air became a reality, the analysis of enemy usage of the electromagnetic spectrum led to airborne SIGINT. SIGINT is traditionally understood to constitute the analysis of ELINT, as well as the interception of adversary communications, facilitating COMINT. In addition, collection of signals between a missile being tested and the controlling ground station, known as telemetry, is considered foreign instrumentation and signals intelligence (FISINT), which also falls under SIGINT. In the context of maritime airborne surveillance, intelligence collection on submarine sound signatures has played a crucial role in the art of ASW, as will be discussed in the final chapters. ACINT is thus an important discipline for reconnaissance and surveillance in the maritime arena. It should be noted that airborne and space-based sensors are (still) not able to penetrate water, which means that they cannot conduct underwater search for submarines from up high. For this, one is still dependent on sensors in the water, such as sonars mounted on a ship, or sonobuoys dropped into the water.

This list is clearly not exhaustive, but it portrays the most common intelligence *products* delivered by airborne sensors. It is important to note that the need for analysing and interpreting sensor data before it can be put to use varies from sensor to sensor. EO camera imagery, for example, is assumed by many to be the most intuitive data to interpret and understand. However, the notion that a "picture is a picture," as some laymen seem to think, can have serious consequences for decisions in targeting situations. This factor will be further discussed below.

Tier 3: Processing and delivery of intelligence products

Having described the two lower tiers of airborne ISR, we have now arrived at the crux of airborne ISR: The intersection between intelligence and operations.[27] Turning to history again, as well as lessons from modern warfare, there is a clear link between the level of interaction with ground and/or surface units and the level of warfare intelligence assets are serving: *Tactical* intelligence collection – close interaction with ground units; *strategic* intelligence collection – no interaction with ground troops. The type of information gathered is also linked to the level of warfare intelligence assets are serving: Information on enemy troop movements – tactical; targeting information for war plans – strategic. In the absence of war, the utilization of airborne intelligence assets was more or less exclusively strategic. The operational requirements that are at the forefront of planning and operational discussion drive investment to service those requirements. In between wars, this means strategic, future requirements. During wars, the ongoing fighting on the ground and at sea shaped urgent developments of intelligence and weapons systems. In order to stay true to the core and purpose of airborne intelligence collection – the support of operations based on decision makers' intent – we must dare to point to the very intersection between airborne intelligence products and operations being supported.

The level of warfare being supported can be expressed in a more principled manner: *Time*. If the intelligence collector is supporting the development and adjustment of war plans, it is considered to be support of a strategic nature, and the intelligence *takes time* to develop and contextualize. Also, this type of strategic intelligence will have time to mature in the sense that it can be supported by other sensors and intelligence sources for the sake of corroboration and verification. The interaction with ground units with specific regard to strategic intelligence does not usually take place – that type of interaction becomes almost irrelevant.

On the other end of the scale we find the real-time dissemination of sensor products – for instance, streaming of live video – from a collecting asset down to ground and surface units. At the lowest level, this is dissemination of a raw, unprocessed sensor product to the consumer. It is picked up by the respective sensor, and the only delay in transmission is whatever delay would be inherent in the electronic system converting data from a sensor format to transmittable information.[28] This is common practice at the tactical level.

The element of time is thus closely linked to the level of warfare that the ISR assets are serving. It is further connected to the type of targets the surveillance and reconnaissance efforts are covering. One important factor in the assessment of the level of warfare being supported is the ability and willingness to share the sensor-products, processed or unprocessed, with the joint force or other national entities.

- *Combat information and tactical intelligence*[29]: Sensor products which are transmitted in order to build an understanding of the target in real time, should be considered *combat information*. Combat information is not processed and contextualized by an analyst at all, but used by the receiving unit in direct support of tactical actions. Airborne ISR here supports immediate and local action, limited in time and space. *Tactical intelligence* is information that is not utilized or acted upon immediately, but facilitates the planning and execution of tactical actions that will take place within minutes, hours, or days after receiving the information.[30]

- *Operational intelligence*[31]: Sensor products that are analysed and provide a basis for how the operational level distributes its forces should be considered operational. Any airborne ISR product that builds a foundation for near-term planning approximately 3–30 days out, at the joint force level, should be considered *operational* ISR.
- *Strategic intelligence*[32]: U-2 aircraft flying over Soviet missile sites during the Cold War and maritime patrol aircraft monitoring enemy ballistic missile test firings from submarines were gathering information on enemy strategic capabilities. The intelligence products inform own weapons and systems development and deliver crucial information to war planners at the strategic level. The level of warfare being supported in these cases is therefore *strategic.*[33]

Any and all sensor data are interpreted at some point, either by a soldier that sees the imagery or other raw data or by a trained analyst that processes, interprets, and possibly collates the data and places it in a tactical, operational, or strategic context. The importance of this aspect is just this: Imagery is not necessarily as easily interpreted as one would think, and there are trained analysts for every type of sensor information. Some ISR aircraft transmit their data in real time, leaving it to the recipient to understand what is being seen on the screen. Other aircraft, with sensor operators on board, can conduct basic interpretation on board and provide tactical and operational updates to the force based on on-board processing. Others yet again bring raw sensor data back to their home base, where the sensor data are thoroughly analysed and contextualized. This spectrum of sensor analysis is shaping *how* intelligence products are utilized, and *at what level* they are being utilized. As we shall see, these elements are also crucial for any discussion on autonomous PED to begin.

Historians discussing airborne ISR conducted during the Cold War highlight the sporadic shift between strategic and tactical focus for the ISR community.[34] This shift back and forth is, if not problematic, challenging both for operators and analysts and the organization as a whole. With the uncertainties that exist on the nature of future conflicts, an ISR organization must be prepared to service the *entire* spectrum of operations. There is a comparison to be made between the Cold War shifts in focus and contemporary ISR units that have been accustomed to a tactical focus for two decades, and that might face strategic challenges in the future. Commentators have also warned of too much tactical focus and what they perceive as "neglect of strategic intelligence," calling for a holistic approach to strategic intelligence as fundamentally necessary to encompass all other intelligence activities.[35] This warning is closely linked to the aforementioned "fetishisation of speed and tacticisation of strategy." The historical chapters of this book have shown airborne ASW to be an excellent case study of the spectrum in sensor analysis, ranging from real-time processing and immediate use of sensor data for their own targeting in a tactical setting, to post-flight, deep analysis of sensor data, facilitating platform, weapons, and sensor development in a strategic sense. Another historical example is the effort during WWII that went into introducing the new ELINT tools to modern warfare – the actual benefits from these new operations were questionable.[36] Several factors played their part: The lack of understanding among commanding officers regarding intelligence as a tool; the inability to change perspective on operations and plans based on new intelligence that contradicts the baseline perspective; and, of course, the organizational challenges that echo to this day – a warfighting organization that does not necessarily know or understand how to process and effectively and efficiently disseminate the information that comes out of the intelligence processes.[37]

In any case, the sensor data being collected are either analysed on board for own use (e.g. weapons drop); transmitted as raw data in real-time, correlated with a database, and then transmitted; they are interpreted by a sensor operator before transmission; or they are analysed post-flight and then disseminated. This relates to the second axis of intelligence processing in support of effects and intent: *The maturity of the intelligence product.* It is useful to distinguish between tactical and strategic processing of raw sensor data. Data with *tactical maturity* refers to sensor data that has been analysed sufficiently to be acted upon in a tactical situation. For acoustic data, it is usually the direction or range to the submarine from the sonobuoy in the water, in other words, positional data. Even in this category there are sublevels of maturity, for example, if the sensor operator is too uncertain about a frequency line for it to be entered into the tactical plot of the aircraft. But if further analysis brings more clarity and certainty, the information can be taken into the plot and acted upon. Once it becomes an active part of the tactical calculations and assessment, we can term the sensor data tactically mature. The other end of the scale of maturity is that of a strategic nature. In terms of acoustic data, this is a level of maturity that comes after time-consuming analysis post-mission, where analysts can dwell on certain frequency lines to identify more and new characteristics of the submarine itself. When information from several sonobuoys is analysed together, analysts are able to find out details on propulsion and propeller blades, machinery internal to the submarine, equipment external to the submarine, details of the gear box technology, and more. This information can in turn be used to describe the actual capabilities of the submarine, what types of missions she can be expected to perform, and other types of information of strategic nature. It provides insight into submarine deployments and behaviour, which, in turn, can help provide insight into enemy operations and strategy. This type of information has been pivotal in NATO's understanding of Soviet and later Russian strategy and operating concepts and capabilities. This level of intelligence can be termed *strategic maturity*, but it is too time-consuming and complex to be performed in-flight. Strategically mature sensor data play a crucial role in many processes, such as the development of own submarines, new sensors, and the improvement of existing ones, the development of underwater weapons and counter-measures, and, most importantly, the improvement of tactics and search parameters for sensor operators to track adversary submarines in the future.

The hierarchy of tenets for airborne ISR can thus be depicted as follows:

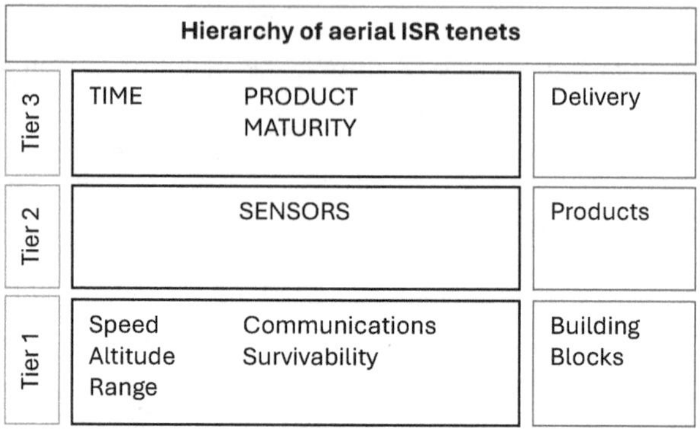

Figure 3.1 Hierarchy of aerial ISR tenets

The key point of this overview is that any discussion on airborne ISR requires a holistic approach. This means that discussing single tenets, such as speed or sensors alone, will not give us any understanding of what the requirements are, nor what products in what format the respective asset is capable of delivering. In order to clarify the requirements for an airborne ISR service, we cannot stop at describing the basic building blocks (Tier 1) – we must go on to describe the actual intelligence product we require (Tier 2), and then in what format and with what urgency we expect this to be integrated with the force and the intelligence community (Tier 3).

A visualization of the top tier in a different way assists us in framing contemporary discussions on ISR in the same overview as potential future technological capabilities:

Figure 3.2 shows the axes of time and product maturity, and the traditional, linear connection between the two. As a rule-of-thumb, the lesser the time between the interception of a signal by a sensor and the transmission of the information by the airborne asset, the less mature the product essentially is. If it is transmitted in real time or near-real-time, the sensor data have not been processed [A]. This means that the responsibility to analyse and contextualize the data rests with the recipient. It is most often the case that this type of data is used during combat and on the tactical level. If, however, a significant amount of time passes before the data are disseminated to the user, it is usually because the asset with the sensor cannot or will not transmit in-flight, and the sensor data will be thoroughly analysed post-mission. This further means that the data usually are obsolete for any tactical use and thus is meant to support the operational or strategic levels [B][C].

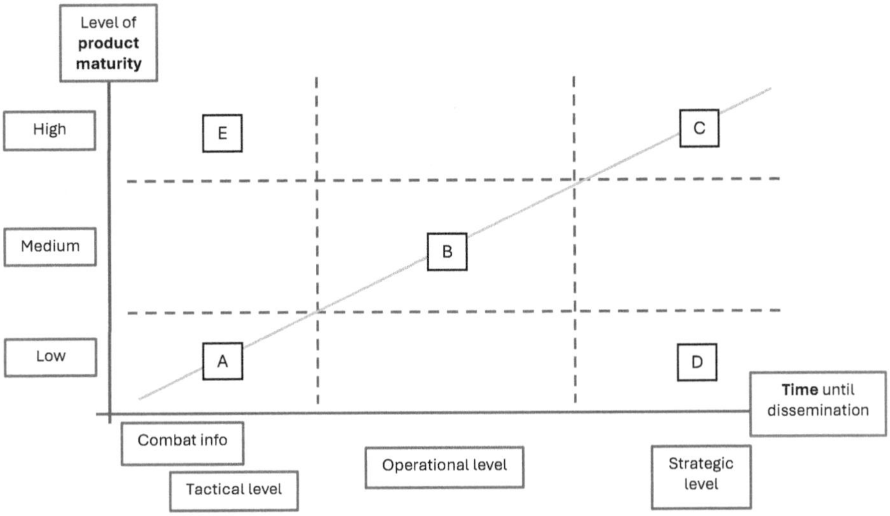

Figure 3.2 Time–product maturity matrix

Both [D] and [E] represent extreme cases, due to different rationales. In the case of [D], immature intelligence, or rather improperly processed and analysed intelligence, is being utilized at the strategic level. This can be either improper use of intelligence or incompetence in the strategic analysis process. Both situations can happen, however, one would hope that any responsible nation would strive to avoid them. The case of [E] is much more interesting, as this situation requires processed information to be disseminated in near-real-time. Some would say that this is a contradiction in terms, but others would argue that this is the next step for autonomous sensor analysis. In the case of [E], the airborne asset is able to remotely sense whatever it is capable of intercepting, and immediately process and analyse the information as it is gathered. Artificial Intelligence (AI) will likely, in the not-too-distant future, be able to process information and sensor data in a manner and speed that humans simply are incapable of doing, and it is AI technology that will facilitate the extreme case of [E]. This discussion is made even more relevant when we know that future unmanned aircraft, or drones, will base their decision making on what they see themselves, including the release of weapons.[38] However, the dissemination of raw data, as in the situation of [A], raises the question of both relevance and utility of the data being transmitted. Who interprets the data on the ground? Are there sensor data where analytical interpretation is not even required? There are important questions to be answered when people make decisions based on unprocessed and uncontextualized data, as is taking place with real-time dissemination of sensor data.[39] The following figure is yet another way of depicting the relationship between time to dissemination and product maturity:

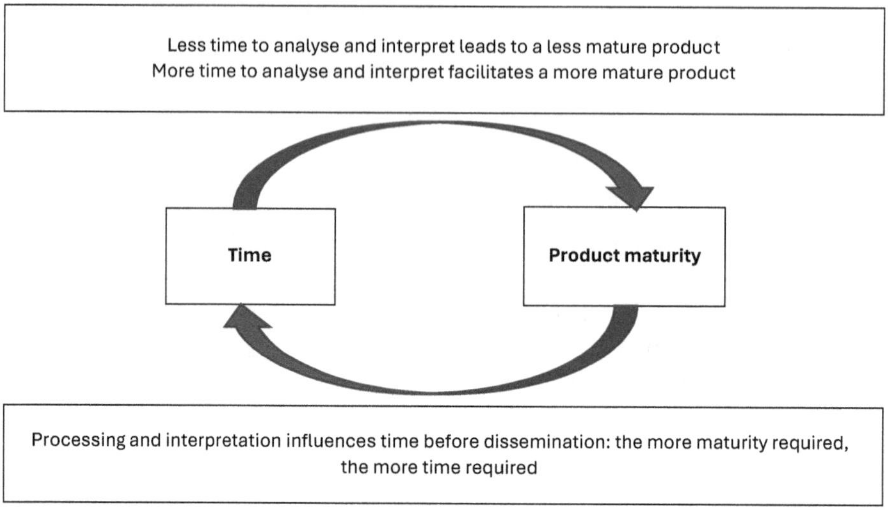

Figure 3.3 Time–product maturity symbiosis

Armed reconnaissance

The previous historical chapters of this book have shown how ASW and acoustic intelligence have constituted a significant part of airborne ISR in the High North since the end of WWII. MPA on patrol are, at least in wartime, prepared to engage the submarines they are hunting with weapons. This means that for maritime patrol, engagement and targeting play important parts. Armed reconnaissance must briefly be discussed for two reasons. Most importantly, it will place ASW in the context of this theoretical framework. Second, it will contextualize the ongoing debate on lethal autonomous weapons systems (LAWSs) in the sense that it provides more clarity to the basis for LAWS' decisions to apply military force.

The theoretical framework presented above also helps to explain the underlying factors facilitating targeting performed by aircraft. Targeting solutions can, in the context of airborne assets, be achieved either by an ISR asset providing targeting information to other assets carrying weapons, or an airborne asset providing targeting for its own weapons, often characterized as *armed reconnaissance*. The mission execution phase of targeting in military operations, as described in NATO doctrine, consists of the elements of find, fix, track, target, engage, exploit, and assess.[40] *Finding* the target relies on the basic intelligence support to operations, and an intelligence planning assessment of what elements in theatre constitute relevant targets for the force. *Fixing* is the process of geolocating and identifying the target, often based on cross-cueing of sensors and multiple intelligence disciplines. *Tracking* means that ISR assets are monitoring and following the target over time. *Targeting* is the process of acquiring attack criteria, and assessing collateral damage, the necessity of the strike, the relevance to the prioritized targeting list, and ultimately receiving final approval to engage the target. *Engaging* the target means releasing a weapon in order to achieve the desired effect. *Exploitation* in this context means taking advantage of short- and long-term opportunities that the immediate engagement presents. *Assessment* is an evaluation of the results of the engagement, assessing whether the desired effects have been achieved.[41] The elements can be structured as has been done in the following:

Table 3.2 NATO targeting sequence.

Function	ISR				Fires	Plans	ISR
Action	Find	Fix	Track	Target	Engage	Exploit	Assess

It is clear that ISR is the foundation for any type of targeting – we require detailed information on a target before any lethal force is applied. This information has traditionally been gathered by designated aircraft, such as airborne ISR platforms, and the information corroborated by other types of intelligence, for example human intelligence (HUMINT).[42] As technology has evolved, more and more armed aircraft are capable of generating their own targeting solutions, or at least the final "piece of the puzzle," with other intelligence sources providing an important foundation before the aircraft arrives on the scene. Essentially, this means that the aircraft is carrying sensors that can bring the firing solution up to the required attack criteria for the respective mission and fire weapons based on that sensor information. Three examples are worth looking at briefly in order to contextualize this discussion.

Example 1. A fighter aircraft is usually carrying an imaging sensor in support of the aircraft's targeting.[43] The fighter aircraft is then able to find and track an enemy tank, or an enemy anti-air artillery battery on the ground, and fire weapons based on this sensor information.

Example 2. An MPA is capable of finding and tracking submarines. During operations, MPA will carry torpedoes, which enable them to attack the enemy submarines being hunted. This means that they can engage their targets in addition to finding and tracking them. This is slightly more complicated than the previous example, as there is a high probability that the enemy submarine will be submerged during the hunt. This further means that there is a good chance that the aircraft crew will drop a weapon on a target they cannot see. The crew has, however, achieved the required confidence level for a targeting solution based on their interpretation of sensor information. ASW requires relatively complex interpretation of sensor data before weapon release.

Example 3. A drone flies an armed reconnaissance mission, searching for a specific human target. The person targeted has been under surveillance for weeks, and several other pieces of intelligence have been accumulated before the drone mission. The aircraft finds a person that meets the parameters of the pre-mission briefing, but the imagery is hard to interpret for the sensor operator, among other things, because of the altitude the aircraft is operating at to avoid detection. Together with the drone pilot and the mission commander, they declare attack criteria to be met and release the weapon on the person as he is driving away in a car.

Example 1 is the least controversial and the least demanding. Military targets are fairly easily distinguished from their surroundings by a trained operator. Example 2 is much more complex, as the sensor data are harder to interpret and the attack criteria are more demanding. Example 3 is a relatively new form of operating armed aircraft, and one that has attracted controversy.[44]

This book will not discuss the important question of the decision to fire a weapon *per se*, including rules of engagement, moral and ethical considerations on maintaining meaningful human control of autonomous weapons, and more. However, it is important to highlight that the *underlying foundation* for making that decision rests on the *interpretation of sensor data*. And it is with intepretation that the maturity and reliability of the sensor data come into play: The level of maturity of the intelligence product that facilitates tactical, operational, and strategic decisions.

Working towards autonomous processing and analysis of sensor data will eventually lead to in-flight, near-real-time processing of sensor data, and further near-real-time dissemination of intelligence products. This will facilitate new ways of operating ISR aircraft and potentially shorten the targeting sequence.

The three categories indicated in Figure 3.4 are relevant when we discuss what we require from the respective ISR asset vis-à-vis a targeting context. A category I delivery is not directed at targeting in the traditional sense, but rather supporting tactical, operational, and strategic levels of warfare with intelligence products.

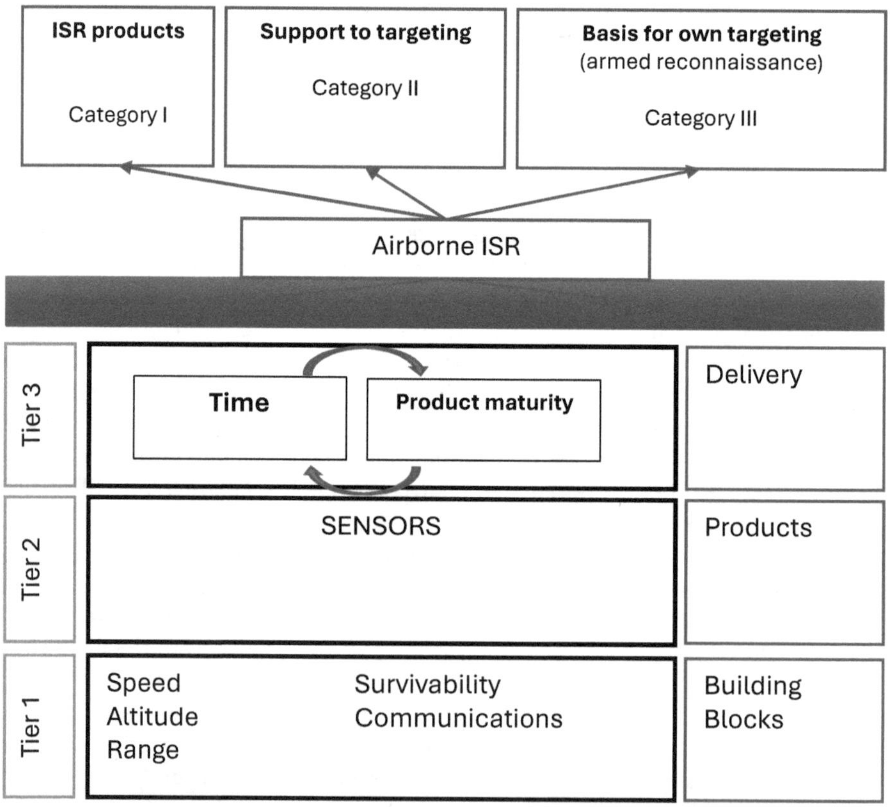

Figure 3.4 Airborne ISR as basis for targeting

A category II delivery is targeting information delivered to a third party, which is actually capable of engaging the target. This can be a drone giving detailed vectors to a frigate firing a torpedo or a missile. Finally, a category III delivery is a platform acquiring attack criteria and being able to engage the target itself. A classic example is an MPA dropping a torpedo on a submerged target based on its own sensor information.

Airborne ISR and the hunt for submarines

In accordance with what we have just discussed, one would require the following for hunting submarines: The elements of Tier 1, to varying degrees; acoustic and other ASW sensors (Tier 2); and the ability to disseminate this information or to act on it by oneself (Tier 3). In discussing the airborne ISR requirements for ASW, some of the above elements stand out more than others.

First, there is an overlap between ASW and ISR. ISR includes general surveillance, more specific reconnaissance, and very specific tracking of targets.

ASW includes finding the target, fixing it, tracking it, targeting it, and then, possibly, engaging it. We know from our definition of ASW that one can execute efficient ASW, not necessarily by *sinking* the adversary submarine, but by *denying* area access, *suppressing* its operations, and more. This overlap between ISR and ASW is illustrated in the figure above. When the NATO Group of Experts in the late 1950s were considering what aircraft they required, they were considering a long-range ASW aircraft, which, given the operational characteristics put forward, would be able to conduct other types of surveillance as well as ASW. When the Norwegian Chief of Defense in 2015 proposed to phase out the P-3 *Orion* from the Norwegian inventory and replace the MPA fleet with drones, he effectively removed the ability to conduct long-range airborne ASW. Instead, he focussed on the ability to conduct *surface* surveillance and reconnaissance with unmanned systems. No unmanned aircraft flying today are capable of conducting searches

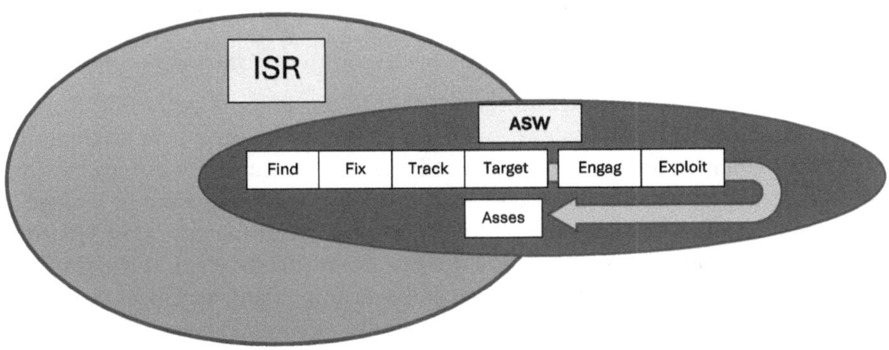

Figure 3.5 The overlap between ISR and ASW

underwater, nor are drones today capable of carrying weapons that can neutralize adversary submarines. This is, however, expected to change in the future.

Second, the factor of *range* is crucial for any credible maritime airborne ISR platform. To begin with, the aircraft must have a minimum range to reach an adversary's operating areas. In addition, one would preferably be able to monitor that area for some time, not only fly out and back. This creates a need for range, depending on how precisely the target's location is known. If one is able to provide cueing to the aircraft, the aircraft would spend less time and fuel to search a given area. Another key factor that played an important role during the Cold War for NATO MPA was the sheer predictability of Russian submarine deployments, as we have seen. The aircraft operators were not only cued by SOSUS, but they knew the general direction of the submarine, broadly their patrol areas, and the approximate duration of their deployment. The very distances involved demanded long endurance, but if such knowledge had not been available, the search for enemy submarines would have taken even longer. And if cueing from other sensors is not available, one would be relying on historical *modus operandi*, intelligence reports, and the sensors of the aircraft itself in the initial search. Thus, the longer the range, the more time an aircraft has to search, conduct general surveillance, and build general situational awareness in an area.

Third, *sensor interpretation* is key to ASW. Acoustic intelligence and in-flight processing of acoustic data have always been challenging. The acoustic environment in which the operators on board an aircraft have to manoeuvre to find and track a submarine is constantly complicated by other sound sources in the water and by increasingly silent submarines. Many of these interpretation processes have been automated compared to the early days of ASW; however, they have not been automated completely. This means that even in 2020, as new P-8 *Poseidon* MPA are operationalized, there is still a requirement for human operators on board to interpret acoustic sensor data. This further means that if one is to track a submarine in a complex acoustic environment, computers without human input or interpretation will likely fall short, even in 2020. We should expect, however, AI to assume this type of interpretation in the not-too-distant future, enabling autonomous drones, surface vessels, and submarines to conduct acoustic intelligence and ASW. Autonomous PED will thus significantly influence Tier 3 aspects mentioned above, and alter what unmanned aircraft and vessels are able to do in a communications-degraded environment.

Finally, if the airborne platform is not able to engage a target, the ISR asset will be somewhat of a lame duck. With the distances involved, for example in the Atlantic, there is little to no time for a flying asset to call for weapons support from another asset when it has found an enemy combatant that must be engaged. In the context of ASW, with the relatively short acoustic ranges involved from state-of-the art submarines such as the Russian *Severodvinsk* class, it is becoming increasingly difficult to find and keep track of enemy submarines. Once a targeting solution has been achieved, that targeting solution is very easily lost. This means that the target submarine will likely not be engaged at all if it is not engaged by the ISR asset tracking it. This, too, will in future factor into targeting and engagement by unmanned aircraft. Autonomous PED will likely facilitate unmanned systems engaging targets based on acoustic attack criteria, altering dynamics and facilitating new possibilities in modern ASW.

The discussion in this chapter has sought to identify and nuance the elements that comprise airborne ISR. The discussion on the basic building blocks can at times overshadow important elements such as processing capability of an ISR asset. In discussing the requirements for surveillance and reconnaissance, one must not only bring out the respective factors of the three tiers, but also carefully consider the most efficient and effective means of satisfying those requirements. Is manned airborne ISR necessarily the answer to one's requirements? How much of the information required can be provided by space-based services? How much can be solved by unmanned systems? What does one gain and lose by shifting from manned to unmanned systems? What type of sensor data still requires human interpretation? These questions can be answered in the context of the presented framework, and it is the main objective of this book to bring clarity to any evaluation of modern airborne ISR requirements, particularly in the context of ASW.

Notes

1 See Deptula and Brown, "A House Divided."
2 Wilson, *Network Centric Warfare*, CRS-2.
3 Ferris, "Netcentric Warfare, C4ISR, and Information Operations," 199–225.
4 John Luddy, *The Challenge and Promise of Network-Centric Warfare*, 3.
5 See for example Ferris, "Netcentric Warfare, C4ISR, and Information Operations," 201.
6 Birkeland, "The Potential of LIDAR as an Anti-Submarine Warfare Sensor," 54–55; Perkins, *Alliance Airborne Anti-Submarine Warfare*, 45–46.
7 Deptula and Brown, "A House Divided."
8 For discussions on the intersection between COIN and intelligence, see for example Zeytoonian mfl., "Intelligence Design: COIN Operations and Intelligence Collection and Analysis," *Military Review*, October 2006.
9 Daniel P. Shibilski, "Future of Air Force Intelligence," *United States Naval Institute Proceedings* 132, no. 2 (February 2006).
10 Ferris, "Netcentric Warfare, C4ISR, and Information Operations," 204.
11 See also Jaylan M. Haley, "An Evolution in Intelligence Doctrine – The Intelligence, Surveillance, and Reconnaissance Mission Type Order," *Air & Space Power Journal* 26, no. 5 (Fall 2012),
12 Examples of articles discussing these factors in modern ISR support to operations are A. Denis Clift, "Intelligence in the Internet Era – From Semaphore to Predator," *Studies in Intelligence* 47, no. 3 (2003); William B. Danskine, "Aggressive ISR in the War on Terrorism – Breaking the Cold War Paradigm," *Air & Space Power Journal* 19, no. 2 (Summer 2005); John Kriendler, *NATO Intelligence and Early Warning* (Swindon, England: Conflict Studies Research Centre, Defense Academy of the United Kingdom, March 2006); Jeff S. Hinrichs, "Education in Intelligence, Surveillance, and Reconnaissance Beyond the 'Green Door'," *Air & Space Power Journal* 22, no. 2 (Summer 2008); Deptula and Brown, "A House Divided"; Deptula and Fransisco, "Air Force ISR Operations"; Dagvin R.M. Anderson, "A Holistic Approach to Intelligence, Surveillance and Reconnaissance," *Air & Space Power Journal* 25, no. 4 (Winter 2011); Haley, "An Evolution in Intelligence Doctrine"; Jason M. Brown, "Airpower Is Intelligence, Surveillance, and Reconnaissance," *Air & Space Power Journal* 27, no. 4 (August 2013); Matthew J. Martin, "Unifying Our Vision – Joint ISR Coordination and the NATO Joint

ISR Initiative," *Joint Force Quarterly (JFQ)*, no. 72 (Quarter 2014); Nicholas P. Cowan, *Rethinking Command and Control of Intelligence, Surveillance and Reconnaissance (ISR)* (U.S. Air Force, July 2015).

13 For an outstanding account of the evolution, shortfalls and potential of the ATO Cycle, see Robert P. Winkler, "The Evolution of the Joint ATO Cycle," (Master of Science in Joint Campaign Planning and Strategy, Joint Advanced Warfighting School, 2006), https://dtic.mil/cgi-bin/GetTRDoc?AD=ADA451239.

14 Michael L. Downs, "Rethinking the Combined Force Air Component Commander's Intelligence, Surveillance, and Reconnaissance Approach to Counterinsurgency," *Air & Space Power Journal* 22, no. 3 (Fall 2008).

15 Cowan, *Rethinking Command and Control of Intelligence, Surveillance and Reconnaissance (ISR)*; Carl Rhodes, Jeff Hagen, og Mark Westergren. *A Strategies-to-Tasks Framework for Planning and Executing Intelligence, Surveillance, and Reconnaissance (ISR) Operations* (RAND, Project Air Force, 2007).

16 Adam B. Young, "Employing Intelligence, Surveillance, and Reconnaissance – Organizing, Training, and Equipping to Get It Right," *Air & Space Power Journal* 28, no. 1 (January–February 2014).

17 See for example Steven Maceda, "Control of Theater Intelligence, Surveillance, and Reconnaissance for the Ground Commander," *Air & Space Power Journal* 22, no. 4 (Winter 2008).

18 For more details, see for example Slava Frayter and Koen Willems, *Increased Efficiency or Beyond Line-of-Sight in Airborne ISR Operations* (National Geospatial Agency (NGA), 2013), proceedings.spiedigitallibrary.org/proceeding.aspx?articleid=1688649; David Bottom, *Overcoming ISR Data Challenges* (National Geospatial Agency (NGA), 2013), spie.org/Publications/Proceedings/Paper/10.1117/12.2019033; USAF Chief Scientist, *Technology Horizons – A Vision for Air Force Technology 2010–2030* (Maxwell AFB, Montgomery, AL: Air University Press, 2010); Mike Fowler, "The Future of Unmanned Aerial Vehicles," *Global Security and Intelligence Studies* 1, no. 1 (Fall 2015); Jenny R. Holzer and Franklin L. Moses, "Autonomous Systems in the Intelligence Community: Many Possibilities and Challenges," *Studies in Intelligence* 59, no. 1 (2015); Puong F. Yeh, "The Case for Using Robots in Intelligence Analysis," *Studies in Intelligence* 59, no. 4 (2015).

19 The Royal Norwegian Air Force, *Forsvarets doktrine for luftoperasjoner* (The Norwegian Armed Forces, 2018), 17.

20 See for example Ferris, "Netcentric Warfare, C4ISR, and Information Operations," 220–21. There has also been a range of articles and books on mistakes done by drone operators in assessing a potential target for a drone attack, leading to the killing of innocent people, see for example Gregoire Chamayou, *Drone Theory* (New York: Penguin Random House, 2015); Ed Pilkington, "Life as a Drone Operator: 'Ever Step on Ants and Never Give It Another Thought?'" *The Guardian*, November 15, 2015; Ryan Devereaux, "The Drone Papers – Manhunting in the Hindu Kush," *The Intercept*, October 15, 2015, https://theintercept.com/drone-papers/manhunting-in-the-hindu-kush/. A good article with a defence of drone operations is T. Mark McCurley, "I Was a Drone Warrior for 11 Years: I Regret Nothing," *Politico*, October 18, 2015, www.politico.com/magazine/story/2015/10/drone-pilot-book-213263. Regardless of viewpoint, the lowered threshold for acting militarily on imagery is on display in the mentioned material.

21 Geoffrey Oxlee, *Aerospace Reconnaissance* (London: Brassey's, 1997), 88; Gordon, *Tactical Reconnaissance in the Cold War*, 181.

22 John L. Trefz, "From Persistent ISR to Precision Strike – The Expanding Role of UAVs" (Master's thesis, Naval War College, 2003), 12–14.

23 Thor A. Simensen, "Link-11 Communication" (Master's thesis, Naval Postgraduate School (NPS), 1992), 8–9.

24 Cowan, "Rethinking Command and Control of Intelligence, Surveillance and Reconnaissance (ISR)," 12–15.

25 For a detailed account of Western surveillance and reconnaissance aircraft that were shot down during the Cold War, see Tart and Keefe, *The Price of Vigilance.*
26 IMINT is part of Geospatial Intelligence (GEOINT), where the purpose is to "*describe, assess, and visually depict physical features and geographically referenced activities on Earth.*" See JP 2-0 "Joint Intelligence" for details.
27 Michael I. Handel, "Intelligence and Military Operations," *Intelligence and National Security* 5, no. 2 (1990); Ferris, "Netcentric Warfare, C4ISR, and Information Operations"; Anderson, "A Holistic Approach to Intelligence, Surveillance and Reconnaissance"; Deptula and Brown, "A House Divided"; Brown, "Airpower Is Intelligence, Surveillance, and Reconnaissance."
28 For an interesting discussion on the development of this capability in the early 1990s, see Marshall, *Near-Real Time Intelligence on the Tactical Battlefield.*
29 NATO defines the "tactical level" of war as "*the level at which activities, battles and engagements are planned and executed to accomplish military objectives assigned to tactical formations and units.*" (AAP-06)
30 For a discussion on the challenges of near-real time distribution of sensor products and intelligence following the Gulf War in 1991, see Marshall, *Near-Real Time Intelligence on the Tactical Battlefield*, 47–78.
31 NATO defines the "operational level" of war as "*the level at which campaigns and major operations are planned, conducted and sustained to accomplish strategic objectives within theatres or areas of operations.*" (AAP-06)
32 NATO defines the "strategic level" of war as "*the level at which a nation of a group of nations determines national or multinational security objectives and deploys national, including military, resources in order to achieve them.*" (AAP-06)
33 For an introduction to the different levels of war and their relationship to intelligence, see Handel, "Intelligence and Military Operations," 26–27.
34 Furquhar, *A Need to Know*; Morton, "Ears and Eyes in the Sky."
35 For more on this discussion, see John G. Heidenrich, "The State of Strategic Intelligence: The Intelligence Community's Neglect of Strategic Intelligence," *Studies in Intelligence* 51, no. 2 (2007); Deptula and Brown, "A House Divided"; Harrison Donnelly, "Q&A – Lt. Gen. John C. Koziol, Deputy Under Secretary of Defense, Joint and Coalition Warfighter Support, Director DoD ISR Task Force," *Geospatial Intelligence Forum* 8, no. 6 (September 2010); Deptula and Fransisco, "Air Force ISR Operations."
36 John Ferris og Michael I. Handel, "Clausewitz, Intelligence, Uncertainty and the Art of Command in Military Operations," *Intelligence and National Security* 10, no. 1 (1995): 17–30.
37 For writings on general intelligence theory, see for example Michael Herman, *Intelligence Power in Peace and War* (Cambridge: Cambridge University Press, 1996); David Kahn, "An Historical Theory of Intelligence," *Intelligence and National Security* 16, no. 3 (2001); Lock Johnson, "Bricks and Mortar for a Theory of Intelligence," *Comparative Strategy* 22, no. 1 (2003); Arthur S. Hulnick, "What's Wrong with the Intelligence Cycle?" *Intelligence and National Security* 21, no. 6 (2006); Michael Herman, J. Kenneth McDonald, og Vojtech Mastny, "Did Intelligence Matter in the Cold War?" *Forsvarsstudier* 1 (2006).
38 For an assessment on this, see John O. Birkeland og Gjert Lage Dyndal, "Fremtidig autonom droneteknologi og -konsepter," in *Når dronene våkner*, redigert av Tor Arne Berntsen, Gjert Lage Dyndal, og Sigrid Redse Johansen (Oslo: Cappelen Damm Akademiske, 2016).
39 Former drone operators from modern drone warfare have revealed significant uncertainty with regards to the basis, that is the sensor data and information, upon which decisions of weapons release have been made. See for example Pilkington, "Life as a Drone Operator"; Chamayou, *Drone Theory.*
40 NATO, *Allied Joint Doctrine for Joint Targeting* (NATO Standardization Agency, April 2016), 2-4–2-6.

41 NATO, 2-5–2-6.
42 HUMINT is a category of intelligence derived from information collected and provided by human sources. For details, see Joint Publication (JP) 2-0, *Joint Intelligence* (United States Department of Defense, October 22, 2013), B-4–B-5.
43 Newer aircraft, such as the F-35 Lightning II, have built-in sensors for targeting, and are not carrying targeting pods.
44 See for example Human Rights Watch, "Heed the Call – A Moral and Legal Imperative to Ban Killer Robots," *Human Rights Watch*, August 21, 2018, www.hrw.org/report/2018/08/21/heed-call/moral-and-legal-imperative-ban-killer-robots; PAX, "Killer Robots – What Are They and What Are the Concerns?" *PAX*, 2018, www.paxforpeace.nl/media/files/pax-booklet-killer-robots-what-are-they-and-what-are-the-concerns.pdf.

Bibliography

Anderson, Dagvin R.M. "A Holistic Approach to Intelligence, Surveillance and Reconnaissance." *Air & Space Power Journal* 25, no. 4 (Winter 2011).

Birkeland, John O. "The Potential of LIDAR as an Anti-Submarine Warfare Sensor." M. Phil (R), War Studies, University of Glasgow, 2009. http://theses.gla.ac.uk/1252/1/2009birkelandmphil.pdf.

Birkeland, John O., og Gjert Lage Dyndal. "Fremtidig autonom droneteknologi og -konsepter." In *Når dronene våkner*, redigert av Tor Arne Berntsen, Gjert Lage Dyndal, og Sigrid Redse Johansen. Oslo: Cappelen Damm Akademiske, 2016.

Bottom, David. *Overcoming ISR Data Challenges*. National Geospatial Agency (NGA), 2013. https://spie.org/Publications/Proceedings/Paper/10.1117/12.2019033.

Brown, Jason M. "Airpower Is Intelligence, Surveillance, and Reconnaissance." *Air & Space Power Journal* 27, no. 4 (August 2013).

Chamayou, Gregoire. *Drone Theory*. New York: Penguin Random House, 2015.

Clift, A. Denis. "Intelligence in the Internet Era – From Semaphore to Predator." *Studies in Intelligence* 47, no. 3 (2003).

Cowan, Nicholas P. *Rethinking Command and Control of Intelligence, Surveillance and Reconnaissance (ISR)*. U.S. Air Force, July 2015. https://dodccrp-testorg.squarespace.com/s/094-i2ln.pdf.

Danskine, William B. "Aggressive ISR in the War on Terrorism – Breaking the Cold War Paradigm." *Air & Space Power Journal* 19, no. 2 (Summer 2005).

Deptula, David A., og R. Greg Brown. "A House Divided – The Indivisibility of Intelligence, Surveillance and Reconnaissance." *Air & Space Power Journal* 22, no. 2 (Summer 2008).

Deptula, David A., og Michael Fransisco. "Air Force ISR Operations – Hunting Versus Gathering." *Air & Space Power Journal* 24, no. 4 (Winter 2010).

Devereaux, Ryan. "The Drone Papers – Manhunting in the Hindu Kush." *The Intercept*, 15. October 2015. https://theintercept.com/drone-papers/manhunting-in-the-hindu-kush/.

Donnelly, Harrison. "Q&A – Lt. Gen. John C. Koziol, Deputy Under Secretary of Defense, Joint and Coalition Warfighter Support, Director DoD ISR Task Force." *Geospatial Intelligence Forum* 8, no. 6 (September 2010).

Downs, Michael L. "Rethinking the Combined Force Air Component Commander's Intelligence, Surveillance, and Reconnaissance Approach to Counterinsurgency." *Air & Space Power Journal* 22, no. 3 (Fall 2008).

Ferris, John. "Netcentric Warfare, C4ISR, and Information Operations: Towards a Revolution in Military Intelligence?" *Intelligence and National Security* 19, no. 2 (2004): 199–225.

Ferris, John, og Michael I. Handel. "Clausewitz, Intelligence, Uncertainty and the Art of Command in Military Operations." *Intelligence and National Security* 10, no. 1 (1995): 1–58.

Fowler, Mike. "The Future of Unmanned Aerial Vehicles." *Global Security and Intelligence Studies* 1, no. 1 (Fall 2015).

Frayter, Slava, og Koen Willems. *Increased Efficiency or Beyond Line-of-Sight in Airborne ISR Operations*. National Geospatial Agency (NGA), 2013. https://proceedings.spiedigitallibrary.org/proceeding.aspx?articleid=1688649.

Furquhar, John T. *A Need to Know: The Role of Air Force Reconnaissance in War Planning, 1945–1953*. Maxwell AFB, Montgomery, AL: Air University Press, 2005.

Gordon, Doug. *Tactical Reconnaissance in the Cold War*. Barnsley: Pen & Sword Aviation, 2006.

Haley, Jaylan M. "An Evolution in Intelligence Doctrine – The Intelligence, Surveillance, and Reconnaissance Mission Type Order." *Air & Space Power Journal* 26, no. 5 (Fall 2012).

Handel, Michael I. "Intelligence and Military Operations." *Intelligence and National Security* 5, no. 2 (1990): 1–95.

Heidenrich, John G. "The State of Strategic Intelligence: The Intelligence Community's Neglect of Strategic Intelligence." *Studies in Intelligence* 51, no. 2 (2007).

Herman, Michael. *Intelligence Power in Peace and War*. Cambridge: Cambridge University Press, 1996.

Herman, Michael, J. Kenneth McDonald, og Vojtech Mastny. "Did Intelligence Matter in the Cold War?" *Forsvarsstudier* 1 (2006).

Hinrichs, Jeff S. "Education in Intelligence, Surveillance, and Reconnaissance Beyond the 'Green Door'." *Air & Space Power Journal* 22, no. 2 (Summer 2008).

Holzer, Jenny R., og Franklin L. Moses. "Autonomous Systems in the Intelligence Community: Many Possibilities and Challenges." *Studies in Intelligence* 59, no. 1 (2015).

Hulnick, Arthur S. "What's Wrong with the Intelligence Cycle?" *Intelligence and National Security* 21, no. 6 (2006): 959–79.

Human Rights Watch. "Heed the Call – A Moral and Legal Imperative to Ban Killer Robots." *Human Rights Watch*, August 21, 2018. www.hrw.org/report/2018/08/21/heed-call/moral-and-legal-imperative-ban-killer-robots.

Johnson, Lock. "Bricks and Mortar for a Theory of Intelligence." *Comparative Strategy* 22, no. 1 (2003): 1–28.

Joint Publication (JP) 2-0. *Joint Intelligence*. United States Department of Defense, October 22, 2013.

Kahn, David. "An Historical Theory of Intelligence." *Intelligence and National Security* 16, no. 3 (2001).

Kriendler, John. *NATO Intelligence and Early Warning*. Swindon, England: Conflict Studies Research Centre, Defense Academy of the United Kingdom, March 2006. https://www.files.ethz.ch/isn/39988/06_Apr.pdf.

Luddy, John. *The Challenge and Promise of Network-Centric Warfare*. Arlington, VA: Lexington Institute, February 2005.

Maceda, Steven. "Control of Theater Intelligence, Surveillance, and Reconnaissance for the Ground Commander." *Air & Space Power Journal* 22, no. 4 (Winter 2008).

Marshall, James P. *Near-Real Time Intelligence on the Tactical Battlefield*. Research Report. Maxwell AFB, Montgomery, AL: Air University, January 1994.

Martin, Matthew J. "Unifying Our Vision – Joint ISR Coordination and the NATO Joint ISR Initiative." *Joint Force Quarterly (JFQ)* 72 (1st Quarter 2014). http://ndupress.ndu.edu/Portals/68/Documents/jfq/jfq-72/jfq-72_54-60_Martin.pdf?ver=2014-03-13-152411-000.

McCurley, T. Mark. "I Was a Drone Warrior for 11 Years: I Regret Nothing." *Politico*, October 18, 2015. www.politico.com/magazine/story/2015/10/drone-pilot-book-213263.

Morton, Tyler. "Ears and Eyes in the Sky: The Evolution of Manned Airborne ISR." Master's thesis, School of Advanced Air and Space Studies, 2012. www.dtic.mil/get-tr-doc/pdf?AD=AD1019401.

NATO. *Allied Joint Doctrine for Joint Targeting*. NATO Standardization Agency, April 2016.

Oxlee, Geoffrey. *Aerospace Reconnaissance*. London: Brassey's, 1997.
PAX. "Killer Robots – What Are They and What Are the Concerns?" *PAX*, 2018. www.paxforpeace.nl/media/files/pax-booklet-killer-robots-what-are-they-and-what-are-the-concerns.pdf.
Perkins, William. *Alliance Airborne Anti-Submarine Warfare – A Forecast for Maritime Air ASW in the Future Operational Environment*. Joint Air Power Competence Centre (JAPCC), June 2016.
Pilkington, Ed. "Life as a Drone Operator: 'Ever Step on Ants and Never Give It Another Thought?'" *The Guardian*, November 15, 2015.
Rhodes, Carl, Jeff Hagen, og Mark Westergren. *A Strategies-to-Tasks Framework for Planning and Executing Intelligence, Surveillance, and Reconnaissance (ISR) Operations*. RAND, Project Air Force, 2007.
The Royal Norwegian Air Force. *Forsvarets doktrine for luftoperasjoner*. The Norwegian Armed Forces, 2018.
Shibilski, Daniel P. "Future of Air Force Intelligence." *United States Naval Institute Proceedings* 132, no. 2 (February 2006).
Simensen, Thor A. "Link-11 Communication." Master's thesis, Naval Postgraduate School (NPS), 1992.
Tart, Larry, og Robert Keefe. *The Price of Vigilance – Attacks on American Surveillance Flights*. New York: Ballantine Books, 2001.
Trefz, John L. "From Persistent ISR to Precision Strike – The Expanding Role of UAVs." Master's thesis, Naval War College, 2003.
USAF Chief Scientist. *Technology Horizons – A Vision for Air Force Technology 2010–2030*. Maxwell AFB, Montgomery, AL: Air University Press, 2010. www.defenseinnovationmarketplace.mil/resources/AF_TechnologyHorizons2010-2030.pdf.
Wilson, Clay. *Network Centric Warfare: Background and Oversight Issues for Congress*. Washington, DC: Congressional Research Service, June 2, 2004.
Winkler, Robert P. "The Evolution of the Joint ATO Cycle." Master of Science in Joint Campaign Planning and Strategy, Joint Advanced Warfighting School, 2006. https://dtic.mil/cgi-bin/GetTRDoc?AD=ADA451239.
Yeh, Puong F. "The Case for Using Robots in Intelligence Analysis." *Studies in Intelligence* 59, no. 4 (2015).
Young, Adam B. "Employing Intelligence, Surveillance, and Reconnaissance – Organizing, Training, and Equipping to Get It Right." *Air & Space Power Journal* 28, no. 1 (February 2014).
Zeytoonian, Dan, Laura Geldhof, Maureen Green, Remi Hajjar, Chris Litwhliler, Christine Locke, James Myers, David Perrine, og Cameron Weathers. "Intelligence Design: COIN Operations and Intelligence Collection and Analysis." *Military Review*, October 2006.

4 The formative years, 1945–1960

The history of submarines is closely intertwined with the history of the forces that have been tasked to neutralize them. There are many entities that must function well in concert with each other in order to "find, fix, and finish" a submarine, and, over the course of the Cold War, one of the most prominent assets for conducting ASW were MPA. The ASW efforts in the Atlantic Ocean in the twentieth century were termed the "Battles of the Atlantic."[1] The first battle aimed to secure allied civilian and military shipping during World War I. The second battle involved the safeguarding of the SLOC across the Atlantic from German U-boats during World War II. The third battle was about holding the Soviet submarines at risk – those that posed the threat of an initial nuclear strike on the United States and Europe and aimed to ensure the Soviets a second-strike capability during the Cold War.

The Northern Fleet on the Kola Peninsula expanded significantly in the 1940s and 1950s, and the role of Russian submarines grew in importance. In NATO, ASW technology was steadily advancing based on a momentum carried on from WWII. The 15 years between the end of WWII and 1960 stand out as formative due to early evolution in nuclear capabilities, the foundation of NATO, the consequences of the Korean War to naval strategy on both sides of the Iron Curtain, and the general development in subsurface technology, both for the hunter and the hunted.

The Norwegian Government was, in the first few years after WWII, reluctant to become too dependent on great power associations in either direction, but were approaching 1950 increasingly engaged in strengthening the Atlantic connection.[2] The single most central factor of Norwegian security politics became a reality in April 1949, when Norway became a founding member of NATO. The years following WWII saw the development of the Norwegian basing policy and the balancing act in the North: Integration towards the West and thus deterrence towards the East, and screening towards the West through restricting Allied movements and presence on Norwegian soil and thus reassurance towards the East.

In the 1940s, the lack of Norwegian capabilities to monitor the military developments in the Barents Sea and on the Kola Peninsula became evident through American surveillance aircraft flying from Norwegian bases into the area of intelligence interest. From the mid-1950s and onwards, Norwegian authorities enforced more explicit restrictions on Allied movements and presence in Norway due to a raised awareness of the requirement for low tensions in the North. This political

DOI: 10.4324/9781003626015-4

move would increase the focus on maritime surveillance in the High North. In the aftermath of WWII, the focus of the Norwegian Government, particularly in the North, was on rebuilding the nation, rather than establishing and improving the ability to follow the military build-up on the Kola Peninsula. A growing paradox became evident: An increase in Allied interest in the military developments in the High North and a decreasing Norwegian capability to monitor them. Maritime air operations in the High North would, starting in the 1950s, become a central factor in the bilateral relation between Norway and the United States. The obvious lack of Norwegian capabilities in all aspects of maritime air operations led, in the late 1950s, to efforts to bring in new MPA to the Norwegian inventory.[3]

This chapter provides an historical backdrop to the evolution of maritime airborne ISR in the North Atlantic. Not so much a discussion of the basic tenets of airborne ISR, it provides essential background for the more detailed discussion that follows later. The chapter shows the growing importance of submarines in the Soviet Union's strategic posture, and it shows that the most common response in NATO to the submarine threat, to the Third Battle of the Atlantic, was the establishment of a credible MPA force.

The Northern Fleet build-up

The build-up of the Soviet Northern Fleet on the Kola Peninsula started after WWII. However, Stalin's grandiose vision stemming from the 1930s of an ocean-going navy had to contend with the traditional Russian defensive focus, in addition to the obvious, harsh economic realities that followed the war. Plans to build aircraft carriers and battleships were, therefore, discarded in favour of submarines and heavy cruisers.[4] After the Korean War, two fundamental factors stood out for the development of submarines for the Soviet Union. First, the Korean War itself cemented the place of the US Navy's aircraft carriers in the evolving nuclear strategy as a launch platform for nuclear strikes against the Soviet Union.[5] The Soviets could no longer focus on defending their national and regional sea spaces: The threat had to be met at sea, long before the carriers came within striking distance of the Soviet mainland. This was a task for a robust and capable submarine service. Second, the death of Stalin in 1953 enabled a general review of Soviet strategy in general, and naval development specifically. The new head of the Navy, Admiral Gorshkov, emphasized the value of single units with strategic impact, in particular submarines, and oversaw a robust expansion of the submarine fleet in his first years as head of the navy.[6]

Gorshkov's massive submarine-building programme had the overall aim of disputing sea control with the NATO navies.[7] It was becoming increasingly clear for the Soviet Union that any contest of control of the Atlantic would be fought by the Northern Fleet, which had access to the only guaranteed year-round ice-free ports of the Russian Navy.[8] Western intelligence also picked up a significant increase in the level of submarine training, and a general extension of peacetime patrol areas, far beyond traditional Russian patrol areas and close to their own coastal waters.[9]

The Potsdam Agreement after WWII had ensured that the Soviet Navy received several Axis submarines as a share of the war spoils, including several advanced German *Type XXI* submarines.[10] After the reconstruction of war-damaged Soviet shipyards in the first 5 years following the war was completed, new submarine construction commenced. Several classes of diesel-electric propulsion were constructed in the 1950s. The benefits of nuclear over conventional propulsion, however, were recognized early: Increased transit speed, range, and unprecedented endurance. The Soviet Navy thus began developing a reactor for nuclear propulsion in 1953. The development of the *November*-class submarine was given high political priority and was rushed into production to such a degree that the prototype stage was bypassed altogether. This led to a leaky and unreliable propulsion system. Although the *November* only carried torpedoes, her commissioning in 1958 was a significant milestone. The submarine worked in tandem with the *Foxtrot* class in scouting far at sea.[11]

In 1954, the US Director of Central Intelligence (DCI) assessed through his National Intelligence Estimate (NIE) that the Soviets "*almost certainly*" would have a requirement for submarine-launched nuclear missiles for attacks on NATO. It was assessed as inevitable that nuclear missiles with an increasing range fired from submarines would pose a substantial threat by 1958.[12] Both the United States and the Soviet Union were already in possession of nuclear weapons, with the latter having detonated their first bomb in 1949. Highlighting this threat, the main advantage of submarine-launched missiles was the remote likelihood of interception of the launch platform. As a final recommendation, the NIE stated that the "primary defensive reliance would have to be placed on anti-submarine warfare."[13] The assessment was that the effectiveness of the Soviet submarine force would increase significantly.[14] Initial investigations into the feasibility of placing ballistic missiles on Soviet submarines date back to the year of the first Soviet nuclear detonation in 1949. As missile and rocket technology matured, test firings were carried out on submarines that were converted and configured for the new mission.[15] The *Zulu V* became the first Soviet nuclear ballistic missile submarine (SSBN) in 1956. This was a major step forward for the submarine service, although the *Zulus* had to fire their SS-N-4 missiles (range: 300–350 nautical miles) from the surface.[16]

The construction of the first purpose-built strategic ballistic missile submarine, the *Golf* class, was initiated in 1958. With a diesel-electric propulsion and carrying three SS-N-4 missiles, the *Golf* had to traverse the Atlantic and surface fairly close to the US coastline in order to fire her missiles. This made her extremely vulnerable to adversary anti-submarine forces. The Soviet Navy continued to develop a submarine that could launch her missiles from underneath the waterline and at a greater distance.[17] Meanwhile, the US Navy deployed the *Polaris A-1* missile with the *George Washington*-class submarine in 1960, which was capable of underwater launch and had a range that was four times that of the Soviet SS-N-4. Underwater launch became a reality for the Soviets in the early 1960s, with retrofitted *Golf*-class submarines with the D-4 system enabling underwater launch from a depth of 40–50 m.[18] The nuclear-powered *Hotel* class was set in production a year after the *Golf* class, essentially constituting a *Golf* class with a nuclear reactor. The *Hotel* was also retrofitted to carry the D-4 launch system, enabling it for underwater launch of the SS-N-4 missile.[19] The *Hotel* still had to manoeuvre close to the US coastline.

Experimentation with Soviet cruise missiles did not gain traction as early as it did in the United States, but was initiated nonetheless in the mid-1950s. The SS-N-3C *Shaddock* land-attack cruise missile (LACM) had a range of 400 NM and was armed with a nuclear warhead. It was fitted on *Whiskey*-class submarines in 1956–1957.[20] This was the precursor to the modern Russian submarines in the 2000s with significant cruise missile capabilities.

In 1956, the Soviet Naval High Command determined that the submarine fleet was to begin long-range patrols, in order to establish a Soviet blue water presence across the globe.[21] Sporadic and experimental in nature, the missions seem to have been more for establishing an institutional foundation for later missions than for actual operational effect. Regular patrols did not take shape until 10 years later. Missions along the European coastlines, interfleet transfers of submarines, and exercises in the Baltic Sea and the North Atlantic were more common.

Western strategy and anti-submarine warfare

A growing concern over the general Soviet naval expansion among NATO countries led to the establishment of the Atlantic Command under one supreme commander in 1952, SACLANT.[22] In the mid-1950s, the intelligence focus of the Americans and NATO started to include assessments of Soviet Northern Fleet expansion and operations in the North Atlantic.[23] The Soviet Navy was assessed to have "the greatest submarine capability in the history of naval warfare," and the primary capability was directed "against NATO sea communications."[24] This was also reflected in the NATO's strategic concept *MC 14/2* of October 1956. Although best known for outlining the overarching strategy of "massive retaliation," the strategic concept detailed the importance of SLOC and the crucial role that the North Atlantic region would play for anti-submarine forces due to the expanding Soviet submarine force and the increasing importance of the Northern Fleet.[25] Emphasis was given to the importance of constituting a credible deterrent, as well as the "ability to carry out an instant atomic counteroffensive by all available means."[26] As NATO strategy progressed, there was an increasing requirement to monitor not only the development of adversary weapon systems, but also their method of delivery. The growing submarine threat prompted discussions in NATO on the Alliance's ability to monitor and, if necessary, neutralize Soviet submarines. The Alliance's assessments of its own measures to meet the threat highlighted a particular lack of MPA.[27]

As World War II ended and a great portion of ASW dedicated vessels were decommissioned, the Americans in particular had channelled resources into improving ASW technology to replace dwindling assets. Between 1943 and 1950, developments included improved ASW radars, sonars, weapons, and tactics. In addition, more revolutionary technology was developed, such as passive acoustic arrays and an entirely new vessel, the ASW submarine.[28]

It was recognized during WWII that significant numbers of aircraft were required for surveillance in the Atlantic, and that these aircraft could not work alone. During WWII, the German submarines were hunting Allied shipping vessels, the latter giving a clear indication of where the enemy submarines would likely be operating. During the Cold War, the search for Soviet submarines required some sort of cueing in order for aircraft to track them down. This cueing would come in the shape of permanently stationed listening devices on the seabed.[29]

Passive acoustic technology had been under development during the 1940s, but it was the programme initiated in 1950 by the US Navy's Office of Naval Research (ONR) on the SOSUS that would change the ASW scene for decades to come. The new technology facilitated detecting hostile submarines passively at ranges unheard of until then. The low-frequency sound propagation phenomenon of *convergence zones* led to sound traveling deep down into the abyss, only to be bent upwards again towards the surface, leading to sensors being able to detect a submarine at approximately 30 NM. Up until the early 1950s, the best ranges experienced were around 10,000 yards, or approximately 5 NM. The new ASW submarines thus became a credible defensive asset in detecting approaching enemy submarines, and the network of SOSUS arrays on the seabed was able to detect enemy submarines at even greater ranges. The post-war *modus operandi* of MPA searching for submarines was to focus on cueing from SIGINT platforms that had picked up submarine communications, then use the radar to vector in on snorkelling or surfaced submarines, and subsequently get in position for an attack with the use of homing torpedoes.[30] Now, ASW submarines together with static, passive acoustic sensors of the SOSUS provided the MPA community with essential cueing for their much narrower search areas, enabling them to save fuel and resources, and arrive at a position closer to the target more rapidly than before. After WWII, the development of passive sonobuoys had nearly stopped, but the success of the SOSUS system gave sonobuoy development a necessary push forward. Sonobuoys were developed based on SOSUS passive acoustic principles and improved on-board processing capabilities.

Probably the most distinct feature of airborne ASW is the dropping of sonobuoys in order to search for frequencies emitted by the submarine.[31] Submarines are mechanically driven and emit sounds from their machinery in almost all circumstances. The sounds come from external units such as the propeller(s) and hydroplanes, or they come from internal equipment such as electrical generators. The aircraft drops disposable sonobuoys into the water, which function as floating microphones listening for sounds from the ocean. After dropping the buoys, several factors complicate matters. Sea life, shipping traffic, surface winds, and seismic activity are examples of what is referred to as *ambient noise*.[32] One of the challenges is to distinguish between ambient noise and the frequencies emitted by the submarine. Submarine commanders, meanwhile, will try to "hide" the sounds of the submarine among these ambient frequencies. Oceanographic factors such as salinity, temperature, and pressure impact how sound moves through the water. They differ according to time of day, proximity to lakes and fjords, littoral waters as opposed to the open ocean, seasonal changes to temperatures, and seasonal changes to currents and winds.

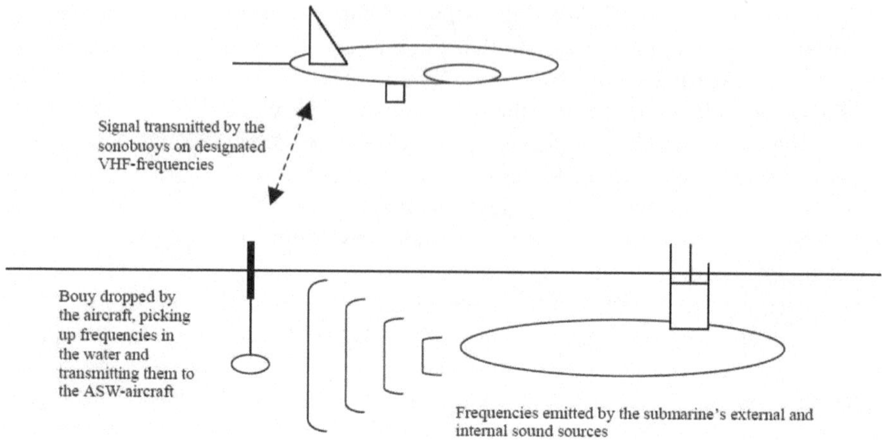

Figure 4.1 Acoustic sensors[33]

If a sonobuoy is only able to listen for submarine frequencies and other sounds in the water, it is usually referred to as a *passive* buoy. A buoy that transmits a sonar sound around it functions almost as an underwater radar that sends out an electronic signal in order to receive an echo of the emitted sound returned by the submarine. These buoys are referred to as *active* sonobuoys. With knowledge of the local water conditions, and the speed of sound through it, the distance and possibly the direction of the submarine from the buoy can be instantly calculated, thereby accelerating target localization.[34] The downside to an active "hit" like this is that the submarine will be warned that it has been detected.

The continued development of sonobuoys meant that MPA could fly to the last known position of the submarine, known as a *datum*, provided by the SOSUS system, and continue the passive search with the use of sonobuoys. Acoustic techniques required high-level skills by the operators, and it took a long time to gain proficiency in operating the equipment. Extensive experimentation was thus undertaken in the late 1950s in order to improve the equipment, operator techniques, and integrated tactics with submarines and the SOSUS community.[35] As a nuclear submarine, the *November* class was able to defeat traditional ASW tactics that were based on radar and active sonar (due to the speed of the nuclear submarine, the sonars of the surface ships were not able to stay focused on her position), but she was vulnerable to the new passive acoustic technology. The Soviets did not focus much on quieting measures for their submarines, which made Soviet nuclear submarine classes very much detectable by passive means. The process of quieting submarines was costly, took time and resources, and considerable investment in research. Quieting of emitted noise was also based on a significant attention to detail during the manufacturing process, something that worked against the usual Soviet rush to field new systems, as exemplified by the *November* class.[36] The Soviets were more concerned about water resistance, which detracted from speed, than they were about noise from water flow and machinery. By the time Soviet nuclear submarines initiated their operations in the North Atlantic and the Pacific in the late 1950s and early 1960s, the Americans and their Allies had developed an extensive ASW network through a triad of ASW submarines, the SOSUS, and MPA.[37]

American P2V *Neptune* MPA operations were initiated in 1947. The aircraft was very advanced both in range and in sensor suite, but it was also costly to operate. The *Neptune* required extensive training to operate, and an elaborate ground support organization. In 1952, the first US Navy *Neptune* squadron deployed to Naval Air Station Keflavik on Iceland, in order to be closer to the operating areas of the growing Soviet submarine fleet.[38] Tracking of Russian submarines by ASW units from NATO nations in the North Atlantic was commonplace by the late 1950s. The first few years operating out of Keflavik required familiarization with the harsh weather environment in the North Atlantic, familiarization with new ASW equipment, and perfecting ASW tactics.[39] In addition, the American *Neptune* aircraft flew extensively in the North Atlantic, not only based in Iceland but also periodically based in Norway.[40]

After WWII, the United States recalled the lend-lease agreement on the B-17s and B-24s that formed the backbone of the RAF's Coastal Command inventory. This caused the RAF to order nearly 30 aircraft of a maritime version of the Avro *Shackleton* in 1949. Although the first *Shackletons* arrived in 1951 and initiated operations and exercises on a small scale the year after, it would take several years before the fleet was operational. Thus, between 1952 and 1956, the British also flew *Neptunes* in the maritime patrol capacity. The *Shackleton* became sufficiently operational in 1956 for the remaining *Neptunes* to be returned to the United States.[41] Through similar arrangements as the Americans, the British also flew out of Norwegian airfields in the 1950s to cover the expanding Soviet naval operations.[42]

After WWII, the Norwegian maritime surveillance aircraft returned to Norway after serving under the British Coastal Command. The *Catalina* flying boats were utilized for everything but maritime surveillance and reconnaissance, such as postal flights and general transportation sorties, flights in support of fishery surveillance, and in support of clearing mines from WWII.[43] Adding to this, the *Catalinas* were put on readiness for search-and-rescue (SAR) tasks, which had the Norwegian maritime surveillance and reconnaissance community drift further away from their wartime tasks: Early warning and ASW operations. During a period of significant development of the Royal Norwegian Air Force, maritime air operations were, throughout the 1940s, put aside to the benefit of the fighter aircraft community.[44]

After WWII, the ideas and functions that formed the basis for the allied fusion of OPINTEL sources perfected during the war were enhanced by new technology.[45] In 1948, the Americans and the British had signed an agreement on intelligence sharing with specific regard to OPINTEL sharing of naval intelligence, building on broader Anglo-American intelligence-sharing agreements.[46] This sharing was crucial for a seamless integration of resources put into detecting and tracking Soviet submarines during the Cold War. NATO did not have any intelligence assets of its own and was dependent on national intelligence efforts feeding into the Alliance's operations and the planning. Arrangements for the bilateral sharing of intelligence functioned to the benefit of NATO in the sense that they raised the quality of the intelligence products. They were not, however, driven and organized by NATO *per se*.[47]

When the nuclear-powered submarine *USS Nautilus* was commissioned in 1954, the speed and lack of requirement to surface for air provided an indication of how difficult it would be to track future enemy nuclear submarines. But it was also discovered that nuclear submarines were quite noisy, even when they were submerged, because of the reactor coolant pumps, which operated continuously, and the reduction gears, which reduced the steam turbine shaft output revolutions.[48] This knowledge led to the Western ASW posture of combining dedicated ASW submarines, MPA, and the SOSUS as the dominant tools in the search and tracking of adversary submarines. Thus, the US Navy Chief of Naval Operations (CNO) gave orders in 1960 to track every Soviet missile-carrying submarine that was operating within range of the United States.[49] And in order to stay ahead of Soviet submarine developments, NATO established an ASW Research Centre in 1959. The centre was proposed by SACLANT in order to streamline Alliance research and development for the advancement of ASW capabilities.[50] Submarines and ASW were thus taking centre stage in day-to-day operations, in intelligence discussions, and doctrinal development early in the Cold War.

The Norwegian balancing act – integration and screening

There were few explicit national requirements in Norway for maritime surveillance in the High North in the late 1940s. NATO intelligence focused on a potential Soviet invasion of the European heartland on the continent.[51] And with the lack of explicit tasking, the maritime airborne surveillance capability was organizationally, technologically, and with regard to competence and skills, in free fall.[52] The United States and the United Kingdom, however, were increasingly concerned about the Soviet build-up in the High North, and the lacking Norwegian surveillance efforts in the north prompted increasing surveillance activity by the Americans and the British in the late 1940s. The increased Allied activity in general shone a light on the Norwegian geopolitical and military-strategic placement. The Norwegian position in geopolitics has been a balancing act between integration towards the West and assurance towards the East. This duality produced a policy vis-à-vis allied partners of *integration/deterrence* and *screening/reassurance*. Norway would integrate with the West and NATO, in order to assure her own security, and NATO membership would constitute a deterrent effect in itself. But, at the same time, Norway would screen her partners in the sense of limiting foreign activity and basing on Norwegian soil, thereby also reassuring the Soviets. Norway thus hoped to raise the threshold for any Soviet aggression through Western integration. At the same time, restrictions were placed on NATO presence and activity in order to maintain low tensions in the High North. The formal Norwegian basing policy was explicitly proclaimed in February 1949, where the government stated that Norway "will not open her bases to foreign powers on Norwegian territory as long as Norway is not subject to attack or a threat thereof."[53] The intention of the policy was to keep foreign powers from *permanently* basing forces on Norwegian soil; exercises would thus be allowed. Only 2 months later, Norwegian foreign and security policy reached its most significant historical milestone when Norway signed the North Atlantic Treaty. Now, the friendly aircraft flying surveillance missions in the High North were alliance partners, and the Norwegian Government saw the need to strengthen the restrictions it had put in place through the basing policy, as we shall see.

It was not until the outbreak of the Korean War in 1950, however, that the Norwegian Government initiated its own regular surveillance missions in the High North. Russian involvement in the Korean War was, by some, suspected to be a distraction in order to tie up Western military resources in the Far East, and NATO closely monitored Russian movements elsewhere in order to notice the run-up to what could be the main event on the European continent.[54] The newly initiated Norwegian surveillance missions aimed to monitor Soviet movements on and outside the Kola Peninsula. However, the squadron and the maritime air operations community had been marginalized due to prioritization of the Norwegian fighter aircraft capability, in addition to other national and organizational factors.[55] The MPA were old; there was a significant lack of personnel; and those that were available were poorly trained. The *Catalinas* had functioned in ASW operations during World War II mostly through visual sightings and torpedo attacks against surfaced submarines. Although the Norwegian *Catalina* aircraft contributed to surveillance also after WWII, the role of early warning had to be performed almost exclusively by American and British aircraft in the North Atlantic.[56] In parallel to Norwegian operations up north, the British established routine monitoring of the Baltic Sea outlets, where the situation was not to be underestimated.[57] In April 1950, Soviet Union fighter aircraft shot down a US Navy PBY4–2 *Privateer* MPA over the Baltic Sea.[58] Two years later, a Soviet fighter aircraft shot down a Swedish SIGINT aircraft, killing everyone on board. Only days later, a Swedish *Catalina* flying boat searching for the missing aircraft was shot at by yet another Soviet fighter aircraft and had to make an emergency landing.[59] A defected Polish pilot later explained that the Soviet fighter pilots were under orders to shoot down any aircraft that did not obey orders to land, whether or not they had opened fire.[60]

The level of Norwegian surveillance efforts, meanwhile, were deemed unacceptable with regard to both regularity and capabilities, and NATO proposed in 1951 to deploy two maritime surveillance squadrons to Norway in order to increase the surveillance capability in the North.[61] The Norwegians rejected the proposal on the basis of the aforementioned basing policy.[62] Norwegian authorities considered acquiring the advanced MPA P2V *Neptune* for surveillance in the High North, but the *Neptune* demanded highly skilled and experienced personnel and was expensive to operate and maintain.[63] The Norwegians were lacking in the former, and could not afford the latter. Surveillance missions thus continued with the outdated *Catalinas*.

During the establishment of the command structure in 1951, Norway was placed under the command area for the Supreme Allied Commander *Europe* (SACEUR) and not SACLANT. The fact that Norway was under the geographical command of SACEUR underlines that the strategic emphasis of the Alliance was on defending against a Soviet thrust through mainland Europe. In the late 1950s, the North Atlantic and the Arctic had not yet received as much strategic planning attention and were considered a part of the Northern Flank to the European mainland theatre of war. Only in the late 1960s would the naval threat from the Soviet Northern Fleet manifest itself to such a degree that the High North came to be considered a theatre of war in its own right.[64] SACLANT had, however, been given the responsibility for allied reconnaissance in the Atlantic. With military cooperation through the mechanisms of the Alliance came increased NATO activity in Norway: More exercises; more equipment flowing into the country through the Mutual Defence Assistance Program (MDAP), and a general increase in foreign presence. The weak Norwegian contribution to maritime surveillance, in particular, led to an increase in Allied maritime surveillance activity in the High North. The Allied desire to conduct surveillance missions out of Norwegian airfields towards the East, in combination with the Norwegian requirement for stability and predictability in the High North, led to bi- and trilateral arrangements for surveillance.

In spite of limited resources, the Norwegian Intelligence Service (NIS) was active and provided credible and sought-after services, particularly in the High North. This led to the NORUSA Agreement of 1954, where the Norwegians provided intelligence to the Americans and the latter provided equipment to the former in exchange.[65] The agreement initially concerned COMINT but expanded to more fields of intelligence cooperation in the following years. This was the starting point for the close relationship between American and Norwegian intelligence agencies. By the end of the 1950s, the NIS had established itself as a reliable intelligence provider to the Norwegian Government as well as to the Allies. To an increasing degree, the NIS worked to keep Allies out of areas where it could handle the information gathering tasks by itself.[66]

In 1955, the Americans requested the Norwegian Government to operate their much more capable P2V *Neptune* out of bases in Northern Norway, since their home base of Keflavik, Iceland, was too far away to give any meaningful time in the operations area up north. Among the requested airbases were Andøya, which then just recently had been built with funding through the NATO infrastructure programme for several reasons.[67] The Soviet build-up caused the SACLANT to demand increased surveillance in the North, in general. Also, he required a dedicated base for conducting ASW in the North, and finally, he required a base for offensive strikes into the Soviet Union in the North.[68] The Americans also saw, from a national perspective, that the requirements for information on Soviet activities north of the Kola Peninsula had grown significantly. The Norwegians, however, still did not have the capacity to meet the requirements for early warning, updated targeting information, and an updated foundation for war plans for the High North.[69] The arrangement that was sought was to fly American aircraft to Norway in order to familiarize the aircrews with bases and topography, as they would have to operate out of these bases in wartime. It was, however, understood that the arrangement would meet general surveillance requirements for the Americans as well. The flights to and out of Norway would take place approximately every three weeks. The arrangement was approved by the Norwegian Government in 1955, and was called *NEPTUNE JOURNEY*.[70] This bilateral agreement was to become the cornerstone for American–Norwegian cooperation with regard to airborne maritime surveillance. Radio communication procedures improved significantly, as did flight safety and tactical procedures for flying in the High North.[71]

Several NATO exercises in the North Atlantic in the early 1950s operated with a 24°E longitude delineation as the easternmost restriction for naval vessels and aircraft, based on concerns about reactions from the Soviet Union to increased Western activity in the High North.[72] The 24° restriction grew into an accepted standard for tasking foreign aircraft in Norway, but the final push for a formal restriction was the American request for a simplified authorization procedure for foreign air operations in Norway. The restriction was finally declared a national policy on 21 January 1959.[73] The 24°E restriction was thus established as the easternmost limitation for any surveillance carried out by allied partners flying out from Norwegian soil, including *NEPTUNE JOURNEY* flights.[74] Also, Norwegian bases were not to be used by Allies conducting operations that involved overflights of Soviet territory. This coordination mechanism was to become a practical example of the balancing act in Norwegian security policy between integration and screening.[75] Inadvertently, the Norwegian Government transformed airborne maritime surveillance efforts into a key instrument of security policy in the High North.[76]

Where the *NEPTUNE JOURNEY* arrangement was meant to meet the general requirements for surveillance, there was a growing requirement for specific coordination with regard to surveillance of specific Soviet exercises when they took place. The three most active nations in this regard were the United States, the United Kingdom, and Norway. Early in 1959, coordination culminated in a trilateral agreement *EXERCISE POPEYE*. The orders divided the ocean areas that were

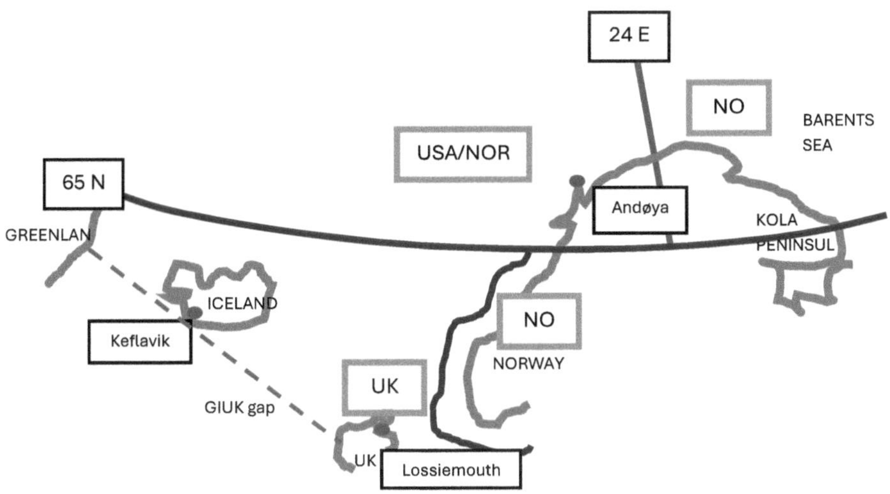

Figure 4.2 Delineation of geographical responsibility

mostly used by the Soviets for exercises into three parts: The Norwegians would be responsible for the Skagerrak basin along the coast up to 65°N and extending 60 nautical miles from the shore; the Norwegians and the Americans would share responsibility for the Norwegian Sea and the Barents Sea; and the British would be responsible for the Norwegian Sea south of 65°N, beyond 60 nautical miles from the Norwegian shore.[77] *EXERCISE POPEYE*, together with *NEPTUNE JOURNEY*, provided the foundation for bi- and trilateral cooperation of surveillance missions in the High North throughout the Cold War.

It was clear to everyone involved in monitoring the High North that the Norwegian *Catalina* aircraft were wholly inadequate for the task due to the lack of capable sensors on board. By the end of the 1950s, the number of American *Neptune* flights increased from two to three a month to approximately eight missions a month covering the North Atlantic. In addition, the Americans flew the RB-47 SIGINT surveillance aircraft over the Barents Sea, along the Soviet coast of the Kola Peninsula, and Novaya Zemlya. The RB-47 aircraft initially flew from the United Kingdom, but soon also flew from Alaska and Greenland, in order to reach further into and along Soviet territory. The RB-47 were supported by tanker aircraft facilitating the long sorties.[78] Due to the poor Norwegian capabilities, early warning missions in the High North and in the Norwegian Sea were, by the end of the 1950s, in their entirety flown by British *Shackleton* and American *Neptune* aircraft.[79] As for Norway and many other alliance partners, there was a dire need for a new aircraft to conduct airborne maritime ISR.

The Norwegian balancing act in the High North led to formal policy, where no foreign power was allowed to permanently place troops, aircraft, or equipment on Norwegian soil. And for air operations specifically, the 24°E restriction kept foreign surveillance aircraft from taking off from Norwegian bases and flying east of 24°E. The intent of these restrictions was to maintain a predictable interaction and low tensions between NATO and the Soviet Union in the Northern Fleet's backyard. As we shall see, this put pressure on the Norwegian Government to acquire the appropriate tools to monitor Soviet naval movements and operations. In keeping track of the Soviet submarine threat, the West deployed an extensive array of ASW measures. The seabed-deployed SOSUS low-frequency arrays acted as cueing mechanisms for more flexible assets such as MPA, which were becoming an increasingly capable tool. Arrangements were in place for Allied temporary deployments to Norway in order to streamline the maritime airborne surveillance and reconnaissance effort.

The 15 years that followed WWII were formative for the two main arguments of this book, and the two lines of thought are closely intertwined. First, the strategic argument is underpinned by the growing importance of submarines with regard to the naval and nuclear threat from the Soviet Union. Second, technological aspects emerged to the fore through the innovative environment that existed in the aftermath of WWII. Research on acoustic technology had gained momentum during the war and was able to maintain traction in the years that followed. Based on experience from WWII and the technological development that took place after the war, the notion established itself in NATO of MPA as the *go-to* platform for effective long-range ASW. MPA were the natural solution to the strategic submarine challenge, and this seems to have shaped the strategic-technological dynamics within ASW for the remainder of the Cold War.

Notes

1 Cote, "The Third Battle – Innovation in the U.S. Navy's Silent War Struggle with Soviet Submarines," 2; Foggo and Fritz, "The Fourth Battle of the Atlantic."
2 Olav Njølstad, *Jens Chr. Hauge – Fullt og helt*, 4. utg. (Oslo: Aschehoug, 2008), 405–25; Klevberg, "Maritime Surveillance in the North," 150.
3 Klevberg, "Maritime Surveillance in the North," 139, 157–67.
4 Gjert Lage Dyndal, "The Northern Flank and High North Scenarios of the Cold War" (Peripherie oder Kontaktzone? Die NATO-Flanken 1961 bis 2013, Zentrum für Militärgeschichte und Sozialwissenschaften der Bundeswehr, Potsdam, Germany, 2013), 8.
5 John Jordan, *Soviet Submarines – 1945 to the Present* (London: Arms & Armour Press, 1989), 46; Jan S. Breemer, *Soviet Submarines – Design, Development and Tactics* (Surrey, UK: Jane's Information Group Ltd., 1989), 103.
6 Natalia Yegorova, "Stalin's Conception of Maritime Power: Revelation from the Russian Archives," *Journal of Strategic Studies* 28, no. 2 (2005): 158–59; Sergei Chernyavskii, "The Era of Gorshkov: Triumph and Contradictions," *Journal of Strategic Studies* 28, no. 2 (2005): 287–93; Jordan, *Soviet Submarines – 1945 to the Present*, 41.
7 Dyndal, "The Northern Flank and High North Scenarios of the Cold War," 10.
8 SG 161/8 (Part II), "The Soviet Bloc Strength and Capabilities 1955–1959," NATO, March 18, 1955, NATO, Online Archive, 124–30.
9 Mats Berdal, "Forging a Maritime Alliance – Norway and the Evolution of American Maritime Strategy 1945–1960," *Forsvarsstudier* 4 (1993): 24–41.
10 For excellent introductions to the Whiskey, Zulu, Quebec, Romeo and Foxtrot classes, see Norman Friedman, *Submarine Design and Development* (London: Conway Maritime, 1984), 100–1; Jordan, *Soviet Submarines – 1945 to the Present*, 22; Breemer, *Soviet Submarines – Design, Development and Tactics*, 80; Defense Intelligence Agency (DIA), *Russia Military Power – Building a Military to Support Great Power Aspirations* (Defense Intelligence Agency, 2017).
11 Jordan, *Soviet Submarines*, 45.
12 NIE 11-6-54, "Soviet Capabilities and Probable Programs in the Guided Missile Field," National Intelligence Estimate, Director of National Intelligence, October 5, 1954, CIA, Online Archive, 3, 12.
13 NIE 11-6-54, 24.
14 NIE 11-3-55, "Soviet Capabilities and Probable Soviet Courses of Action Through 1960," Director of National Intelligence, May 17, 1955, CIA, Online Archive, 2–3.
15 Podvig, *Russian Strategic Nuclear Forces*, 237; Friedman, *Submarine Design and Development*, 103.
16 Jordan, *Soviet Submarines – 1945 to the Present*, 57; Podvig, *Russian Strategic Nuclear Forces*, 283–86.
17 Friedman, *Submarine Design and Development*, 103; Podvig, *Russian Strategic Nuclear Forces*, 286–90; Jordan, *Soviet Submarines – 1945 to the Present*, 61–62.
18 Podvig, *Russian Strategic Nuclear Forces*, 237–38, 317.
19 Friedman, *Submarine Design and Development*, 103; Jordan, *Soviet Submarines – 1945 to the Present*, 62–63; Podvig, *Russian Strategic Nuclear Forces*, 290–93.
20 Jordan, *Soviet Submarines – 1945 to the Present*, 66.
21 Gary E. Weir og Walter J. Boyne, *Rising Tide – The Untold Story of the Russian Submarines That Fought the Cold War* (New York: Basic Books, 2003), 43.
22 Berdal, "Forging a Maritime Alliance," 13.
23 See for example Øystein Espenes and Gjert Lage Dyndal, "The Changing Focus of NATO-Intelligence – From Southern Scandinavia to the 'High North' at the End of the 1950s," *Tidsskrift (Kungliga Krigsvetenskapsakademien)* 2 (June 2013), for an

extensive discussion on increased intelligence emphasis given to the Northern Flank in the time period around 1956/57.

24 SG 161/10, "The Soviet Bloc Strength and Capabilities 1957–1961 – Part I," March 20, 1957, Electronic Document Folder: IMS/MC-MCM-SG/1953–1970, NATO Archives, Brussel, 19.

25 MC 14/2, "The Overall Strategic Concept for the Defence of the NATO Area," October 14, 1956, Electronic Document Folder: IMS/MC-MCM-SG/1953-1970. NATO Archives, Brussel, 14, 16, 25, 26.

26 MC 14/2, 12.

27 AR(56)GENERAL-WP/1, "Report on the 1956 Annual Review," November 23, 1956. Electronic Document Folder: IMS/1955–1959. NATO Archives, Brussel, 2.

28 Cote, "The Third Battle – Innovation in the U.S. Navy's Silent War Struggle with Soviet Submarines," 14.

29 Cote, 24.

30 Owen, *Anti-Submarine Warfare*, 164–67; Cote, "The Third Battle – Innovation in the U.S. Navy's Silent War Struggle with Soviet Submarines," 31.

31 Birkeland, "The Potential of LIDAR as an Anti-Submarine Warfare Sensor," 15.

32 Birkeland, 15.

33 Birkeland, 16.

34 R.A. Mason, *Air Power – An Overview of Roles* (London: Brassey's Defence Publishers, 1987), 104.

35 Podvig, *Russian Strategic Nuclear Forces*, 276–77; Cote, "The Third Battle – Innovation in the U.S. Navy's Silent War Struggle with Soviet Submarines," 31–33.

36 Jordan, *Soviet Submarines – 1945 to the Present*, 54; Cote, "The Third Battle – Innovation in the U.S. Navy's Silent War Struggle with Soviet Submarines," 39, 45.

37 Cote, "The Third Battle – Innovation in the U.S. Navy's Silent War Struggle with Soviet Submarines," 38–40.

38 "Command History Report (CHR) for VP-7 (1952)," 7; Reade og Harper, "Naval Air Station Keflavik, Iceland," 11; Borgeld, "Keflavik – from the Past to the Present," 39; Michael D. Roberts, *Dictionary of American Naval Aviation Squadrons*, vol. 2 (Washington, DC: Naval Historical Center, 2000), 2:62, 71, 81, 760.

39 "Command History Report (CHR) for VP-10 (1955)," US Navy, 1955, FOIA case number: DON-NAVY-2018-00574, Naval History and Heritage Command; "Command History Report (CHR) for VP-21 (1959)," US Navy, 1959, FOIA case number: DON-NAVY-2018-00574, Naval History and Heritage Command.

40 For an elaborate account of the development and service life of the P2V *Neptune* Maritime Patrol Aircraft, see Wayne Mutza, *Lockheed P2V Neptune – An Illustrated History* (Atglen, PA: Schiffer Publishing Ltd., 1996).

41 Several aircraft of the original 52 *Neptunes* delivered had been crashed or written off. For a detailed overview of the development and operationalization of the Shackleton MPA, see Ashworth, *Avro's Maritime Heavyweight*.

42 This book will not discuss the American background for acquiring the *Neptune* nor the British background for acquiring the *Shackleton*. Suffice it to say, they were flying operationally and were middle-aged aircraft not requiring immediate replacement when other alliance partners started collectively looking for replacement aircraft for their increasingly obsolete aircraft in the mid- and late 1950s. This collective effort will be discussed in the next chapter.

43 Klevberg, "Maritime Surveillance in the North," 139–43.

44 Klevberg, 147–55.

45 Christopher Ford og David Rosenberg, *The Admiral's Advantage – U.S. Navy Operational Intelligence in World War II and the Cold War* (Annapolis, MD: Naval Institute Press, 2005), 34–35.

46 Richard J. Aldrich, *GCHQ – The Uncensored Story of Britain's Most Secret Intelligence Agency* (London: Harper Press, 2011), 89; Ford og Rosenberg, *The Admiral's Advantage*, 38.

47 For articles on intelligence sharing within NATO, see for example Adriana N. Seagle, "Intelligence Sharing Practices Within NATO: An English School Perspective," *International Journal of Intelligence and Counterintelligence* 28, no. 3 (2015); Jan Ballast, "Merging Pillars, Changing Cultures: NATO and the Future of Intelligence Cooperations within the Alliance," *International Journal of Intelligence and Counterintelligence* 31, no. 4 (2018).
48 Cote, "The Third Battle – Innovation in the U.S. Navy's Silent War Struggle with Soviet Submarines," 21.
49 Michael A. Palmer, *Origins of the Maritime Strategy – The Development of American Naval Strategy, 1945–1955*, reissue ed. (Annapolis, MD: Naval Institute Press, 1990), 92–93.
50 AC/137-R/3, "Science Committee – Summary Records of Meetings Held at the Palais de Chaillot, Paris, January 5–8, 1959," 30–33; C-M(59)39, "Standing Group – SACLANT Anti-Submarine Warfare Centre," NATO, April 14, 1959, CD: IS-AC-0385. NATO Archives, Brussel.
51 Espenes og Dyndal, "The Changing Focus of NATO-Intelligence," 32–33.
52 Svein Duvsete, "Fra Luftforsvar til Strategisk Angrep. Norsk Luftmilitær Doktrine 1945–1955," *Forsvarsstudier* 2 (1998): 97–98.
53 Tamnes, "Integration and Screening," 60–61.
54 Gaddis, *Strategies of Containment. A Critical Appraisal of Postwar American National Security Policy*, 114.
55 Klevberg, "Maritime Surveillance in the North," 158–61.
56 Klevberg, 180.
57 Aldrich, *GCHQ – The Uncensored Story of Britain's Most Secret Intelligence Agency*, 117–19.
58 Aldrich, 116.
59 Aid og Wiebes, *Secret of Signals Intelligence During the Cold War and Beyond*, 234; Polmar and Bessette, *Spyplanes*, 69.
60 Aldrich, *GCHQ – The Uncensored Story of Britain's Most Secret Intelligence Agency*, 125–27.
61 Tamnes, *The United States and the Cold War in the High North*, 117.
62 Tamnes, 117.
63 Klevberg, "Maritime Surveillance in the North," 194, 221.
64 For an explanation of the evolution of the Northern Flank in NATO strategy, see Gjert Lage Dyndalm, "How the High North Became Central in NATO Strategy: Revelations from the NATO Archives," *Journal of Strategic Studies* 34, no. 4 (2011).
65 Olav Riste og Arnfinn Moland, *Strengt Hemmelig – Norsk etterretningstjeneste 1945–1975* (Oslo: Universitetsforlaget, 1997), 132–37.
66 Tamnes, *The United States and the Cold War in the High North*, 173–74; Riste og Moland, *Strengt Hemmelig – Norsk etterretningstjeneste 1945–1975*, 129–55.
67 Vengstad, "Fleksibilitet og slagkraft. Luftforsvaret og NATOs infrastrukturprogram 1950–1957," 70–72.
68 Vengstad, 70–72.
69 Hopkins, *U.S. Strategic Aerial Reconnsaissance and the Cold War, 1945–1961*, 41–42.
70 Tamnes, *The United States and the Cold War in the High North*, 118; Klevberg, "Maritime Surveillance in the North," 175.
71 Klevberg, "Maritime Surveillance in the North," 177–78.
72 Tamnes, "Integration and Screening," 60–61; Klevberg, "Maritime Surveillance in the North," 168; Knut E. Eriksen og Helge Ø. Pharo, *Kald Krig og Internasjonalisering, 1945–1965*, Bd. 5. Norsk utenrikspolitisk historie (Oslo: Universitetsforlaget, 1997); Knut Egeland Moen, *Selvpålagte restriksjoner i nord, 1945–65* (Oslo: Institutt for Forsvarsstudier, 1998), 44. Enforced during exercise *Lifeline, Sea Enterprise, Air Enterprise* and *Fox Paw in 1955.*
73 Knut E. Moen, "Norway's Self-Imposed Restrictions in the High North, 1945–65," *Forsvarsstudier* 5 (1998): 44–49.

74 Klevberg, "Maritime Surveillance in the North," 177.
75 Egeland Moen, "Selvpålagte restriksjoner i nord, 1945–65," 44–49; Tamnes, *The United States and the Cold War in the High North*, 175; Klevberg, "Maritime Surveillance in the North," 181–82.
76 For an outstanding discussion on Norwegian internal politics and dynamics in the development of airborne maritime surveillance while simultaneously standing up the Norwegian Air Force post-World War II, see Klevberg, "Maritime Surveillance in the North."
77 Klevberg, 178–79.
78 Hopkins, *U.S. Strategic Aerial Reconnaissance and the Cold War, 1945–1961*, 53–54; DUPI, *Grønland under den Kolde Krig. Dansk og amerikansk sikkerhedspolitik 1945–68* (Copenhagen: Dansk Udenrigspolitisk Institut, 1997), 315–16; Klevberg, "Maritime Surveillance in the North," 175–81.
79 Tamnes, *The United States and the Cold War in the High North*, 116–17.

Bibliography

AC/137-R/3. "Science Committee – Summary Records of Meetings Held at the Palais de Chaillot, Paris, January 5–8, 1959." North Atlantic Treaty Organization (NATO), January 19, 1959. CD: IS-AC-0385. NATO Archives, Brussel.

Aid, Matthew M., og Cees Wiebes, reds. *Secret of Signals Intelligence During the Cold War and Beyond*. London: Frank Cass, 2001.

Aldrich, Richard J. *GCHQ – The Uncensored Story of Britain's Most Secret Intelligence Agency*. London: Harper Press, 2011.

AR(56)GENERAL-WP/1. "Report on the 1956 Annual Review." November 23, 1956. Electronic Document Folder: IMS/1955–1959. NATO Archives, Brussel.

Ashworth, Chris. *Avro's Maritime Heavyweight: The Shackleton*. Bourne End: Aston Publications Ltd., 1990.

Ballast, Jan. "Merging Pillars, Changing Cultures: NATO and the Future of Intelligence Cooperations Within the Alliance." *International Journal of Intelligence and Counterintelligence* 31, no. 4 (2018).

Berdal, Mats. "Forging a Maritime Alliance – Norway and the Evolution of American Maritime Strategy 1945–1960." *Forsvarsstudier* 4 (1993).

Birkeland, John O. "The Potential of LIDAR as an Anti-Submarine Warfare Sensor." M. Phil (R), War Studies, University of Glasgow, 2009. http://theses.gla.ac.uk/1252/1/2009birkelandmphil.pdf.

Borgeld, Warner. "Keflavik – from the Past to the Present." *Maritime Patrol Aviation*, 2004.

Breemer, Jan S. *Soviet Submarines – Design, Development and Tactics*. Surrey, UK: Jane's Information Group Ltd., 1989.

Chernyavskii, Sergei. "The Era of Gorshkov: Triumph and Contradictions." *Journal of Strategic Studies* 28, no. 2 (2005): 281–308.

C-M(59)39. "Standing Group – SACLANT Anti-Submarine Warfare Centre." NATO, April 14, 1959. CD: IS-AC-0385. NATO Archives, Brussel.

"Command History Report (CHR) for VP-7 (1952)." US Navy, 1952. FOIA case number: DON-NAVY-2018-001782. Naval History and Heritage Command.

"Command History Report (CHR) for VP-10 (1955)." US Navy, 1955. FOIA case number: DON-NAVY-2018-00574. Naval History and Heritage Command.

"Command History Report (CHR) for VP-21 (1959)." US Navy, 1959. FOIA case number: DON-NAVY-2018-00574. Naval History and Heritage Command.

Cote, Owen R. Jr. "The Third Battle – Innovation in the U.S. Navy's Silent War Struggle with Soviet Submarines." *Naval War College Newport Papers* 16 (2003).

Defense Intelligence Agency (DIA). *Russia Military Power – Building a Military to Support Great Power Aspirations*. Defense Intelligence Agency, 2017.

DUPI. *Grønland under den Kolde Krig. Dansk og amerikansk sikkerhedspolitik 1945–68*. Copenhagen: Dansk Udenrigspolitisk Institut, 1997.

Duvsete, Svein. "Fra Luftforsvar til Strategisk Angrep. Norsk Luftmilitær Doktrine 1945–1955." *Forsvarsstudier* 2 (1998).

Dyndal, Gjert Lage. "How the High North Became Central in NATO Strategy: Revelations from the NATO Archives." *Journal of Strategic Studies* 34, no. 4 (2011): 557–85.

Dyndal, Gjert Lage. *The Northern Flank and High North Scenarios of the Cold War*. Potsdam, Germany: Zentrum für Militärgeschichte und Sozialwissenschaften der Bundeswehr, 2013.

Egeland Moen, Knut. *Selvpålagte restriksjoner i nord, 1945–65*. Oslo: Institutt for Forsvarsstudier, 1998.

Eriksen, Knut E., og Helge Ø. Pharo. *Kald Krig og Internasjonalisering, 1945–1965*. Bd. 5. Norsk utenrikspolitisk historie. Oslo: Universitetsforlaget, 1997.

Espenes, Øystein, og Gjert Lage Dyndal. "The Changing Focus of NATO-Intelligence – From Southern Scandinavia to the 'High North' at the End of the 1950s." *Tidsskrift (Kungliga Krigsvetenskapsakademien)* 2 (June 2013). http://kkrva.se/hot/2015:2/HoT_2_2015.pdf.

Foggo, James, og Alanik Fritz. "The Fourth Battle of the Atlantic." *United States Naval Institute Proceedings* 142, no. 6 (June 2016).

Ford, Christopher, og David Rosenberg. *The Admiral's Advantage – U.S. Navy Operational Intelligence in World War II and the Cold War*. Annapolis, MD: Naval Institute Press, 2005.

Friedman, Norman. *Submarine Design and Development*. London: Conway Maritime, 1984.

Gaddis, J.L. *Strategies of Containment. A Critical Appraisal of Postwar American National Security Policy*. Oxford: Oxford University Press, 1982.

Hopkins, Robert S. III. *U.S. Strategic Aerial Reconnaissance and the Cold War, 1945–1961*. University of Virginia, 1998.

Jordan, John. *Soviet Submarines – 1945 to the Present*. London: Arms & Armour Press, 1989.

Klevberg, Håvard. "Maritime Surveillance in the North: 333 Squadron in Norwegian Security Politics." PhD, History, University of Oslo, 2011.

Mason, R.A. *Air Power – An Overview of Roles*. London: Brassey's Defence Publishers, 1987.

MC 14/2. "The Overall Strategic Concept for the Defence of the NATO Area." October 14, 1956. Electronic Document Folder: IMS/MC-MCM-SG/1953-1970. NATO Archives, Brussel.

Moen, Knut E. "Norway's Self-Imposed Restrictions in the High North, 1945–65." *Forsvarsstudier* 5 (1998).

Mutza, Wayne. *Lockheed P2V Neptune – An Illustrated History*. Atglen, PA: Schiffer Publishing Ltd., 1996.

NIE 11-3-55. "Soviet Capabilities and Probable Soviet Courses of Action Through 1960." Director of National Intelligence, May 17, 1955. CIA, Online Archive.

NIE 11-6-54. "Soviet Capabilities and Probable Programs in the Guided Missile Field." National Intelligence Estimate. Director of National Intelligence, October 5, 1954. CIA, Online Archive.

Njølstad, Olav. *Jens Chr. Hauge – Fullt og helt*, 4. utg. Oslo: Aschehoug, 2008.

Owen, David. *Anti-Submarine Warfare – An Illustrated History*. Annapolis, MD: Naval Institute Press, 2007.

Palmer, Michael A. *Origins of the Maritime Strategy – The Development of American Naval Strategy, 1945–1955*, reissue ed. Annapolis, MD: Naval Institute Press, 1990.

Podvig, Pavel, red. *Russian Strategic Nuclear Forces*, 2004 paperback ed. Cambridge, MA: MIT Press, 2001.

Polmar, Norman, og John F. Bessette. *Spyplanes – The Illustrated Guide to Manned Reconnaissance and Surveillance Aircraft from World War I to Today*. Minneapolis: Quarto Publishing Group Inc., 2016.

Reade, David, og Bob Harper. "Naval Air Station Keflavik, Iceland." *Airborne LOG* (Fall 1994): 8–11.

Riste, Olav, og Arnfinn Moland. *Strengt Hemmelig – Norsk etterretningstjeneste 1945–1975*. Oslo: Universitetsforlaget, 1997.

Roberts, Michael D. *Dictionary of American Naval Aviation Squadrons*, vol. 2. Washington, DC: Naval Historical Center, 2000.

Seagle, Adriana N. "Intelligence Sharing Practices Within NATO: An English School Perspective." *International Journal of Intelligence and Counterintelligence* 28, no. 3 (2015).

SG 161/8 (Part II). "The Soviet Bloc Strength and Capabilities 1955–1959." NATO, March 18, 1955. NATO, Online Archive.

SG 161/10. "The Soviet Bloc Strength and Capabilities 1957–1961 – Part I." March 20, 1957. Electronic Document Folder: IMS/MC-MCM-SG/1953-1970. NATO Archives, Brussel.

Tamnes, Rolf. "Integration and Screening – The Two Faces of Norwegian Alliance Policy." *Forsvarsstudier* 6 (1987).

Tamnes, Rolf. *The United States and the Cold War in the High North*. Oslo: Ad Notam, 1991.

Vengstad, K. "Fleksibilitet og slagkraft. Luftforsvaret og NATOs infrastrukturprogram 1950–1957." *Forsvarsstudier* 3 (2006).

Weir, Gary E., og Walter J. Boyne. *Rising Tide – The Untold Story of The Russian Submarines That Fought the Cold War*. New York: Basic Books, 2003.

Yegorova, Natalia. "Stalin's Conception of Maritime Power: Revelation from the Russian Archives." *Journal of Strategic Studies* 28, no. 2 (2005): 157–86.

5 A common NATO maritime patrol aircraft

MPA had been established during WWII as a key factor in combating enemy submarines, and procuring the required aircraft inventory for the Alliance was considered to be of crucial importance. By the mid-1950s, discussions on the Soviet submarine threat had been ongoing for some time, and the 1956 Annual Review confirmed that NATO capabilities were not robust enough to address the threat in case of war. The Alliance recognized that there was an "overall deficiency of maritime patrol aircraft" in NATO European countries, and most nations relied on the MDAP to "bridge the gap between NATO requirements and aircraft availabilities."[1]

In the 1950s, NATO conducted several major exercises in the North Atlantic and in Norway, as a part of the containment strategy towards the Soviet Union. These exercises further led to the explicit restrictions on Allied troops and aircraft in the northernmost parts of Norway, with the overarching political aim to maintain low tensions in the High North. Simultaneously, there was a mounting pressure on Norwegian authorities to acquire the appropriate tools to conduct maritime surveillance on the Northern Flank. The organizational understanding for the requirement in the armed forces was lacking, however, as was the political will to prioritize surveillance on behalf of the Alliance. It was from a political standpoint more important to ensure stability in the North than to venture off with a new and expensive capability based on external pressure, and without an internal desire to build the capability up. The weapons assistance programme from the United States would be the releasing factor in this dynamic, as we shall see.

The NATO 1956 Annual Review

Although there had been several reviews preceding it, the 1956 Annual Review of NATO's force structure and capabilities relative to that of the Soviet Bloc painted a grim picture.[2] Among the shortfalls depicted, the only military capability that is *specifically* mentioned in the document's main body is the large reductions and shortfalls in MPA, with no improvement in sight before 1959 or 1960.[3] From a defence planning standpoint, the Alliance acknowledged the "insidious changes in Soviet tactics" together with a "growing awareness of the mounting costs of modern forces" for the Alliance members. The NATO leadership thus began to emphasize

DOI: 10.4324/9781003626015-5

the requirement to handle defence challenges "collectively and with a sense of urgency."[4] It was underlined that the "problems of the development and production of modern weapons are far beyond the capacity of most NATO countries to resolve single-handed."[5] International cooperation should be explored at every opportunity for the development of modern military equipment. In November 1956, NATO military commanders recommended that the North Atlantic Council (NAC) consider international collaboration for MPA procurement.[6] Building on the military comments, the report from the Defence Planning Committee on the 1956 Annual Review suggested that "common action might also be most helpful in overcoming the deficiencies in maritime patrol aircraft."[7]

Establishment of the Group of Experts

The NAC recognized in December 1956 the need to replace the P2V *Neptune* MPA that were in service with several European nations, due to the age of the airframe and its sensors. Based on challenges related to logistics and operational cooperation for MPA during WWII, there was a desire to seek a common aircraft type across the Alliance. If NATO was to operate in a unified manner, common solutions and standardized equipment would be necessary.[8] Based on efforts within the field of general anti-submarine research, the French Delegation proposed to establish a Group of Experts to study the feasibility of providing "suitable maritime patrol aircraft for the NATO forces."[9] It was widely accepted that it would be too costly for a single nation to develop and procure maritime surveillance aircraft in the relatively small quantities required.[10] The proposal was agreed by the Defence Planning Committee on 5 March 1957.

An aircraft with operational versatility, capable of operating in all conditions of weather, was in itself considered to be a complex system through its wide array of sensors and equipment. Based on experience from civilian manufacturing of aircraft, a lead time of approximately 10 years was to be expected, an expectation later revised to 8 years.[11] This lead time, together with the costs associated with the research and development (R&D) of entirely new systems, led to discussions in the group on whether it was feasible to initiate an entirely new project to meet operational requirements. An alternative could be a study of already existing aircraft covering the same roles, which might prove more productive and economical.[12] The possibility of fulfilling NATO operational and national requirements by procuring an aircraft already in production was indeed examined, but it was decided that no existing aircraft fully met the initially agreed-upon operational requirements from August 1957.[13] It is worth noting that at the time of these investigations, the US Navy P-3 *Orion* programme was underway, but had not yet been firmly established. However, the American requirements that led to the development of the P-3 *Orion* were considered to be in excess of European needs, and consequently "uneconomical for operation by European governments."[14] The decision was made in December 1958 to go ahead with a unique design, and the Group of Experts encouraged the initiation of design studies within the respective nations.

A question of range

The starting point for the discussion on the numbers of aircraft was a questionnaire that the nations had responded to in relation to the 1956 Annual Review.[15] The questionnaire was specifically related to the Alliance's force goals for 1957. The requirements for MPA were as follows:

Table 5.1 Initial requirements based on ARQ(56).[a]

Country	*Long-range aircraft*	*Short-range aircraft*
Denmark	4	–
France	60	–
Germany	–	10
Italy	40	–
Netherlands	8	8
Norway	24	–
Portugal	24	–
United Kingdom	88	–
Total	248	18

[a] This overview is depicted in the minutes of "AC/126-D/1 (REV)" and shows the national answers to the questionnaire. The overview shows only European requirements, and it is unclear why neither Canadian nor American requirements are portrayed. One suggestion is that the intention of the survey was to examine the potentially different requirements with regards to long- and short-range aircraft, where it would be plausible to assume that the Canadians and Americans only required long-range aircraft. They would thus fall outside the purview of the survey. Then again, the responses suggest, as is clear from the discussions within the Group of Experts as well, that the delineation between long and short range was not at all defined by the collective.

The explicit difference between long-range and short-range aircraft was not clear at the point of answering the questionnaire, although existing airframes provided a guide. Short-range ASW aircraft of modern design were the French *Alize*, the British *Fairey Gannet* and *Short Seamew*, the American *Tracker*, and the Italian *P155-AS*. The endurance of these aircraft varied between 4 and 9 h, with the ability to cover between 750 NM and 1,400 NM, depending on the speed of the aircraft.[16] In the context of short-range ASW, the use of helicopters operating from aircraft carriers and from the shore was being examined, and there seemed to be a relatively wide choice of modern aircraft to fill this particular role. It was therefore suggested to defer the considerations of short-range aircraft pending the completion of the long-range ASW project.[17]

The available long-range maritime patrol and reconnaissance aircraft in the late 1950s were the Canadian *Argus*, the American *Marlin*, the American *Neptune*, and the British *Shackleton*. The endurance of these aircraft spanned from 13 to 19 h, with a range between 2,000 NM and 5,300 NM.[18] Operational commanders were invited to provide specific operational requirements and specifications to the Group of Experts.[19] Based on early discussions between the NATO commands, there emerged a general NATO requirement for a *medium*-range MPA.[20] This was supported by neither Britain nor the United States, both of whom saw the emerging threat in the North Atlantic as requiring longer range for any NATO MPA.

It seems clear that nations forwarded their views based on their own geographical situation. Britain was adamant that the MPA required significant range in order to cover their areas of responsibility in the Atlantic Ocean. Italy saw no need for any range on those scales and advocated early on a shorter range for Mediterranean operations. The Soviet build-up in the North and increasing Soviet submarine activity in the Atlantic meant that any Alliance MPA would necessarily require more than the minimum range necessary for Mediterranean operations, in order to bring some credibility to meet the submarine threat elsewhere as well.[21] This balancing act between national and Alliance requirements meant that the Group of Experts opted to focus on "medium-range" requirements.

In order to further distil the requirements for the nations, two different types of missions were foreseen. First, a type of mission where the MPA would fly 700 NM to an operating area, operate for 3 h, and have a minimum of 10 h of endurance at sea level. Second, a mission where the MPA would fly 1,000 NM and operate for 4 h, with a minimum endurance of 14 h at sea level.[22] The discussion on range requirements included strong stands for both an "Atlantic type" (long-range) and a "Mediterranean or enclosed-water type" (intermediate/short-range), with all delegations from an economic standpoint discarding the idea to develop two separate aircraft in order to service the two missions.[23]

In Mediterranean terms, these distances mean that an MPA based in Sicily would be able to operate just south of the southern coast of Turkey for 3 h in the

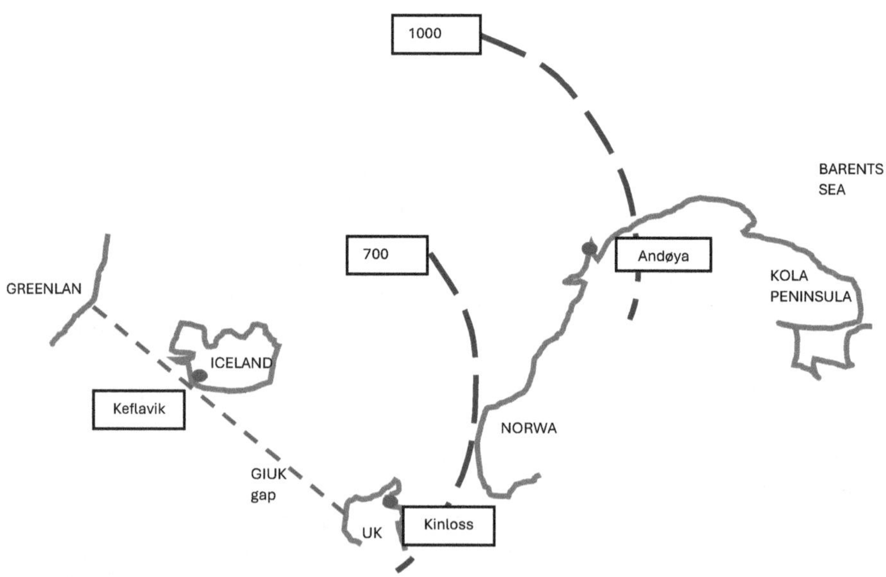

Figure 5.1 Ranges for a common MPA based on Iceland

first instance, and off the coast of Syria in the second. In North Atlantic terms, the missions proposed meant that an MPA based in Iceland would be able to operate off the Western coast of Norway in the first instance and outside the northern coast of Norway in the second. The group settled, with the exception of Italy, on a "medium-range aircraft."[24] The medium range was set to 8 h of patrol at 600 NM, or, alternatively, 4 h of patrol at 1,000 NM.[25]

Roles and missions

It was emphasized initially that the *sole mission* for the aircraft would be ASW.[26] Any additional missions that would include reconnaissance, mining, and other types of warfare should only be performed to the extent that such missions could be executed with the ASW equipment and sensors that were already installed.[27] However, in the proposed requirements forwarded on 11 October 1957, the role of the aircraft was slightly widened. The Group of Experts saw that it would be inefficient use of resources to send such capable aircraft hundreds and at times thousands of miles out to sea in order to perform a single mission (ASW).[28] Additionally, the financial discussions demanded that "in the interests of economy, it should also have a transport capability."[29] However, the American delegation, which found itself in a fundamentally different situation with regard to their national economy than did the Europeans, did not agree to this added capability. The Americans, even after it became clear that they would not procure the aircraft, rejected the notion that an added role would be transportation. The American delegation would permit "no compromise of function of ASW to provide defensive and transport capabilities," and the Americans took a strong stand as they provided significant funding for the project, regardless of US procurement.[30] In the final specifications of March 1958, the role of the NATO MPA would read:

> The aircraft will be used for the detection, the localisation and the destruction of submarines, and for general maritime reconnaissance.[31]

However, the Europeans were able to squeeze in several secondary roles. Four were specifically mentioned: Search-and-rescue; shadowing-and-attack (which, in modern terms, can be translated into Anti-Surface Warfare [ASuW]); minelaying; and finally, light transport. The transport mission was meant to include a number of ground crew, a certain amount of ground equipment, and a spare engine, for what is termed self-deployment.[32]

Operational characteristics and specifications

High *speed* for the aircraft was recognized as important for "early development of a contact," in other words to get to the position of the submarine and accumulate sufficient information on the target for an attack. The minimum cruising speed was set to 225 knots at sea level, and a cruising speed of "not less than 300 knots" in general.[33] The aircraft was to have a service ceiling of at least 20,000 feet, and good manoeuvrability at 5000 feet at low speeds. Performance for take-off and landing also indirectly set the fundamental requirements for the weight of the aircraft.

There were few discussions on the *sensor* suite of the NATO MPA. The required sensors were those of the standard suite of the time: Underwater acoustic sensors; a good radar (primary detection method of the previous war); electronic emission sensors (referring to the electronic warfare that played a key role during WWII); exhaust-sniffing equipment (searching for exhaust from submarines running their diesel engines on the surface); and a powerful search light for surface searches at night and in poor weather.[34] All the sensors noted were representative of a fully equipped MPA in the late 1950s. A robust inventory of sonobuoys would take the NATO ASW aircraft into the nuclear age with regard to sensors, which now would have to search for submarines with nuclear propulsion that were submerged for extended periods of time. Both the United States and the Soviet Union had operationalized nuclear propulsion in submarines in 1958. A strict criterion was established when it came to weight and growth potential. The American delegation provided advice based on their own extensive aircraft production experience, stating a maximum weight limitation due to growth potential and general performance.[35]

In addition, the disposable stores to be carried were torpedoes, depth charges, and air-to-ground guided weapons. Pertaining to attack capabilities, the requirement was for the MPA to be able to make two separate attacks using depth bombs or homing weapons, to include atomic weapons.[36] The initial requirements developed on operational characteristics also stated that armaments, which in this context means self-protection, would not be necessary. However, the German representative to the group emphasized that the "particular circumstances in which the German Air Force would operate such aircraft" would pose specific requirements with regard to armament. The Baltic and the North Sea areas were "particularly vulnerable to shore based aircraft and it was obviously necessary to provide at least a minimum degree of armament as a protection against attack from hostile units."[37] In view of the shoot-downs of Western intelligence flights by Soviet fighter aircraft over the Baltic Sea during the early 1950s, the German concerns were understood by all.[38] The Germans recommended that the aircraft armament should include guided air-to-air missiles. In the end, missiles were not included. Instead, wiring would be in place for the respective nation to fit the aircraft with such weapons if deemed necessary.[39] In all the documents detailing the Group of Experts' efforts, the discussion on armaments seems almost anecdotal. Nations other than Germany showed less concern over armaments, presumably because operating far into the Atlantic or Mediterranean posed less of a direct threat to the aircraft than did missions over the Baltic Sea.

The Group of Experts also did not spend much time discussing the requirements for *communications* equipment. Satellite communications were years from achieving operational capability, and the use of data link was still a decade or so away from being operational. The first mention of requirements for communications capabilities occurred when proposing the initial, desired operational characteristics in August 1957. The Group of Experts emphasized that the aircraft would normally operate individually, "but provisions are required for communications and co-ordinating with surface, air, shore and submarine units. A desired feature is the ability to act as an inflight area co-ordinator at a scene of action."[40] Although not stated explicitly, this would require several tactical radios in the VHF and ultra-high-frequency (UHF) range in order to coordinate efficiently between several assets simultaneously. It would also require at least one long-range radio covering the HF range, in order to enable aircraft-to-shore communications when operating far out to sea. The ability to transmit detection data was also brought forward, although the Group of Experts did not elaborate on what type of transmission this was meant to be. The final operational requirements demanded that "the aircraft is to be fitted with radio equipment for communicating with ships, submarines and shore bases at long range."[41]

The final decision

The letter to industry was sent out in March 1958, with a deadline for submissions of design proposals of 21 June 1958.[42] After evaluating the 16 proposals received from industry across the Alliance, the working group unanimously decided on one design, namely the French Breguet 1150, later called *Atlantic*.[43] One of the deciding factors was that the Breguet 1150 offered more working space and room for expansion for future equipment.[44] Other important factors were flexibility in weapons load and a high capacity for sonobuoy storage.

As a next step, the national requirements for number of aircraft were further adjusted. As mentioned above, Britain, Canada, and the United States would not procure the NATO MPA. The United States did, however, contribute funds to the development of the aircraft.[45] In the final specification letter sent to the aircraft industry, the potential contractors had been asked to plan for the production of anything between 100 and 200 aircraft. And in order to ensure international cooperation, the production plans had to include "existing or proposed plans for international co-operation with at least one firm in another country."[46] The final required numbers established by the Group of Experts would read:[47]

Table 5.2 Final requirements established by the Group of Experts.

Country	*Number*	*Country*	*Number*
Belgium	6	Italy	0
Canada	0	Netherlands	20
Denmark	0	Norway	6
France	70	Portugal	12–24
Germany	18	United Kingdom	0
USA	0		

For the maritime nation of Norway, the number six stands out as low. The number of six aircraft was a continuation of the number of *Catalina* aircraft, and the

Table 5.3 The Minimum Essential Force Requirements, 1958–1963 (issued January 1958).[a]

NATO Command	*SACLANT*		*SACEUR*		*CHANNEL*		*TOTAL*	
Year	*1958*	*1963*	*1958*	*1963*	*1958*	*1963*	*1958*	*1963*
Belgium	–	–	–	–	0	0	0	0
Canada	28	40	–	–	–	–	28	40
Denmark	0	0	0	0	–	–	0	0
France	23	32	24	30	4	11	51	73
Germany	–	–	0	24	–	–	0	24
Greece	–	–	0	0	–	–	0	0
Italy	–	–	18	27	–	–	18	27
Netherlands	0	0	–	–	8	12	8	12
Norway	–	–	9	9	–	–	9	9
Portugal	18	24	–	–	–	–	18	24
Turkey	–	–	0	0	–	–	0	0
UK	72	72	16	16	16	17	104	105
USA	216	261	54	54	–	–	270	315

[a] Notice that Norway falls under the responsibility of SACEUR (European continent), and not yet under SACLANT (the Northern Atlantic), underlining the focus of a mainland invasion by the Soviets, making Scandinavia a flank in the North. See further discussion in previous chapter.

Norwegian Defense Staff did not see how they could afford to raise their ambition in this field.[48]

The SACLANT representative to the Group of Experts underlined that there was an urgent military requirement for this aircraft to become operational in quantity by 1964 in order to meet the overall NATO operational requirements in MC-70.[49] The force requirements for MPA laid out in MC 70 in early 1958 were as depicted in the force requirements for 1958–1963.[50] In the end, France, (West) Germany, the Netherlands, Italy, Portugal, and Belgium actually procured the NATO-developed MPA *Atlantic*.[51]

The Norwegian decision to acquire Albatross MPA

The Norwegian *Catalina* fleet was in dire need of replacement. In 1952, the Norwegian Government, through the MDAP process, sought to procure American *Albatross* MPA as a replacement, an amphibious aircraft with long endurance able to support a wide array of tasks.[52] The *Albatross* was not, however, configured for ASW. Overhauled *Catalinas* were lent to Norway in 1952 as a temporary solution to the capability issues in the north while discussions continued for a long-term replacement. The S-2 *Tracker* was an early candidate for a replacement aircraft. It was small, but constructed specifically for ASW operations. It was made for flying off US aircraft carriers and was able to both find and attack enemy submarines. Being based on a carrier, it was constructed with limited space in mind, and at 1,250 NM in range it had significantly shorter endurance than the *Neptune*

(3,450 NM) and other, larger aircraft. Regardless, it was made available for Norway through the MDAP.[53]

British aircraft were considered, but neither the *Short Seamew* nor the *Fairey Gannet* met the overarching requirements for range, and they were too expensive compared to any MDAP solution. By 1955, the pressure was mounting from NATO operational entities as well as from the Americans for modernizing the Norwegian maritime air surveillance fleet. The ageing *Catalinas* were spending more time in the hangar being serviced than they were flying, in addition to not being equipped with the appropriate modern sensors required for maritime surveillance in the High North. The intelligence services that the *Catalinas* provided were thus inadequate.[54] The pressure from NATO for increased military surveillance by the Norwegians collided with the Norwegian perception of their maritime air assets as *utility* aircraft supporting a wide array of tasks, to include a heavy focus on maritime SAR. In a case of intra-service rivalry, the navy argued for an aircraft with ASW capabilities, and the air force leadership argued against ASW (they thought ASW should be left to more capable and knowledgeable allies).[55]

The *Neptune* aircraft was considered for its range, but it was assessed as much too complicated to operate for Norwegian aircrews, and too demanding for Norwegian infrastructure, organization, and logistics to handle. It was also important to the Norwegian Defence Staff to procure an aircraft that would not affect the resources required to run the fighter aircraft service. In an attempt to gather loose threads in the discussion between NATO and national requirements, the Air Force declared in 1956 that the main task for Norwegian maritime surveillance aircraft was "to conduct surveillance on the Norwegian coast and adjacent areas, and to early detect a seaborne invasion force."[56] The operational requirement with regard to range was thus set to 900 NM. The *Tracker* was unable to meet this requirement.[57] The 900 NM requirement was established in parallel to the NATO MPA project and the establishment of the bi- and trilateral agreements for maritime airborne surveillance mentioned earlier.

The common procurement project for an MPA in NATO came at the perfect time, so it is worth commenting briefly on the Norwegian interaction with the project, or lack thereof. Norwegian delegations were present for the first meeting in April 1957 and took part in the discussion on how to move forward to establish a common requirement.[58] But as discussions with the Americans progressed in the context of the MDAP where aircraft would be provided at no cost, Norwegian participation in the NATO project ended. The final meeting that the Norwegian delegates took part in was that of 23 October 1957. At this meeting, the Norwegian delegates were unwilling to take part in the discussion on range requirements for the new aircraft, due to the "not yet established Allied and national operational requirements for Norwegian areas of interest."[59] This reflected both the chaotic national process in Norway with leadership shying away from decision making, but also a noticeable lack of competence in the field. It should be noted that this was also the time that Norwegian authorities were in the discovering phases of their own "areas of interest" in the High North, and a Norwegian awakening to the emerging military threat on the Kola Peninsula.

The Norwegian exit from the NATO common MPA procurement programme seems to rest on three factors. First, economic constraints meant that the replacement aircraft would have to come through the MDAP.[60] Second, the 900 NM range requirement put forward by the Norwegian Air Force in 1956 meant that the final 700 NM requirement for what would become the NATO *Atlantic* MPA came up short. Finally, the situation with regard to strategy, ambition, and requirements for the Norwegian maritime air fleet was, in the 1950s, so chaotic, and the individual knowledge in maritime air matters so low, that any further participation in the project was likely futile.

The northern command for NATO, the Commander Allied Forces North (COMAFNORTH), was in charge of all NATO forces operating in and out of the northernmost parts of the Alliance, and was subordinate to SACEUR. The headquarters, situated in Oslo, Norway, produced a report in December of 1957 that addressed the different requirements for maritime surveillance and early warning, and recommended the *Albatross* for Norway, which stood out as a compromise of range, sensors, and economy in maintenance.[61] MDAP required that the aircraft would be American-made. With an ambition to acquire an aircraft that would cost approximately the same to operate as the *Catalinas* did, the *Albatross* seemed to be a viable and sustainable choice for the Norwegian Government.

Further complications arose when the Americans rejected the notion of the Norwegians flying in the High North without an ASW capability; the *Albatross* was a surface surveillance aircraft. The requirement for ASW in the north was also emphasized by SACEUR himself in 1958.[62] After arguing the pros and cons for almost a year, the Norwegians decided to go back to the original proposal for the *Tracker*, in spite of the short range, because of the ASW capability. But, in the winter of 1960, the producer of the *Albatross*, Grumman, notified Norwegian authorities that they were able to modify the *Albatross* to include ASW capabilities. This would give the Norwegians an ASW-capable aircraft with significant range, more capable than the *Tracker*, however, not as capable as the *Neptune*. The Norwegians changed their minds yet again, and decided to procure the *Albatross*, moderately modified for ASW.[63]

The common NATO MPA procurement project facilitated four nations to fulfil their MPA force requirements for the Alliance. The NATO-developed MPA was faster, had longer endurance, and better manoeuvrability than the P2V *Neptune* it was meant to replace. It also seems likely that the group development and procurement was the best way forward for the nations, as it resulted in procurement of the *Atlantic* from an acquisition standpoint. The NATO MPA project achieved what it set out to do; it enabled several nations to meet their force requirements against a significant naval threat to the Alliance. Three big nations pulled out due to national projects of greater magnitude and technological scope, one small nation pulled out because of negotiations on a different aircraft within the MDAP framework, several nations never took part due to the lack of NATO force requirements, and the rest of the nations specifically mentioned in MC 70 ended up procuring the *Atlantic* at some point.

The Alliance discussions in the late 1950s function as a case study of the basic characteristics of airborne ISR: Range, speed, altitude, sensors, communications, and armament (self-protection). We shall discuss these more in detail later in this book. Of these characteristics, the question of *range* was most extensively discussed. This is ultimately the factor that lays the foundation for where, how far away, and for how long a maritime surveillance mission is capable of operating. Nations bordering the Atlantic demanded longer range than the nations bordering the Mediterranean. As we shall see in the coming chapters and the final discussion, range has historically been, and will in the future be, crucial to how an airborne surveillance asset can approach its ISR mission in general, and the hunt for adversary submarines in particular. This timeless tenet played a crucial role in the late 1950s, and continues to do so today.

Notes

1 AC/126-D/1 (REV), "The Group of Experts on Maritime Patrol Aircraft – The Status of Development of Maritime Patrol Aircraft," 1.
2 IPT 098-195, "Military Comments on the 1956 Annual Review – Note by the International Planning Team," 3–10.
3 AR(56)GENERAL-WP/1, "Report on the 1956 Annual Review," 2.
4 AR(56)GENERAL-WP/1, 1.
5 AR(56)GENERAL-WP/1, 4.
6 IPT 098-195, "Military Comments on the 1956 Annual Review – Note by the International Planning Team," 10.
7 AR(56)GENERAL-WP/1, "Report on the 1956 Annual Review," 4.
8 Chris Gibson, *Nimrod's Genesis – RAF Maritime Patrol Projects and Weapons Since 1945* (Manchester: Hikoki Publications, 2015), 96.
9 AC/126-D/1 (REV), "The Group of Experts on Maritime Patrol Aircraft – The Status of Development of Maritime Patrol Aircraft," 1.
10 AC/126-D/1 (REV), 3.
11 AC/126-D/1 (REV), 4.
12 AC/126-D/1 (REV), 4.
13 AC/126-D/5, "Group of Experts on Maritime Patrol Aircraft – Report to the Armaments Committee – 16th December 1958," 4.
14 AC/126-D/5, 4.
15 Annual Review Questionnaire: ARQ(56).
16 Simon Mitchell, red., *Jane's Civil and Military Aircraft Upgrades 1994–95* (Coulsdon, UK: Jane's Information Group Ltd., 1994), 57–58; Owen Thetford, *British Naval Aircraft Since 1912* (London: Putnam, 1978), 190; Jerry Proc, "CS2F Tracker," *Radio Communications and Signals Intelligence in the Royal Canadian Navy* (blog), February 18, 2019, http://jproc.ca/rrp/rrp3/tracker.html.
17 AC/126-D/1 (REV), "The Group of Experts on Maritime Patrol Aircraft – The Status of Development of Maritime Patrol Aircraft," 2.
18 Donald, *The Encyclopedia of World Aircraft*, 118; Stewart Wilson, *Combat Aircraft Since 1945* (Fyshwick, Australia: Aerospace Publications Pty Ltd., 2000), 82; Jones, *Avro Shackleton*, 108. For unknown reasons the Marlin was considered a long-range aircraft, even though it only had an endurance of 8–9 hours.
19 The Standing Group (SG) acted as a link between the Military Council and the Regional Planning Groups and the Defense Planning Commiittee.
20 AC/126-R/1, "Group of Experts on Maritime Patrol Aircraft – Summary of the First Meeting, April 16, 1957," 3–4.
21 See for example AC/137/D/20, "Statement on future trends in anti-submarine warfare made at the Second Meeting of the Science Committee by the SACLANT Representative."
22 AC/126-D/2, "Group of Experts on Maritime Patrol Aircraft – Desired Operational Characteristics," 3.

23 AC/126-R/4, "Group of Experts on Maritime Patrol Aircraft – Summary Record of Meeting on March 10, 1958," 3.
24 AC/126-R/4, 4.
25 AC/126(DS)R/1, "Group of Experts on Maritime Patrol Aircraft – Sub-Group on Design Studies – Summary of Meeting Held on 10–11 March 1958," 8.
26 AC/126-D/2, "Group of Experts on Maritime Patrol Aircraft – Desired Operational Characteristics," 3.
27 AC/126-D/2, 3.
28 AC/126-R/2, "Group of Experts on Maritime Patrol Aircraft – Summary Record of the Meeting Held at Palais de Chaillot, Paris, October 11, 1957," 11.
29 AC/126-R/3, "Group of Experts on Maritime Patrol Aircraft – Summary of Meeting Held on January 16–17, 1958," 15.
30 AC/126-D/3, "Group of Experts on Maritime Patrol Aircraft – Desired Operational Characteristics of the Maritime Patrol Aircraft," Addendum 4, 2; David Reade, *The Age of Orion – The Lockheed P-3 Story* (Atglen, PA: Schiffer Publishing Ltd., 1998), 8, Addendum 4, 2.
31 AC/126(DS)R/1, "Group of Experts on Maritime Patrol Aircraft – Sub-Group on Design Studies – Summary of Meeting Held on March 10–11, 1958," 5.
32 AC/126(DS)R/1, 11.
33 AC/126(DS)R/1, 7, AC/126-D/2, "Group of Experts on Maritime Patrol Aircraft – Desired Operational Characteristics," 3.
34 AC/126-D/2, "Group of Experts on Maritime Patrol Aircraft – Desired Operational Characteristics," 4.
35 AC/126-R/4, "Group of Experts on Maritime Patrol Aircraft – Summary Record of Meeting on March 10, 1958," 10.
36 AC/126-D/2, "Group of Experts on Maritime Patrol Aircraft – Desired Operational Characteristics," 4.
37 AC/126(DS)R/1, "Group of Experts on Maritime Patrol Aircraft – Sub-Group on Design Studies – Summary of Meeting Held on March 10–11, 1958," 9.
38 Lashmar, *Spy Flights of the Cold War*, 42; Hopkins, *Spyflights and Overflights*, 185–89.
39 AC/126(DS)R/1, "Group of Experts on Maritime Patrol Aircraft – Sub-Group on Design Studies – Summary of Meeting Held on March 10–11, 1958," 9.
40 AC/126-D/2, "Group of Experts on Maritime Patrol Aircraft – Desired Operational Characteristics," 3.
41 AC/126-R/3, "Group of Experts on Maritime Patrol Aircraft – Summary of Meeting Held on January 16–17, 1958," 21.
42 AC/126(DS)R/1, "Group of Experts on Maritime Patrol Aircraft – Sub-Group on Design Studies – Summary of Meeting Held on March 10–11, 1958."
43 AC/126-D/5, "Group of Experts on Maritime Patrol Aircraft – Report to the Armaments Committee – 16th December 1958," 6; AC/126(DS)R/4, "Group of Experts on Maritime Patrol Aircraft – Sub-Group on Design Studies – Summary Record of Meeting, July 9–10, 1958," 126.
44 AC/126-D/6, "Group of Experts on Maritime Patrol Aircraft – NATO Multilateral Procurement and/or Production for a Maritime Patrol Aircraft," 3–4.
45 AC/126-R/3, "Group of Experts on Maritime Patrol Aircraft – Summary of Meeting Held on January 16–17, 1958," 3–4.
46 AC/126(DS)R/1, "Group of Experts on Maritime Patrol Aircraft – Sub-Group on Design Studies – Summary of Meeting Held on March 10–11, 1958," 20.
47 AC/126-D/5, "Group of Experts on Maritime Patrol Aircraft – Report to the Armaments Committee – December 16, 1958," 7.
48 Klevberg, "Maritime Surveillance in the North," 198–99.
49 AC/126-R/5, "Group of Experts on Maritime Patrol Aircraft – Summary Record of Meeting October 22–24, 1958," 4.

50 MC 70, "A Report by the Military Committee to the North Atlantic Council on the Minimum Essential Force Requirements, 1958–1963," Annex 2 to Appendix E, 2–4.
51 Dassault Aviation, "Atlantic – Origins and Context," *Dassault Aviation*, Åpnet March 23, 2017, www.dassault-aviation.com/fr/passion/avions/dassault-militaires/atlantic/; Mutza, *Lockheed P2V Neptune*, 222; Gibson, *Nimrod's Genesis*, 106.
52 Klevberg, "Maritime Surveillance in the North," 198–201.
53 Michael Green, *United States Naval Aviation 1911–2014* (Barnsley: Pen & Sword Aviation, 2015), 122; Klevberg, "Maritime Surveillance in the North," 204.
54 Klevberg, "Maritime Surveillance in the North," 198.
55 Klevberg, 198–209.
56 Klevberg, 205.
57 Klevberg, 200–5.
58 AC/126-R/1, "Group of Experts on Maritime Patrol Aircraft – Summary of the First Meeting, April 16, 1957."
59 Klevberg, "Maritime Surveillance in the North," 194.
60 AC/126-R/2, "Group of Experts on Maritime Patrol Aircraft – Summary Record of the Meeting Held at Palais de Chaillot, Paris, October 11, 1957," 8.
61 Klevberg, "Maritime Surveillance in the North," 211–12.
62 Klevberg, 222.
63 Klevberg, 211–29.

Bibliography

AC/126-D/1 (REV). "The Group of Experts on Maritime Patrol Aircraft – The Status of Development of Maritime Patrol Aircraft." June 28, 1957. CD: IS-AC-0385. NATO Archives, Brussel.

AC/126-D/2. "Group of Experts on Maritime Patrol Aircraft – Desired Operational Characteristics." August 22, 1957. CD: IS-AC-0385. NATO Archives, Brussel.

AC/126-D/3. "Group of Experts on Maritime Patrol Aircraft – Desired Operational Characteristics of the Maritime Patrol Aircraft." December 3, 1957. CD: IS-AC-0385. NATO Archives, Brussel.

AC/126-D/5. "Group of Experts on Maritime Patrol Aircraft – Report to the Armaments Committee – 16th December 1958." December 16, 1958. CD: IS-AC-0385. NATO Archives, Brussel.

AC/126-D/6. "Group of Experts on Maritime Patrol Aircraft – NATO Multilateral Procurement and/or Production for a Maritime Patrol Aircraft." November 10, 1958. CD: IS-AC-0385. NATO Archives, Brussel.

AC/126(DS)R/1. "Group of Experts on Maritime Patrol Aircraft – Sub-Group on Design Studies – Summary of Meeting Held on 10–11 March 1958." March 11, 1958. CD: IS-AC-0385. NATO Archives, Brussel.

AC/126(DS)R/4. "Group of Experts on Maritime Patrol Aircraft – Sub-Group on Design Studies – Summary Record of Meeting 9–10 July 1958." July 10, 1958. CD: IS-AC-0385. NATO Archives, Brussel.

AC/126-R/1. "Group of Experts on Maritime Patrol Aircraft – Summary of the First Meeting, 16 April 1957." April 23, 1957. CD: IS-AC-0385. NATO Archives, Brussel.

AC/126-R/2. "Group of Experts on Maritime Patrol Aircraft – Summary Record of the Meeting Held at Palais de Chaillot, Paris, 11 October 1957." October 23, 1957. CD: IS-AC-0385. NATO Archives, Brussel.

AC/126-R/3. "Group of Experts on Maritime Patrol Aircraft – Summary of Meeting Held on January 16–17, 1958." January 17, 1958. CD: IS-AC-0385. NATO Archives, Brussel.

AC/126-R/4. "Group of Experts on Maritime Patrol Aircraft – Summary Record of Meeting on 10 March 1958." March 10, 1958. CD: IS-AC-0385. NATO Archives, Brussel.

AC/126-R/5. "Group of Experts on Maritime Patrol Aircraft – Summary Record of Meeting 22–24 October 1958." October 24, 1958. CD: IS-AC-0385. NATO Archives, Brussel.
AC/137/D/20. "Statement on Future Trends in Anti-Submarine Warfare Made at the Second Meeting of the Science Committee by the SACLANT Representative." NATO, July 17, 1958. CD: IS-AC-0385. NATO Archives, Brussel.
AR(56)GENERAL-WP/1. "Report on the 1956 Annual Review." November 23, 1956. Electronic Document Folder: IMS/1955–1959. NATO Archives, Brussel.
Dassault Aviation. "Atlantic – Origins and Context." *Dassault Aviation*. Åpnet March 23, 2017. www.dassault-aviation.com/fr/passion/avions/dassault-militaires/atlantic/.
Donald, David. *The Encyclopedia of World Aircraft*. Etobicoke, ON: Prospero Books, 1997.
Gibson, Chris. *Nimrod's Genesis – RAF Maritime Patrol Projects and Weapons Since 1945*. Manchester: Hikoki Publications, 2015.
Green, Michael. *United States Naval Aviation 1911–2014*. Barnsley: Pen & Sword Aviation, 2015.
Hopkins, Robert S. III. *Spyflights and Overflights – US Strategic Aerial Reconnaissance – Volume I, 1945–1960*. Manchester: Hikoki Publications, 2016.
IPT 098-195. "Military Comments on the 1956 Annual Review – Note by the International Planning Team." November 8, 1956. Electronic Document Folder: IMS/1955–1959. NATO Archives, Brussel.
Jones, Barry. *Avro Shackleton*. Ramsbury, UK: Crowood Press, 2002.
Klevberg, Håvard. "Maritime Surveillance in the North: 333 Squadron in Norwegian Security Politics." PhD, History, University of Oslo, 2011.
Lashmar, Paul. *Spy Flights of the Cold War*. Phoenix Mill: Sutton Publishing Ltd., 1996.
MC 70. "A Report by the Military Committee to the North Atlantic Council on The Minimum Essential Force Requirements, 1958–1963." January 29, 1958. Electronic Document Folder: IMS/1955–1959. NATO Archives, Brussel.
Mitchell, Simon, red. *Jane's Civil and Military Aircraft Upgrades 1994–95*. Coulsdon, UK: Jane's Information Group Ltd., 1994.
Mutza, Wayne. *Lockheed P2V Neptune – An Illustrated History*. Atglen, PA: Schiffer Publishing Ltd., 1996.
Proc, Jerry. "CS2F Tracker." *Radio Communications and Signals Intelligence in the Royal Canadian Navy* (blog), February 18, 2019. http://jproc.ca/rrp/rrp3/tracker.html.
Reade, David. *The Age of Orion – The Lockheed P-3 Story*. Atglen, PA: Schiffer Publishing Ltd., 1998.
Thetford, Owen. *British Naval Aircraft Since 1912*. London: Putnam, 1978.
Wilson, Stewart. *Combat Aircraft Since 1945*. Fyshwick, Australia: Aerospace Publications Pty Ltd., 2000.

6 Foundation for long-range anti-submarine warfare, 1961–1969

Although the Soviet approach to shipbuilding and naval strategy fluctuated somewhat in the decade and a half following WWII, the importance of submarines was, as a whole, on the rise. The 1960s would see a continued ambivalence as to the primary mission of Russian submarines. However, with missile technology increasingly taking centre stage in the arms race between the superpowers, submarines established themselves further as cornerstones of Soviet naval strategy.

The shoot-down of the U-2 over the Soviet Union and the RB-47 over the Barents Sea in 1960 caused the Norwegian Government to panic. All maritime surveillance missions, Norwegian as well as Allied taking off from Norway, were forbidden for a month, in fear of more incidents. When the missions were resumed again, they were conducted under stricter restrictions, meaning that the willingness of the Norwegian Government to approve Allied surveillance missions taking off from Norway into the Barents Sea was significantly reduced. The situation in the North made both the defence minister as well as the prime minister personally involved in the approval process for several of the surveillance missions.[1] The reluctance on the part of the Norwegian Government to allow allied activities to be conducted from Norwegian soil into the Barents Sea almost by accident turned Norwegian surveillance and intelligence operations in the High North into a security policy commodity.[2] This dynamic would take centre stage when the defence ministers of the United States and Norway both personally negotiated the terms of the Norwegian procurement of the P-3 *Orion*, which became operational late in the decade.

The decade in question also experienced shifts in Western and NATO strategy, resulting in a more defensive naval posture in the North Atlantic. Surface force operations were pulled further back, and an increasing reliance was put on barriers consisting of aircraft and seabed hydrophones. However, the posture of maritime airborne surveillance and reconnaissance units did not decline in tandem with the overall naval posture. In response to the expanding nature of Soviet submarine operations, the main nations handling maritime surveillance in the High North did so largely through the use of MPA. And although the respective MPA had significant range, the sheer size of the patrol areas required close international cooperation. This requirement brought the Norwegian basing policy under increasing pressure in the 1960s. At the same time, developments within the fields of sensors

DOI: 10.4324/9781003626015-6

and communications improved the search and tracking of submarines and facilitated closer collaboration between different aircraft.

Soviet naval strategy and submarine capabilities

At the turn of the decade, Soviet Premier Krushchev was concerned that each branch of the armed services was establishing independent strategic nuclear missions for themselves. This led to the organization of all strategic ballistic missile capabilities under one entity, the Strategic Rocket Forces (SRF) in late 1959.[3] The strategic nuclear mission was thus removed from the submarine force and placed more or less exclusively on land. The ICBM was now the strategic weapon of choice, with the Soviet Union's as well as the world's first ICBM becoming operational in January 1960.[4] The shift toward ICBM created a void in the mission portfolio of Soviet submarines, but the Soviet Navy did not have to search for a new mission for long. The US Navy had, in the years following the Korean War, put into service new supercarriers weighing more than 60,000 tons. These could carry long-range nuclear bomber aircraft such as the A-3 *Skywarrior* with a range of 1,200 NM.[5] The only effective weapon against the supercarriers was, by the Soviets, deemed to be the anti-ship stand-off cruise missile. The only two credible weapon platforms for this missile were considered to be long-range bombers and submarines.[6] The policy change therefore did not degrade the continued value of the submarine insofar as naval strategy was concerned. Naval exercises and submarine operations continued, as did missile development for submarines, although development of strategic ballistic missiles fired from submarines had been temporarily halted.

The first nuclear-powered *Echo*-class submarine was completed in 1960, and it carried SS-N-3 *Shaddock* LACMs.[7] However, the mentioned policy changes in prioritization of weapons development ensured that the anti-carrier mission was to be pursued, leading to the construction of *Echo II* submarines equipped with SS-N-3A *Shaddock* anti-ship missiles becoming operational in 1962. Several of the *Golf*-class SSBNs were converted into *Golf II* multirole cruise missile submarine (SSGN) configurations carrying the new SS-N-5 *Sark* ballistic missile, and the *Hotel*-class SSBNs were converted into *Hotel II*-class SSGNs carrying the same missile.[8] The SS-N-5 *Sark* had a range of 700 NM/1,300 km and was the first Soviet ballistic missile that could be launched from a submerged position. It became operational in 1963, some 3 years after the American *Polaris* missile with similar launch capabilities.[9] But the move to a new mission also brought technical and tactical challenges, the most prominent of which was targeting for the anti-ship missiles. Even with an intricate network of shore-based direction-finding radio stations, airborne sensors such as the *Bear D* long-range maritime reconnaissance and targeting aircraft, and designated targeting submarines such as the *Whiskey Canvas Bag* class, it still would be a challenge to efficiently target adversary carrier strike groups. It took 20–30 min on the surface to prepare the launch of an SS-N-3A missile. This created to problems for the submarine: Firstly, by staying surfaced for that long she was vulnerable to detection by ASW forces. Secondly, by the time the submarine could fire her missiles the position of a carrier group sailing at 25 knots would have changed its position by at least 8 NM.[10] Faster-moving nuclear submarines and targeting sensors in space would aim to remedy such challenges.

The speed and mobility required by submarines hunting fast-moving aircraft carriers were facilitated by nuclear reactors. New weapons systems such as the US Navy *Polaris* submarine pointed to "limited" nuclear wars as more plausible than a complete nuclear exchange.[11] This complicated the Soviet policy decision to rely solely on land-based ICBMs for a nuclear exchange, and thoughts on operating strategic submarines carrying nuclear ballistic missiles re-emerged in the mid-1960s. The Cuban missile crisis had emphasized the requirement for improvements to the naval force.[12] When Brezhnev took over from Krushchev as Soviet premier in 1964, Admiral Gorshkov could, as the head of the Soviet Navy, pursue his ambition of a nuclear role for his submarine force. Although the Soviet Naval High Command also wanted long-range patrols to be initiated as early as 1956, regular patrols by Soviet submarines off the coast of the United States did not commence until 1966, likely due to the combination of the overarching focus given to land-based missiles (ICBM) and the still nascent strategic submarine capability.[13]

In 1967, the new *Yankee*-class SSBN became operational. The *Yankee* carried the new nuclear ballistic missile SS-N-6 *Serb* with an initial range of 1,300 NM/2,400 km, which was a significant improvement. The new SSBN was quieter, moved faster, and had immense firepower on board with a significant range. The SSBNs had become truly strategic.[14] The range of the SS-N-6 meant that although the *Yankees* had to cross the GIUK gap, they could fire their missile much further away from the US coastline. The *Yankee*s patrolled in dedicated areas, holding the US coast at risk, starting in the summer of 1969.[15] Up until then, *Golf*- and *Hotel*-class submarines had patrolled sporadically in the Atlantic. Patrol patterns would consist of one *Yankee* submarine north and one south of Bermuda, and *Hotel* and *Golf* submarines stationed east of Nova Scotia and west of the Azores. Later, this was increased to three *Yankee*-class submarines on station in the Atlantic.[16] Although improved, the range of the SS-N-6 was not sufficient to keep the *Yankees* out of American ASW defences. The *Yankee*/SS-N-6 combination was thus viewed as an interim solution, awaiting more capable systems that would facilitate strategic submarines operating closer to home waters and still able to fire ballistic missiles at the US mainland. This would become reality with the *Delta* class carrying SS-N-8 missiles in 1973.

The *Charlie*-class SSGN also entered service in 1967. The *Charlie* was nuclear-powered and carried the supersonic SS-N-7 *Starbright* anti-ship missile with a range of 30–35 NM. The missile could be fired from beneath the surface, forcing the development of new methods for protecting American carrier groups. The result was ASW helicopters and ASW fixed-wing aircraft on carriers, in addition to new passive acoustic sensors for the surface fleet.[17] The silent, high-speed *Los Angeles*-class attack submarine would clear the waters ahead of the force.[18] The *Charlie* was, however, just as dependent on external targeting data as her predecessors, making her vulnerable to detection as she had to raise mast to receive updated targeting information.[19] The Soviets attempted to engage surveillance satellites for cueing the *Charlies* instead of their MPA force, but technological challenges with the satellites kept the SSGN force dependent on target cueing by land-based aircraft.[20] The *Victor* class carried anti-ship and anti-submarine torpedoes, as well as the SS-N-15 *Starfish* ASW missile.[21] ASW missiles are fired from a torpedo tube that leaves the water and flies a distance in the air, before re-entering the water and functioning like a torpedo, alternatively like a depth charge. The *Victor* was built to support strategic operations as an escort, working to hunt down enemy submarines and protecting the *Yankee* SSBN whilst on patrol.[22]

Evolution of Western naval strategy

After the Second Berlin Crisis in 1958, discussion on engagements short of total war led to the general threat of a massive retaliatory response to any hostile action losing its credibility.[23] But changes to the overarching strategy of massive retaliation were met with resistance, particularly in Europe, where much of the reasoning was that the threat of a massive American retaliatory response was the only language the expansionist communists in Moscow would understand.[24] It was when the French opted out of the NATO military command structure in March 1966 that the door opened to a change in NATO's overarching strategy. The strategy included improved intelligence and early warning, prompt coordinated action in the period of warning time, increased readiness, flexibility, and other elements, and would be known as *flexible response*.[25]

SACLANT initiated two studies in 1965, in order to raise awareness of the expanding Soviet fleet.[26] NATO estimates indicated that approximately 134 submarines, some 80% of the Soviet total operational submarine inventory, would be used for rocket attacks against the US coastline, against the NATO strike fleet or against Western SLOCs in a war. The task of keeping track of them and providing early warning was thus formidable.[27] It was with the operationalization of the *Yankee* class that the NATO intelligence community really woke up to the threat.[28] The nuclear-powered strategic submarines operated with increasing speeds, were equipped with missiles with increasing range, and they were capable of underwater launch. SACLANT continued to warn about the expansionist Soviet Navy in early 1968.[29] The expanded pattern of operations had removed the previous warning time that would be provided by the Soviets to NATO, as the former had been focussing on activities in home waters. Now, SACLANT warned, the Soviets were able to launch attacks with significantly less warning or none at all. Submarines in particular were able to deploy for extended periods of time without any supply or support, patrolling off the US coast with ballistic missiles.[30] The SACLANT studies significantly influenced discussions held 2 years later on a renewed NATO Maritime Strategy. The *Yankee*-class patrols brought about a fundamental shift in perception of the North Atlantic and the Northern Flank of NATO. The Russian Navy now posed a continuous threat to both the European and the American mainland. This prompted SACLANT to establish the Standing Naval Force Atlantic (STANAVFORLANT) in 1968, with Alliance nations rotating contributions of assets. Also, SACLANT pointed to a severe lack of NATO surveillance to monitor the expanding Soviet activities and emphasized multinational measures and intelligence sharing between national commands and NATO commanders.[31] SACLANT's efforts prompted NATO Secretary General Manlio Brosio to call for further studies into the nature of the threat in the North Atlantic.[32] The studies assessed that forces operating in the North Atlantic were necessary for a prolonged conflict, in order to meet the threat of maritime air forces seeking control of the ocean areas in the north. If Soviet forces also operated with submarines in support of their aircraft, the survivability of NATO aircraft carriers would decrease. SACLANT thus claimed it an imperative to contain and destroy Soviet submarines as far forward as possible.[33]

On 1 May 1960, a U-2 surveillance aircraft with plans to land in Bodø, Norway, was shot down over the Soviet Union. Not long thereafter, an American RB-47 reconnaissance aircraft flying from the United Kingdom was shot down over the Barents Sea. These incidents influenced the Norwegian Government's decision to alter the existing restrictions on foreign aircraft flying from Norwegian soil. As tensions did not just rise over land, starting on 25 October 1960, Allied aircraft taking off from Norway were formally no longer permitted to fly beyond 24°E, even over *international waters*, without the explicit consent from the Norwegian Government.[34] However, due to delays in the delivery of the *Albatross* aircraft to the Norwegian Air Force in 1961, there was an increased requirement for support from allied partners to cover surveillance operations in the High North. This support fell under the aforementioned arrangements of *NEPTUNE JOURNEY* and *EXERCISE*

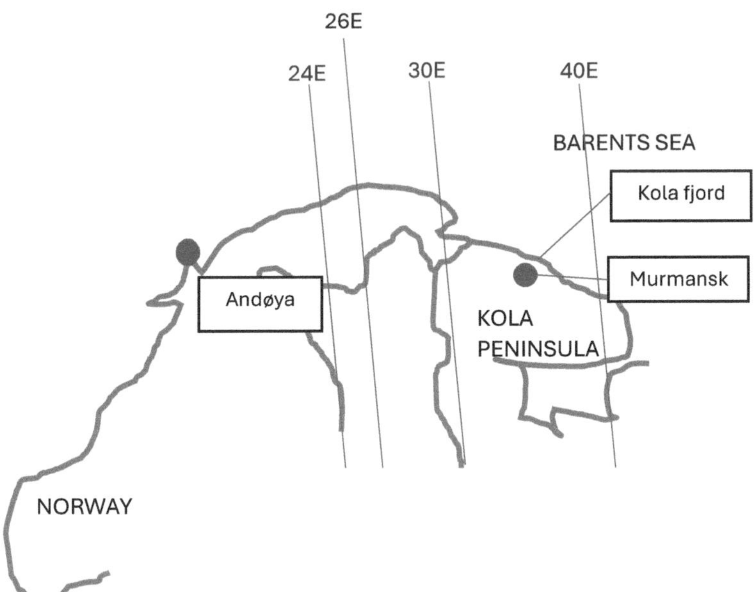

Figure 6.1 Overview of longitudinal delineation of Norwegian restrictions towards allied aircraft

POPEYE. In addition to these two arrangements, yet another regime for trilateral cooperation between the United States, the United Kingdom, and Norway on airborne maritime surveillance came into being. *EXERCISE COLD ROAD* was established in 1961 to assure sufficient coverage in the High North in the transition from *Catalina* to the new *Albatross* for Norway.

The *COLD ROAD* arrangement distinguished itself from the other two in that it represented a more systematic approach to the planning of surveillance missions, where a given number of flights were *planned* well in advance, whereas the other two arrangements functioned as a regime for *approving* allied surveillance flights. In the months between July 1960 and the winter of 1961, the Norwegian *Catalinas* were taken out of service, and the *COLD ROAD* arrangement was implemented. Even though the *COLD ROAD* regime underlined the fact that the 24° restriction remained in effect, the transition period came with relaxations. On several occasions, the British and Americans were allowed to operate to 24°E, and at times even further east. Some Allied missions based in Norway were authorized to 30°E, one even to 40°E, but such missions were subject to high-level authorization.[35] When the *Albatrosses* were operational in Norway, the strict enforcement of the 24° restriction went back into place. Although not fully ASW-capable, the *Albatrosses* provided important surveillance information from the High North.

Norwegian authorities asked for British and American assistance in covering the High North with surveillance flights during the transition to *Albatross* aircraft, and it was the British that conducted most of those missions.[36] British aircrews flew from their bases in the United Kingdom to the North Atlantic and carried out their surveillance missions in the assigned areas, regularly stopping in Bodø in Northern Norway for the night. At times, selected members of the crew would visit the Norwegian Maritime Headquarters (MHQ) for discussions and information exchange.[37] When the British were tasked by their own headquarters, they usually flew further south than the Norwegians normally did. But when they operated out of Norwegian bases on Norwegian tasking orders, they tended to cover areas further north and east, although not further east than 24°.[38] The British could also be requested by headquarters in alliance nations to cover Russian naval activity further to the West between Norway and Iceland, sometimes with several aircraft simultaneously if the naval activity was of significant size and scope.[39]

Yet another arrangement for international coordination and cooperation for aerial maritime surveillance was the *BOX CAR* Agreement. This was put in place during the Norwegian upgrade to the *Albatross* in 1960–1961 in order to structure the landing of allied surveillance aircraft on Norwegian and Danish airfields.[40] The *BOX CAR* Agreement was thus an agreement for surveillance in more *southern* areas of responsibility in the Atlantic, such as the Norwegian Sea and Skagerrak. It was supported significantly by the British for several years after the Norwegian transfer to *Albatross* aircraft, with the new aircraft becoming fully operational in 1963. Operational records depict regular British sorties almost every month under the arrangement.[41]

The British effort in support of *COLD ROAD* surveillance also increased in the years that followed. In the mid-1960s, it became commonplace with routine expansion of the areas covered, especially Westwards. This meant that British aircraft could fly under the *COLD ROAD* arrangement first from the United Kingdom to Keflavik, Iceland, and then from Keflavik to Andøya or Bodø in Norway, before leaving Norway to return to the United Kingdom in a three-day operational roundtrip.[42] At times, the *Shackleton* crews were tasked with more than one mission out of Norway, which meant that they would transit to Norway while carrying out a mission, stay in Norway and typically perform a mission a day for several days, and then return to the United Kingdom on an operational transit.[43] The number of British missions under the *COLD ROAD* arrangements decreased sharply in the fall of 1965, as the Norwegian *Albatrosses* were increasingly capable of covering their own, assigned areas.[44]

With the new aircraft came sensors that were a significant improvement to the *Catalinas*, although not by any means on par with the American *Neptunes* or *Orions*. The combination of the lack of ASW and ELINT capabilities in the Norwegian *Albatrosses* and the self-imposed restrictions for Allied surveillance aircraft brought the Norwegian basing policy under pressure. However, Allies monitored and followed Russian exercises and international deployments, but the strategic, nuclear threat came from ICBMs on *land* after the SRF had been established. It wasn't until 1967 and the *Yankee* patrols that the requirements for continuous ASW really set in in the North Atlantic.

The evolution of allied ASW and MPA capabilities

In anticipation of quieter submarines, there was a steady improvement of Western ASW technology's ability to detect Russian submarines in the 1960s. At the core of the modern ASW operations stood barrier operations, with SOSUS arrays on the seabed providing cueing of approaching submarines to other platforms. The first arrays were deployed from Argentina not far from the US coastline, but the new technology now facilitated a more forward-leaning approach to the search. In 1964, a SOSUS array was placed between Andøya, Bear Island, and Svalbard, covering the entrance to the Atlantic for the Northern Fleet.[45] The Norwegian BRIDGE system was a seabed system similar to SOSUS, and the two worked in tandem, with BRIDGE contributing significantly to the search efforts.[46] In 1965, another SOSUS array was placed between Greenland, Iceland, and the United Kingdom, in the so-called GIUK gap. SOSUS was a cornerstone in the search for and detection of Soviet submarines, but utilization of the acoustic data had to be handled so as not to disclose the SOSUS capabilities and whereabouts. The estimated position of the adversary submarine was revealed indirectly to the aircraft crew through other sources. Some intelligence officials even actively prohibited the use of SOSUS data by ASW aircraft, in fear of disclosing their acoustic capabilities to the Soviets.[47]

The US Navy introduced the P-3 *Orion* operationally in the summer of 1962. The aircraft had the most capable ASW suite available, in addition to a new radar, but, most importantly, the *Orion* featured on-board computers with an unprecedented processing capability. The *Orion* hit the track running with significant involvement in the Cuban Missile crisis of October that year.[48] The new MPA commenced operational sorties in the North Atlantic soon thereafter. With the P-3 *Orion* becoming operational, further integration of SOSUS with MPA assets provided a high level of awareness of Russian submarine movements. This removed the hunter-killer role of ASW-dedicated aircraft carriers in the 1960s, and led to the removal of the ASW-dedicated aircraft carriers altogether in the 1970s.[49]

When the *Yankee* class initiated their strategic patrols, Western ASW forces had embraced a *barrier* approach in full. SOSUS was up and running, the integration with the MPA community was functioning, and forward-deployed ASW-optimized attack submarines were operating in trail of the great Soviet missile platforms. The ASW barrier strategy was the framework within which Alliance ASW forces in the North Atlantic operated. They detected and tracked Soviet strategic submarines more or less continuously as they initiated their patrols in the late 1960s off the US coast.[50]

Towards the late 1960s, the Norwegians initiated negotiations with the Americans for procurement of the P-3 *Orion*. The developments of the 1960s had emphasized the requirement for capable ASW aircraft in the North Atlantic, and the Norwegians were given a significant reduction in procurement costs from the Americans based on an explicit bilateral arrangement for sharing of intelligence gathered by the new aircraft.[51] Norway finally had the right aircraft to conduct maritime airborne ISR in the High North, not just in terms of range, but also in terms of sensors and communications.

The British *Shackletons* were, by the late 1960s, becoming outdated in most respects, and compared to the P-3 *Orion*, the *Shackleton's* ASW performance was poor. However, the British were still capable of conducting ASW operations, as they operated a relatively high number of aircraft.[52] At times, the *Shackletons* operated in cooperation with the much faster *Victor* bomber aircraft, which flew at high altitude and provided radar cover, with the *Shackletons* investigating contacts.[53] In any case, the process to find a new maritime surveillance aircraft for the British was initiated in 1963.[54] Close interaction with Allies also helped prepare the British aircrews for the new equipment that was inbound with their new MPA, the *Nimrod*.[55] The RAF took delivery of the first *Nimrod* in October 1969.[56]

Moreover, the early 1960s also witnessed the launch of the first reconnaissance satellites. The *CORONA* IMINT satellite of August 1960 was predated by more than a month by the US Naval Research Laboratory's ELINT satellite, the *Galactic Radiation and Background* (GRAB) in June 1960.[57] Intended to map Soviet air defence radars that couldn't be reached through long-range ELINT flights along the Soviet border, *GRAB* marked the dawn of what would become a crucial part of modern reconnaissance and surveillance: Satellite systems. In time, ELINT satellites were to eventually provide between 80 and 90 percent of surface vessel OPINTEL for the US Navy.[58] Although more Western SIGINT satellites were launched in the 1960s, they were mostly focussed on activities on the Soviet land mass and not naval activities.[59] In addition, although communications satellites were operationalized in the 1960s, they were in naval terms dedicated for supporting fleet flag ships and carriers.[60] Satellites focussing on naval reconnaissance were, to a larger extent, fielded in the 1970s.

All in all, the 1960s witnessed improvements for the MPA force with regard to computer and signal processing, signal integration and time compression, miniaturization of electronics, and improvements in sonobuoy technology.[61] The automation of sensor products started in the 1960s, and is an important factor for airborne ISR assets in modern surveillance operations. Automated intelligence processing systems were introduced, as were (at the time) high-speed data links between aircraft and between aircraft and headquarters.[62] But most of the automated processing capability took place at processing facilities ashore. Experiences from supporting the war in Vietnam provided important lessons for the fusion of all source data for naval OPINTEL.[63] It was in the late 1960s that the ELINT-specific maritime aircraft EP-3 *Aries* was developed, specializing in collecting ELINT in the maritime domain.[64]

The deployment of allied MPA capabilities

An increase in Soviet operational naval activities in the North Atlantic took place in the middle of the 1960s, leading to a significant increase in reporting on Soviet submarine activity by MPA aircrews.[65] From the mid-1960s, moreover, American aircrews took part in intelligence gathering operations on the suspected air-submarine manoeuvre by Soviet forces against transiting US aircraft carriers.[66] This could be seen as reaction to the fielding of tactical submarines with anti-carrier missiles in the Soviet Navy, which has been discussed above. The primary mission for MPA was ASW and intelligence gathering against the Soviet submarine fleet.[67]

During the Soviet naval exercise in the North Atlantic in August 1963, several Allies flew out of Norwegian bases to collectively cover the Russian activities. *Shackleton* aircraft took part in this surveillance activity, named *Operation Bargold* by the British. By this time, the Norwegian *Albatrosses* were operational and flew together with the British and the Americans. Canadian aircraft took part as well, but these flew from bases outside Norway, as they were not a part of the *COLD ROAD* arrangement. Tasking was efficiently coordinated between the maritime headquarters (MHQ) in Bodø and the UK MHQ in Rosyth, and intelligence reports were shared between the nations after the sorties. Rigid and strict procedures for this sharing of intelligence were yet to be established, and the information flowed relatively seamlessly between the participants.[68]

Many within the ASW community emphasized that the primary search sensor in the hunt for submarines was still radar (in order to find surfaced submarines at periscope depth) and vision. However, increasing activities with more capable submarines in the Northern Fleet initiated discussions both in Norway as well as in NATO on both ASW equipment and national competency levels of modern ASW. During *Operation Bargold* in 1963, the British and the Americans performed well with regard to detecting Russian submarines, but the Norwegian *Albatrosses* were unable to detect any submarines at all.[69] They simply did not carry the appropriate ASW sensors to do the required job. However, with the increasing Soviet activities in the Barents Sea, which occurred to a high degree beyond the approved areas of Allied aircraft, the Norwegian *Albatrosses* were able to deliver much needed surveillance of Soviet surface activities. Even though the strongest feature of the *Albatross* aircraft was *surface* surveillance, the aircraft was at times capable of providing intelligence on submerged submarines as well. One such instance was the collection of acoustic details on the relatively new *Echo II* submarine in 1965. However, the Norwegian ASW shortfalls in the High North were deemed a critical problem to the Alliance in general, and the United States in particular.[70] *Operation Adjutant* in the summer of 1964 was a follow-up of *Bargold* the previous year, and had the objective of shadowing the Northern Fleet as they exercised off the coast of Norway. The British aircrews were also to cooperate with a submarine barrier situated between Bear Island and the Norwegian coast. The *Shackleton* post-mission report was filled with sightings of Russian naval vessels that would normally not be seen further south.[71] Coordinated surveillance operations took place again in 1965, through *Operation Beresford*. British authorities emphasized that

> these exercises give us a unique opportunity to observe Soviet ships, submarines, and aircraft at work in this vital sea area, to assess their capabilities under operational conditions, and to judge their current state and development of their equipment and tactics.[72]

The respective aircraft operated under national control, but coordination between the respective national MHQ's de-conflicted areas among themselves.[73] The operational report claimed that the surveillance was a "marked success," and that the "volume of intelligence gained in the operation is the greatest and most valuable since these operations were instituted."[74]

On an overarching level, the Norwegians, the Americans, and the British came to an understanding for a trilateral coordination of anti-submarine intelligence collection in the Norwegian Sea with emphasis on the exploitation of the underwater sensor at Andøya. This specific coordination of intelligence had been established in 1965, and was termed *Operation Canasta*.[75] More specifically, with the Norwegian acquisition of the P-3B *Orion* as a baseline, the NORCANUSUK Agreement between Norway, Canada, the United States, and the United Kingdom was signed in 1969. The agreement delineated the responsibility to conduct maritime surveillance in the North Atlantic and the High North. Norway was to maintain surveillance of the "north-eastern waters."[76] The NORCANUSUK Agreement reaffirmed that Norwegian aircraft were in charge of the High North east of 24°E. In 1969, however, the British would yet again fill in the gap left in surveillance by the turnover in Norwegian capabilities, this time from the *Albatross* to the *Orion*. British aircrews deployed in August 1969 to Bodø for *Operation Knockabout*, the code for the deployment of coastal command units for shadowing of the Soviet fleet exercises.[77] As the Soviet force moved south, British aircraft based in Scotland took over. As the force moved north-west, American aircraft based on Iceland took over.[78]

The surveillance flights in the early 1960s were mostly coordinated in advance. British aircrews flying *Shackletons* out of Kinloss, Scotland, were tasked by the MHQ in Pitreavie, Rosyth, close to Edinburgh, Scotland. The Headquarters in

Pitreavie also served as the NATO entity responsible for maritime air operations in the North Atlantic, as *Commander Maritime Air Northern Sub-Area*. First answering directly to *Commander-in-Chief Eastern Atlantic Area* (CINCEASTLANT) under SACLANT, maritime air operations came under centralized control of the maritime headquarters in Northwood, London, in the early 1960s.[79] NATO did not control any assets in peacetime, but coordinated and wrote strategy, plans, doctrine, as well as directing exercises. These command and control entities played important roles in shaping the way the maritime air community interacted. Norwegian aircrews flying *Albatrosses* out of Andøya were tasked by the Maritime Air Operations Centre (MAOC), which was run by air force personnel but situated at the Maritime Headquarters in Bodø.[80] The American *Neptunes* were tasked by the *Commander Iceland Sector ASW Group* (CTG 84.1) based in Keflavik, who, in his national chain-of-command, answered to *Commander Task Force (CTF) 84*, one of the subunits of *Commander Submarine Forces Atlantic* in Norfolk, United States.[81] The close interaction at the national level between the Americans at Keflavik, the British in Rosyth and later Northwood, and the Norwegians in Bodø, was executed outside the formal NATO command structure. The bi- and trilateral interaction facilitated the coordination of maritime air operations in the North Atlantic, which was required for maintaining an overview of the Soviet Navy long-range submarine patrols in the Atlantic. Allied aircraft were regularly tasked by the military authorities of another nation, especially British and American aircraft regularly flying out of Norwegian airfields.

There is little evidence of in-flight coordination between aircraft of different nationalities in the early 1960s, coordination that did not seem to take shape before regular Soviet strategic patrols began in the mid- to late 1960s. Even after this, it was the position, course, and speed of the target submarine that was shared in real time, as long as the information was available. The intelligence collection of more detailed nature was kept for national analysis.

The North Atlantic ASW environment in the 1960s was busy. Over the course of 1 month in March 1968, one US squadron at Keflavik flew 91 sorties for a total of 784 h, providing continuous coverage with at least one aircraft over the target submarine for three weeks straight.[82] Over the course of a deployment of 5–6 months to Keflavik in the late 1960s, it was common to be in contact with 15–20 submarines.[83] As the Soviet *Yankee* patrols settled into routine, the Keflavik squadrons worked in tandem with their Allies to cover the Soviet submarines emerging from the North.[84] Similar to the British squadrons, the US Navy squadrons routinely deployed to allied countries such as Norway, the United Kingdom, Denmark, Germany, and the Netherlands in order to support local efforts, to practice operating out of foreign airfields, and often times to get closer to the submarine operating areas.[85] Whether the Americans based on Iceland at times flew beyond 24°E based from their home-base or not was not known by Norwegian authorities.[86] American operational records do not reveal any particular reports that would indicate that they did.[87] Norwegian authorities considered loosening the restrictions in the course of the 1960s, moving the restricted line from 24° to 27°, but the *Albatrosses* managed to support intelligence gathering efforts sufficiently for the 24° restriction to remain in place.[88] This indicates that the British and American surveillance operations further to the West and South were considered sufficient to execute the barrier strategy against the Soviet submarines.

After the Soviet *Sputnik* launch, it was of particular interest to follow Soviet missile development, and British *Comet* ELINT aircraft regularly deployed to Bodø and Andøya in order to fly missions in the Barents Sea, flying with sensors that the Albatrosses didn't have. Approval for this was given through the prime minister's Security Council. The Comets flew along the Kola coastline, then up north along the Novaya Zemlya coastline, before returning to Bodø.[89] After a pause from flying these missions in 1960 and 1961 due to the previously mentioned shoot downs of the U-2 and the RB-47, the British flights resumed until 1967, when they were denied for the foreseeable future.[90] The denial likely came as Norwegian intelligence capabilities were growing in assertiveness, credibility, and general standing. Allied operations in the High North also included RC-135 *Rivet Joint* ELINT operations flown by the Americans.[91] These were, however, flown from the United Kingdom and did not fall under the Norwegian restrictions.

In what can be termed a theme of general expansion in both inventory and operational mission portfolio, the strategy and posture of the Soviet Navy fluctuated somewhat in the 1960s. But even if the Strategic Rocket Forces were established early in the decade, the submarines were given back their mission of strategic nuclear strike by the end of the decade, most evidently represented by the operationalization of the *Yankee* class in 1967. Both the anti-carrier submarines and missiles and the strategic nuclear submarines with their missiles were brought forward by the 1970s. In spite of fluctuations, the submarine force only strengthened its place in Soviet naval as well as grand strategy throughout the 1960s. And through the 1960s, the Allied approach to the Soviet submarine threat was to stay on top and keep track of Russian submarines on patrol, regardless of the overarching NATO strategy.

The 1960s were also formative for airborne maritime ISR in many ways, particularly in the sense of trilateral partnering in the surveillance of Soviet naval operations. MPA had, by then, been established as the weapon of choice to meet the submarine threat. Several of the fundamental tenets of airborne ISR can be identified in the dynamics for MPA development and operationalization in the 1960s. For communications, data link was introduced, facilitating more efficient communication and coordination between both tactical units as well as headquarters ashore. The poor ASW suite on the *Albatrosses* put Norwegian flight restrictions for foreign aircraft operating from Norwegian soil under pressure. A combination of the flight restrictions and the poor ASW capabilities of the *Albatross* seems to ultimately have secured the *Orion* deal for the Norwegians in the late 1960s.

The question of range for MPA is subsumed in the overarching dynamics altogether. In order to get more time on station in areas with the most activity, American and British aircraft sought to operate from Norwegian airfields in the north. This was the basis for the bi- and trilateral coordination between these three nations, as manifested through the respective coordination arrangements: *NEPTUNE JOURNEY*, *EXERCISE POPEYE*, *COLD ROAD*, *BOX CAR*, and the *NORCANUSUK*

Agreement. However, had range not been a factor, the UK and US aircraft could easily have operated from their home bases without being restricted by Norwegian policy in the first place.

Notes

1 Klevberg, "Maritime Surveillance in the North," 183, 252.
2 Klevberg, 250–51.
3 Jan S. Breemer, "Estimating the Soviet Submarine Missile Threat: A Critical Examination of the Soviet Navy's SSBN Bastion Strategy" (PhD, International Relations, University of Southern California, 1987), 61; Friedman, *Submarine Design and Development*, 103–4; Jordan, *Soviet Submarines – 1945 to the Present*, 71.
4 Podvig, *Russian Strategic Nuclear Forces*, 121.
5 Palmer, *Origins of the Maritime Strategy*, 90; Jordan, *Soviet Submarines – 1945 to the Present*, 77–78; Cote, "The Third Battle – Innovation in the U.S. Navy's Silent War Struggle with Soviet Submarines," 60.
6 Jordan, *Soviet Submarines – 1945 to the Present*, 78.
7 Jordan, 68.
8 Jordan, 74; Ford og Rosenberg, *The Admiral's Advantage*, 41.
9 Podvig, *Russian Strategic Nuclear Forces*, 237–38; Weir og Boyne, *Rising Tide*, 109–10; Friedman, *Submarine Design and Development*, 103.
10 Jordan, *Soviet Submarines – 1945 to the Present*, 88–91; Friedman, *Submarine Design and Development*, 104.
11 Jordan, *Soviet Submarines – 1945 to the Present*, 95–96; Breemer, "Estimating the Soviet Submarine Missile Threat."
12 MC/CS 42, Summary Record (December 20, 1968), NATO Archives, quoted in Dyndal, "How the High North Became Central in NATO Strategy."
13 Secondary sources diverged on this point. Pavel Podvig, *Russian Strategic Nuclear Forces*, 239; Cote, "The Third Battle – Innovation in the U.S. Navy's Silent War Struggle with Soviet Submarines," 20; Jordan, *Soviet Submarines – 1945 to the present*, 96; Friedman, *Submarine Design and Development*, 104–5; AIR 27/2978, "Operational Records – 201 squadron – 1961–1963"; AIR 27/2979, "Operational Records – 201 squadron – 1963–1965"; AIR 27/3161, "Operational Records – 201 squadron – 1966–1968." First with *Golf* and *Hotel* class submarines, and then there is a consensus on the commissioning of the *Yankee* class in 1967–69, with operational patrols commencing in 1969.
14 Dyndal, "How the High North Became Central in NATO Strategy," 572.
15 Podvig, *Russian Strategic Nuclear Forces*, 274.
16 Breemer, "Estimating the Soviet Submarine Missile Threat," 81–82; Friedman, *Submarine Design and Development*, 105–6; Jordan, *Soviet Submarines – 1945 to the present*, 96–99.
17 Cote, "The Third Battle – Innovation in the U.S.Navy's Silent War Struggle with Soviet Submarines," 61–63.
18 Friedman, *Submarine Design and Development*, 72; Jordan, *Soviet Submarines – 1945 to the present*, 103–8.
19 Jordan, *Soviet Submarines – 1945 to the present*, 108.
20 Friedman, *Seapower and Space*, 155–72.
21 Jordan, *Soviet Submarines – 1945 to the present*, 108–10; Friedman, *Submarine Design and Development*, 106.
22 Jordan, *Soviet Submarines – 1945 to the present*, 113–15.
23 See for example Brodie, *Strategy in the Missile Age*, 174–77; Schelling, *Strategy of Conflict*, 230–38; Friedman, *Submarine Design and Development*, 99.
24 Pedlow, "NATO Strategy Documents 1949–1969"; Tamnes, *The United States and the Cold War in the High North*, 196–98; Dyndal, "How the High North Became Central in NATO Strategy," 565.

25 NATO, "MC 48/3 – Measures to Implement the Strategic Concept for the Defence of the NATO Area" (North Atlantic Treaty Organization (NATO), December 8, 1969), Electronic Document Folder: IMS/MC-MCM-SG/1953–1970, NATO Archives, Brussel; Pedlow, "NATO Strategy Documents 1949–1969," XXV.
26 Dyndal, "How the High North Became Central in NATO Strategy," 568.
27 Riste og Moland, *Strengt Hemmelig – Norsk etterretningstjeneste 1945–1975*, 220.
28 Podvig, *Russian Strategic Nuclear Forces*, 238–39.
29 C-R(68)5, "Summary record of a meeting of the Council held at the Headquarters, Brussels, on Wednesday, 24th January at 10.15 A.M."
30 C-R(68)5.
31 NATO IMSM-483–68 (rev), Memorandum on 'Status reports' (12 November 1968), and NATO IMSWM-362–68, Memorandum on 'Status Report' for DPC in Ministerial Session (23 December 1968), quoted in Dyndal, "How the High North Became Central in NATO Strategy," 573; Eric Grove og Graham Thomson, *Battle for the Fjords* (Annapolis, MD: Naval Institute Press, 1991), 14.
32 NATO PO/68/117, 13 February 1968, quoted in Dyndal, "How the High North Became Central in NATO Strategy," 575.
33 Grove og Thomson, *Battle for the Fjords*, 14–15, 23.
34 Tamnes, *The United States and the Cold War in the High North*, 183.
35 Tamnes, 184; Klevberg, "Maritime Surveillance in the North," 254–55.
36 Although Canadian aircraft took part in the surveillance efforts in the Atlantic, they were not a part of the Cold Road arrangement.
37 AIR 27/2978, "Operational Records – 201 Squadron – 1961–1963."
38 AIR 27/2978.
39 AIR 27/2978.
40 Aircraft from US, UK, Canada, and eventually West Germany and the Netherlands. See Klevberg, "Maritime Surveillance in the North," 326.
41 AIR 27/2979, "Operational Records – 201 Squadron – 1963–1965"; AIR 27/3161, "Operational Records – 201 Squadron – 1966–1968."
42 AIR 27/2979, "Operational Records – 201 Squadron – 1963–1965."
43 AIR 27/2979.
44 AIR 27/2979.
45 Olav Riste and Arnfinn Moland, *Strengt Hemmelig – Norsk Etterretningstjeneste 1945–1975* (Oslo: Universitetsforlaget, 1997), 215–32; Cote, "The Third Battle – Innovation in the U.S. Navy's Silent War Struggle with Soviet Submarines," 41; Owen Wilkes and Nils Petter Gledistch, *Onkel Sams Kaniner – Teknisk Etterretning i Norge* (Oslo: Pax Forlag, 1981), 125–27.
46 Kjetil Skogrand, Olav Njølstad, og Rolf Tamnes, "Hot Spot – Cold War, the High North, and Grand Strategy," *Institutt for Forsvarsstudier (IFS) Info* 5 (1998): 22–23.
47 Riste og Moland, *Strengt Hemmelig – Norsk etterretningstjeneste 1945–1975*, 219; Ford og Rosenberg, *The Admiral's Advantage*, 37.
48 "Command History Report (CHR) for VP-56 (1962)."
49 Cote, "The Third Battle – Innovation in the U.S. Navy's Silent War Struggle with Soviet Submarines," 50–51.
50 Larry Robideau, "Third Battle for the North Atlantic 1962–1991," *Cold War Times* 6, no. 1 (February 2006), www.coldwar.org/text_files/coldwartimesfeb2006.pdf; Podvig, *Russian Strategic Nuclear Forces*, 240; Cote, "The Third Battle – Innovation in the U.S. Navy's Silent War Struggle with Soviet Submarines," 52.
51 Klevberg, "Maritime Surveillance in the North," 333–35.
52 Gjert Lage Dyndal, "Land Based Air Power or Aircraft Carriers? The British Debate About Maritime Air Power in the 1960s" (PhD, History, University of Glasgow, 2009), 194–98.
53 AIR 27/3162, "Operational Records – 201 Squadron – 1969–1970."
54 Tony Blackman, *Nimrod – Rise and Fall* (London: Grub Street Publishing, 2011), 14.

55 AIR 27/3162, "Operational Records – 201 Squadron – 1969–1970."
56 Blackman, *Nimrod – Rise and Fall*, 18–25.
57 Norris, *Spies in the Sky*, 26.
58 Press Release, "GRAB Satellite Declassified – NRL Built and Deployed First Reconnaissance Satellite System," U.S. Naval Research Laboratory, June 17, 1998, www.nrl.navy.mil/media/new-news-releases/1998/grab-satellite-declassified-nrl-built-and-deployed-first-reconnaissance-satellite-system.
59 Friedman, *Seapower and Space*, 102–10.
60 Friedman, 77.
61 Robideau, "Third Battle for the North Atlantic 1962–1991."
62 Ford og Rosenberg, *The Admiral's Advantage*, 46–47.
63 Ford og Rosenberg, 53–54.
64 Reade, *The Age of Orion – The Lockheed P-3 Story*, 106–7.
65 See for example "Command History Report (CHR) for VP-23 (1965)."
66 "Command History Report (CHR) for VP-44 (1967)."
67 "Command History Report (CHR) for VP-21 (1960–1965)"; "Command History Report (CHR) for VP-5 (1960–1963)."
68 Klevberg, "Maritime Surveillance in the North," 289.
69 Klevberg, 290.
70 Tamnes, *The United States and the Cold War in the High North*, 216; Klevberg, "Maritime Surveillance in the North," 302.
71 AIR 27/2979, "Operational Records – 201 Squadron – 1963–1965."
72 DEFE 24/20, "British Surveillance Efforts Against the Soviet Northern Fleet – Summer Exercises 1965," UK Director of Naval Intelligence, 1965, UK National Archives, Kew.
73 DEFE 24/20, "Operation Beresford – Planning Guide," UK Director of Naval Intelligence, 1965, UK National Archives, Kew, Section 1A, para 7.
74 DEFE 24/20, "Operation Beresford – Letter from Deputy Director of Naval Intelligence to the Ministry of Defence," UK Director of Naval Intelligence, October 5, 1965, UK National Archives, Kew.
75 Riste og Moland, *Strengt Hemmelig – Norsk etterretningstjeneste 1945–1975*, 227; Skogrand, Njølstad, og Tamnes, "Hot Spot – Cold War, the High North, and Grand Strategy," 23.
76 Tamnes, *The United States and the Cold War in the High North*, 213, 216.
77 AIR 27/3162, "Operational Records – 201 Squadron – 1969–1970."
78 AIR 27/3162.
79 NATO, *NATO – Facts About the North Atlantic Treaty Organization – 1962* (Paris: NATO Information Service, 1962); Dave Angus, "Command and Control of MPA Assets on the Northern Flank," *Maritime Patrol Aviation* 2, no. 1 (September 1991).
80 Klevberg, "Maritime Surveillance in the North," 312–15.
81 Robideau, "Third Battle for the North Atlantic 1962–1991."
82 "Command History Report (CHR) for VP-24 (1968)."
83 "Command History Report (CHR) for VP-24 (1969)"; "Command History Report (CHR) for VP-44 (1969)."
84 "Command History Report (CHR) for VP-24 (1968)."
85 "Command History Report (CHR) for VP-24 (1968)."
86 Klevberg, "Maritime Surveillance in the North," 249–50.
87 Operational records for US Navy squadrons deploying to Keflavik from 1952 to 1994 have been surveyed for this thesis: Naval History and Heritage Command, FOIA case number: DON-NAVY-2018-00574.
88 Klevberg, "Maritime Surveillance in the North," 318–19.
89 Blackman, *Nimrod – Rise and Fall*, 69–70.
90 Klevberg, "Maritime Surveillance in the North," 302–4.
91 Tamnes, *The United States and the Cold War in the High North*, 216–17.

Bibliography

AIR 27/2978. "Operational Records – 201 Squadron – 1961–1963." UK Ministry of Defense, 1963–1961. UK National Archives, Kew.

AIR 27/2979. "Operational Records – 201 Squadron – 1963–1965." Royal Air Force, 1965–1963. UK National Archives, Kew.

AIR 27/3161. "Operational Records – 201 Squadron – 1966–1968." Royal Air Force, 1968–1966. UK National Archives, Kew.

AIR 27/3162. "Operational Records – 201 Squadron – 1969–1970." Royal Air Force, 1970. UK National Archives, Kew.

Angus, Dave. "Command and Control of MPA Assets on the Northern Flank." *Maritime Patrol Aviation* 2, no. 1 (September 1991).

Blackman, Tony. *Nimrod – Rise and Fall.* London: Grub Street Publishing, 2011.

Breemer, Jan S. "Estimating the Soviet Submarine Missile Threat: A Critical Examination of the Soviet Navy's SSBN Bastion Strategy." PhD, International Relations, University of Southern California, 1987.

Brodie, Bernard. *Strategy in the Missile Age*, 8. utg. Princeton, NJ: Princeton University Press, 1991.

"Command History Report (CHR) for VP-5 (1960–1963)." US Navy, 1963–1960. FOIA case number: DON-NAVY-2018-001782. Naval History and Heritage Command.

"Command History Report (CHR) for VP-21 (1960–1965)." US Navy, 1965–1960. FOIA case number: DON-NAVY-2018-001782. Naval History and Heritage Command.

"Command History Report (CHR) for VP-23 (1965)." US Navy, 1965. FOIA case number: DON-NAVY-2018-00574. Naval History and Heritage Command.

"Command History Report (CHR) for VP-24 (1968)." US Navy, 1968. FOIA case number: DON-NAVY-2018-001782. Naval History and Heritage Command.

"Command History Report (CHR) for VP-24 (1969)." US Navy, 1969. FOIA case number: DON-NAVY-2018-001782. Naval History and Heritage Command.

"Command History Report (CHR) for VP-44 (1967)." US Navy, 1967. FOIA case number: DON-NAVY-2018-001782. Naval History and Heritage Command.

"Command History Report (CHR) for VP-44 (1969)." US Navy, 1969. FOIA case number: DON-NAVY-2018-001782. Naval History and Heritage Command.

"Command History Report (CHR) for VP-56 (1962)." US Navy, 1962. FOIA case number: DON-NAVY-2018-00574. Naval History and Heritage Command.

Cote, Owen R. Jr. "The Third Battle – Innovation in the U.S. Navy's Silent War Struggle with Soviet Submarines." *Naval War College Newport Papers* 16 (2003).

C-R(68)5. "Summary Record of a Meeting of the Council Held at the Headquarters, Brussels, on Wednesday, 24th January at 10.15 A.M." NATO, February 7, 1968. Electronic Document Folder: IMS/MC-MCM-SG/1953-1970. NATO Archives, Brussel.

DEFE 24/20. "British Surveillance Efforts Against the Soviet Northern Fleet – Summer Exercises 1965." UK Director of Naval Intelligence, 1965. UK National Archives, Kew.

DEFE 24/20. "Operation Beresford – Letter from Deputy Director of Naval Intelligence to the Ministry of Defence." UK Director of Naval Intelligence, October 5, 1965. UK National Archives, Kew.

DEFE 24/20. "Operation Beresford – Planning Guide." UK Director of Naval Intelligence, 1965. UK National Archives, Kew.

Dyndal, Gjert Lage. "How the High North Became Central in NATO Strategy: Revelations from the NATO Archives." *Journal of Strategic Studies* 34, no. 4 (2011): 557–85.

Dyndal, Gjert Lage. "Land Based Air Power or Aircraft Carriers? The British Debate About Maritime Air Power in the 1960s." PhD, History, University of Glasgow, 2009.

Ford, Christopher, og David Rosenberg. *The Admiral's Advantage – U.S. Navy Operational Intelligence in World War II and the Cold War.* Annapolis, MD: Naval Institute Press, 2005.

Friedman, Norman. *Seapower and Space.* Annapolis, MD: Naval Institute Press, 2000.

Friedman, Norman. *Submarine Design and Development*. London: Conway Maritime, 1984.
Grove, Eric, og Graham Thomson. *Battle for the Fjords*. Annapolis, MD: Naval Institute Press, 1991.
Jordan, John. *Soviet Submarines – 1945 to the Present*. London: Arms & Armour Press, 1989.
Klevberg, Håvard. "Maritime Surveillance in the North: 333 Squadron in Norwegian Security Politics." PhD, History, University of Oslo, 2011.
MC 48/3. "Measures to Implement the Strategic Concept for the Defence of the NATO Area." North Atlantic Treaty Organization (NATO), December 8, 1969. Electronic Document Folder: IMS/MC-MCM-SG/1953-1970. NATO Archives, Brussel.
NATO. *NATO – Facts About the North Atlantic Treaty Organization – 1962*. Paris: NATO Information Service, 1962.
Norris, Pat. *Spies in the Sky – Surveillance Satellites in War and Peace*. Chichester, UK: Praxis Publishing Ltd., 2008.
Palmer, Michael A. *Origins of the Maritime Strategy – The Development of American Naval Strategy, 1945–1955*, reissue ed. Annapolis, MD: Naval Institute Press, 1990.
Pedlow, Gregory W. "NATO Strategy Documents 1949–1969." October 1997. NATO, Online Archive. www.nato.int/docu/stratdoc/eng/intro.pdf.
Podvig, Pavel, red. *Russian Strategic Nuclear Forces*, 2004 paperback ed. Cambridge, MA: MIT Press, 2001.
Press Release. "GRAB Satellite Declassified – NRL Built and Deployed First Reconnaissance Satellite System." U.S. Naval Research Laboratory, June 17, 1998. www.nrl.navy.mil/media/new-news-releases/1998/grab-satellite-declassified-nrl-built-and-deployed-first-reconnaissance-satellite-system.
Reade, David. *The Age of Orion – The Lockheed P-3 Story*. Atglen, PA: Schiffer Publishing Ltd., 1998.
Riste, Olav, og Arnfinn Moland. *Strengt Hemmelig – Norsk etterretningstjeneste 1945–1975*. Oslo: Universitetsforlaget, 1997.
Robideau, Larry. "Third Battle for the North Atlantic 1962–1991." *Cold War Times* 6, no. 1 (February 2006). www.coldwar.org/text_files/coldwartimesfeb2006.pdf.
Schelling, Thomas C. *Strategy of Conflict*. London: Oxford University Press, 1970.
Skogrand, Kjetil, Olav Njølstad, og Rolf Tamnes. "Hot Spot – Cold War, the High North, and Grand Strategy." *Institutt for Forsvarsstudier (IFS) Info* 5 (1998).
Tamnes, Rolf. *The United States and the Cold War in the High North*. Oslo: Ad Notam, 1991.
Weir, Gary E., og Walter J. Boyne. *Rising Tide – The Untold Story of the Russian Submarines That Fought the Cold War*. New York: Basic Books, 2003.
Wilkes, Owen, og Nils Petter Gledistch. *Onkel Sams kaniner – teknisk etterretning i Norge*. Oslo: Pax Forlag, 1981.

7 The emergence of the Russian bastion strategy, 1970–1991

As the 1970s began, the SSBNs had taken centre stage in the strategic nuclear stalemate between the United States and the USSR, and the Russian *Yankee* class was conducting more or less continuous patrols in the Atlantic. The *Yankees* were, however, somewhat of an interim solution to the Soviet Union's geography-ASW conundrum. In order to avoid Western ASW defences as a whole, an even more capable platform/missile combination was required. This came through the *Delta*-class SSBN/SS-N-8 missile in 1972. This also shifted the strategic posture of the Soviet Navy, and the emergence of what in the West was perceived as a "bastion strategy" was first heavily debated, and then taken as a foundation for Russian naval posture and strategy by the 1980s. The Soviet bastion approach put the submarines front and centre in assuring their second-strike capability, as well as protecting waters close to vital home-bases in the High North.

From 1970 and onwards the Norwegian threat assessments to an increasing degree focussed on the military build-up on the Kola Peninsula. Although the strategic nuclear submarines based on Kola did not pose a direct threat to Norwegian soil, they certainly posed a threat to the United States and NATO as a whole. Also, the land and air forces on Kola were focussed on seizing land territory in the northern parts of Norway for two reasons. First, achieving control of Northern Norway would deprive the Allies of staging bases for air attacks on the Soviet Union. Second, controlling air bases in Northern and mid-Norway would provide Soviet air forces further reach in their quest for air supremacy in the Northern Atlantic.[1] Through the Soviet Union's establishment of the bastion defence in the mid-1970s, the North Atlantic and the Barents Sea, the value of Norwegian bases increased even more. Norwegian air bases would, in case of a Soviet land grab, become an integrated part of the land-based infrastructure that supported Soviet air operations in denying Allied access through the GIUK gap and freedom of manoeuvre in the North Atlantic.[2]

Through the 1970s and 1980s, Norwegian authorities were therefore engaged in a policy of invitation towards Allied partners. Through actively inviting Allied partners to Norway and northern waters for presence, exercises, and training, Norwegian authorities ensured Allied engagement in the High North.[3] This policy of invitation ensured further Norwegian integration with the Western bloc. However, through upholding the basing policy and self-imposed restrictions, Norway also

DOI: 10.4324/9781003626015-7

was able to demonstrate an ability to develop independent security policies, within the framework of the NATO membership.[4]

The 1970s were characterized by negotiations and a somewhat tumultuous relationship on the political level with the Soviet Union in the High North. The decade saw the establishment of vast exclusive economic zones for many coastal states, which led to peaceful agreements between some nations and forceful disputes between others. The vast ocean areas in the North and their known and potential natural resources had to be handled with care by Norwegian authorities in dialogue with the Russians.[5]

The forward-leaning *Maritime Strategy* of the US Navy and CONMAROPS of NATO, however, caused challenges for the Norwegian Government. Norway and the High North increased their strategic importance, but this made the balancing act in security policy a challenge. The focus of the government was on ensuring the credibility and commitment of American and NATO reinforcements through allowing Allied activities to a certain extent but also maintaining low tensions in the North. A part of the *Maritime Strategy* was to hold Russian submarines at risk before they left home waters, but the Norwegian Government viewed such operations as destabilizing. At the same time, it was important to ensure Allied navies were part of the continuous presence in the maritime domain in the High North.[6]

The more forward-leaning maritime strategy for the US Navy and NATO led to an increase in naval exercises in the Norwegian Sea. The MPA forces of the respective nations maintained their pace of high-intensity surveillance of Russian activities, even when that surveillance required following the Russians closer to their home bases. The established 24°E restriction would, more than ever, manifest the High North as an area for intelligence gathering more or less exclusively by Norwegian surveillance aircraft. Alliance MPA were well set up with capable sensors, communications, and efficient use of range. Towards the end of the 1980s, however, these MPA sensors would be put under significant pressure.

Soviet naval strategy and submarine capabilities

From the early 1970s, the Soviet strategic submarines would once again play a key role in the defence of the Soviet Union. Other units, from attack submarines to surface vessels, were dedicated to the task of defending those strategic submarines. The idea of the SSBNs providing an assured second-strike capability after a potential pre-emptive counterforce strike by their adversaries was by then embraced in full by the Soviets. However, the Russian strategic submarines had to pass through extensive NATO ASW barriers. Starting in the late 1960s, John Walker of the US Navy submarine service began spying for the Soviet Union, revealing just how vulnerable the Russian SSBNs were to NATO ASW defences.[7] It became clear to the Soviets that the strategic situation was unacceptable. In order to avoid Western forces, the range of the next submarine-launched strategic missile would have to be more than three times the range of the SS-N-6 fitted on the *Yankee*. A range of 4,000 NM or more would enable the Russian strategic submarines to remain north of NATO ASW forces in the Atlantic and still constitute a threat to NATO.

By the 1970s, the main focus of Western ASW efforts was to track Soviet strategic submarines on patrol. The *Yankee* class was the main threat, but Western ASW technology was ahead of Soviet silencing efforts, and the *Yankee* class was detectable both by SOSUS and by MPA, although they had to work in tandem. In addition to being detectable by Western acoustic systems, the *Yankees* were sailing on what became somewhat of a regular patrol pattern. This led to a predictability that the bastion defence would later remove. When the revolutionary SS-N-8 missile was developed, it was simply too large for any existing submarine, and its use by submarines demanded the construction of a larger vessel than the *Yankee*. As the Soviets did not want to disturb ongoing submarine production, they developed the new *Delta* class as a modified version of the *Yankee*.[8] The *Delta I*-class SSBN initially carried 12 SS-N-8 missiles with a range of 4,300 NM/7,800 km, and the first submarine was completed in 1972. The *Delta*/SS-N-8 combination had thus more than three times the range of the *Yankee*/SS-N-6 combination. The *Delta II* carried 16 instead of 12 missiles, and the *Delta III* carried the new SS-N-18, with a range of 3,600 NM/6,500 km and 4,400 NM/8,000 km, with multiple- and single-warhead configuration, respectively.[9] The SS-N-18 was the first operational Soviet SLBM with multiple, independently targeted re-entry vehicles (MIRV). The *Delta* carrying the SS-N-8 thereby facilitated a retreat from offensive forward deployments by the *Yankee* class.

By the early 1980s, most analysts had concluded that Soviet naval strategy was indeed focussed on withholding strategic submarines in local waters under the protection of attack submarines, the surface fleet, and maritime air units, in a *bastion*.

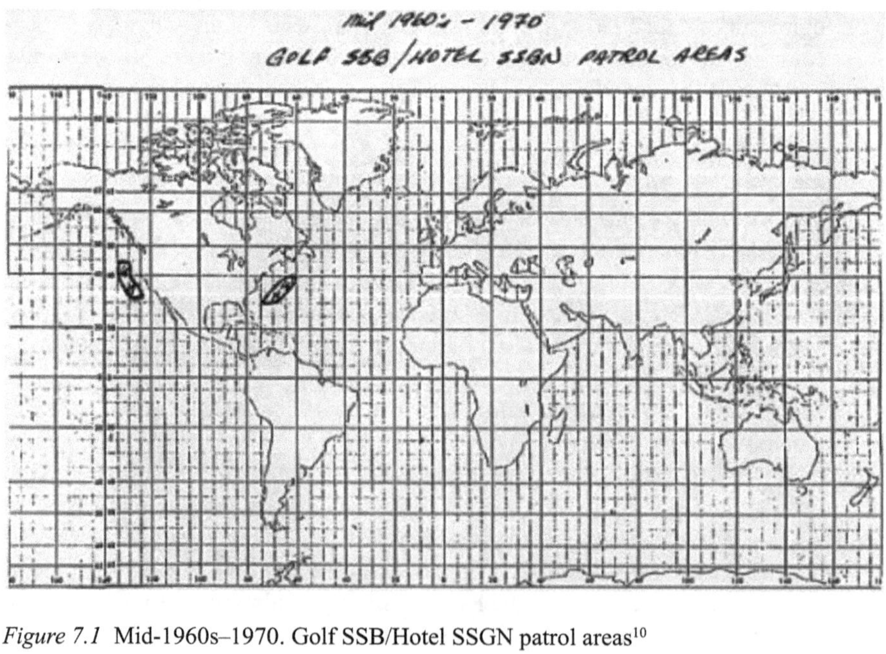

Figure 7.1 Mid-1960s–1970. Golf SSB/Hotel SSGN patrol areas[10]

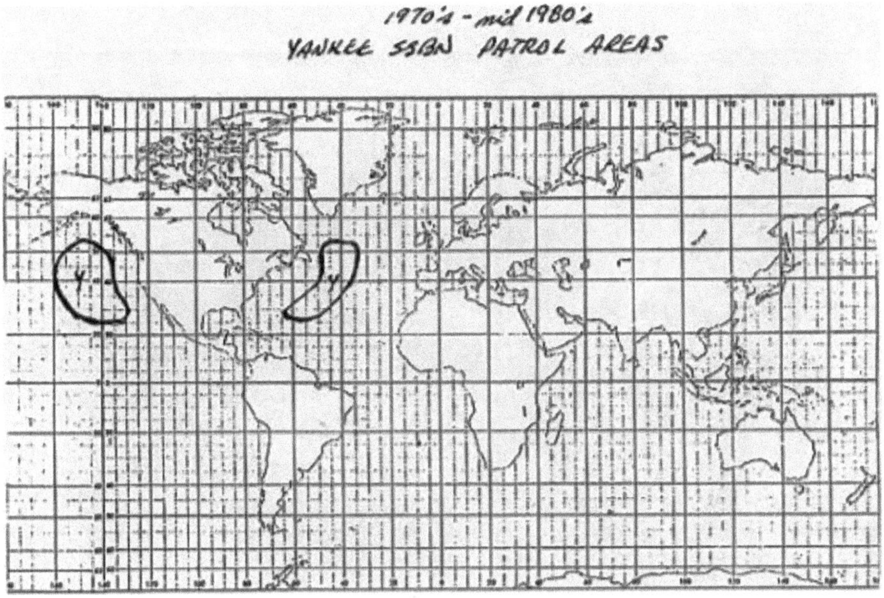

Figure 7.2 1970s–mid-1980s. Yankee SSBN patrol areas

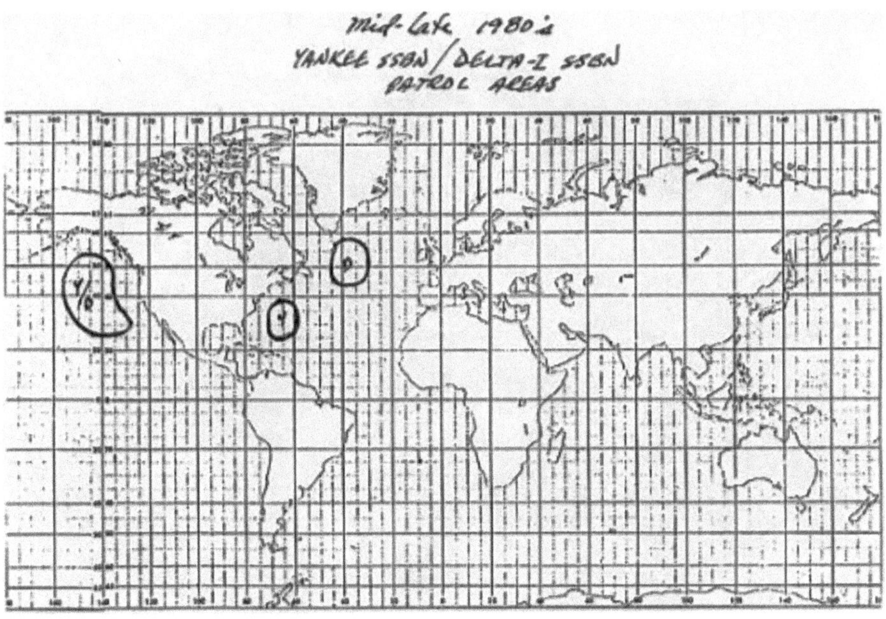

Figure 7.3 Mid-to-late 1980s. Yankee SSBN/Delta I SSBN patrol areas

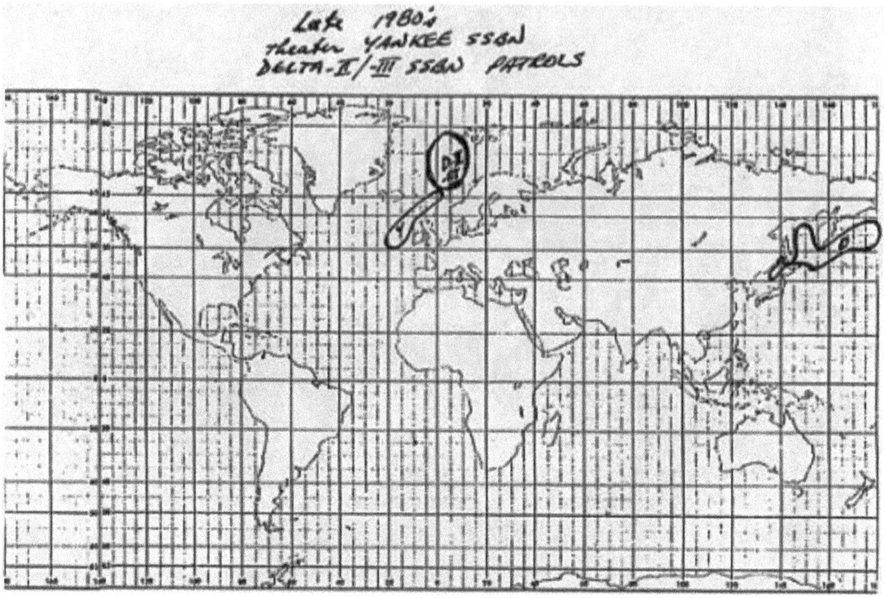

Figure 7.4 Late 1980s. Yankee SSBN/Delta II/III SSBN patrol areas

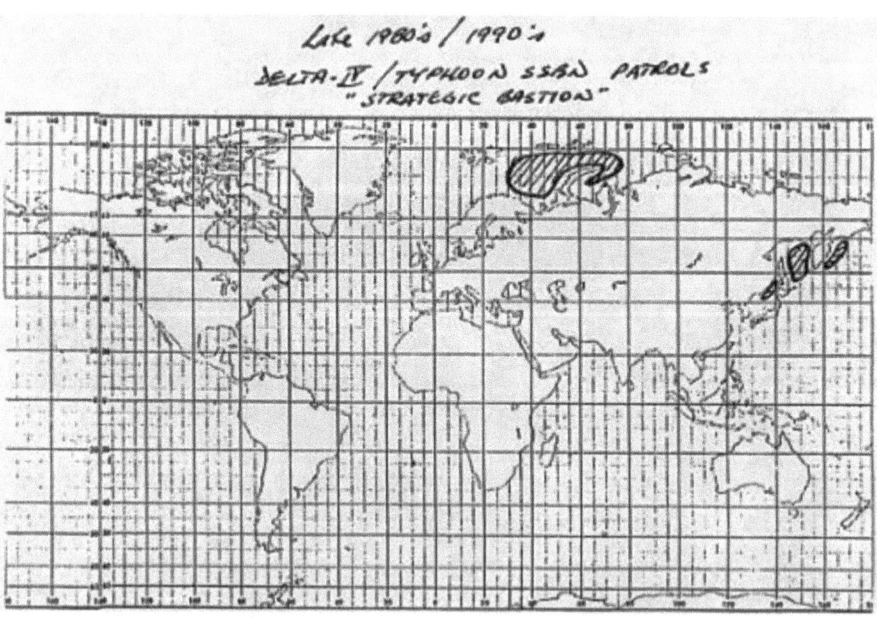

Figure 7.5 Late 1980s/1990s. Delta IV/Typhoon SSBN patrol areas. Strategic Bastion

This shift in Soviet naval posture was to become the strategic foundation for the emerging *Maritime Strategy* of the United States.[11] The *Delta*-class submarines both facilitated and solidified the bastion concept of the Soviet Union's approach to naval strategy. From having to leave home waters and traverse ASW-infested waters all the way to the US coastline, and patrol whilst constantly being held at risk by adversary forces, the assured second-strike assets of the Russians could now operate within the defensive perimeters of national forces. The improved ranges of the missiles even enabled the SSBNs to fire their weapons while in their home ports. The strategic submarines were, in the 1970s and 1980s, gradually retracted from patrolling off the coast of the United States, and, to an increasing extent, sent out on patrol in what is now consistently termed the bastion.

In the early 1970s, new submarines focussed on bastion-related tasks were developed. The *Tango* class brought with it a return to diesel-electric propulsion, which was deemed ideal for defensive operations in the bastions. From 1973, the *Charlie II* carried the SS-N-15 as well as the SS-N-9 *Siren*, an anti-ship missile with a 60 NM range. But NATO ASW development, particularly the S-3 *Viking* carrier-based ASW aircraft and the *Los Angeles* attack submarine, quickly made the *Charlie II* obsolete. The *Oscar* class did not take over until 1982, and will be discussed further below. However, there were developments for Soviet attack submarines as well. In the early 1970s, the *Alfa* and the *Papa* classes were set to sea. Very fast and deep-diving, they surprised the West with their capabilities. However, the production was shut down after only a few units due to high cost.[12] In 1972,

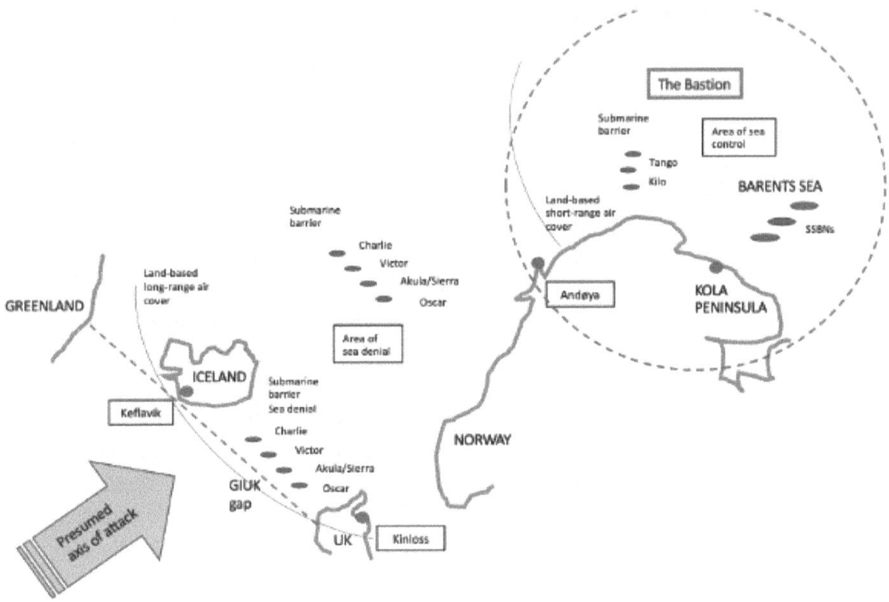

Figure 7.6 The Bastion, with principled placement of Soviet submarine forces

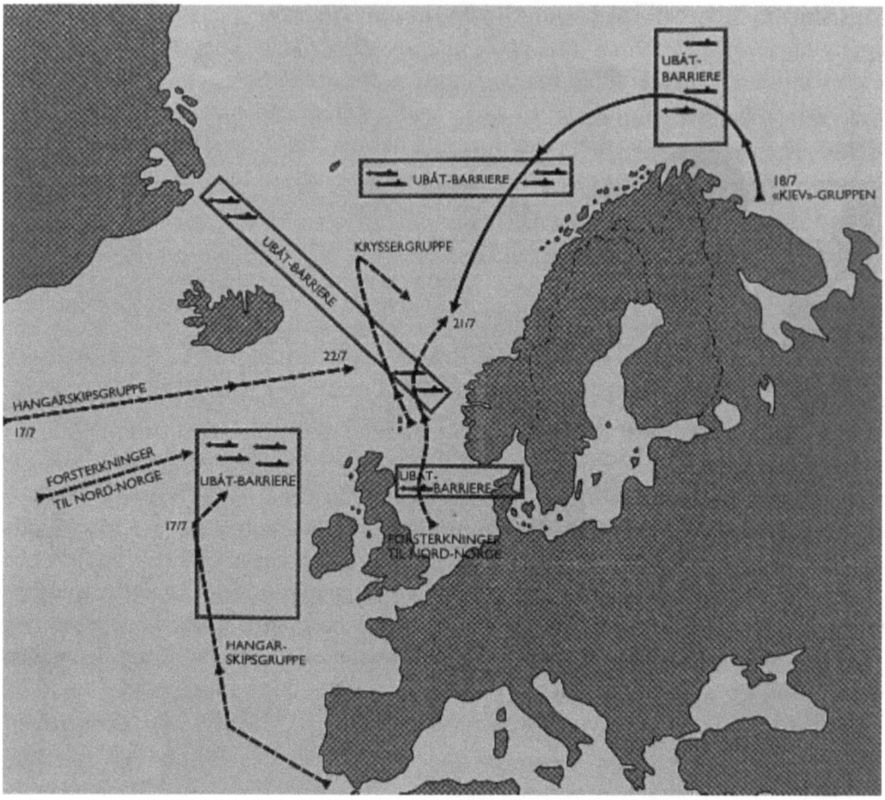

Figure 7.7 Soviet Navy force deployments during the exercise SUMMEREX 1985. The naval exercise SUMMEREX in 1985 was the largest and most complex exercise conducted by the Soviet Navy during the Cold War. The Russians established five submarine barriers in order to meet the simulated NATO threat from the south-west. The objective of the exercise was to establish control of the Norwegian Sea early in the conflict in order to defend the strategic assets in the Barents Sea and the Kola Peninsula.[21]

the *Victor II* entered service, a more capable version of its predecessor. The *Victor II* accommodated the SS-N-15 *Starfish*, which could carry a nuclear-charged depth bomb a little more than 20 NM. Analysts believe that the SS-N-15 armament was indicative of an anti-SSBN mission.[13]

The Soviet Union also finally initiated programmes to silence their submarines, and by the end of the decade their efforts bore fruit. The *Victor III* became operational in the late 1970s, designed to counter forward-deployed adversary attack submarines. She was thus an important feature of the bastion defence in the North. The Russian silencing efforts continued with a helping hand from the Walker spy ring. The stolen information showed the Soviet Union just how vulnerable their submarines were to Western ASW systems, and likely gave impetus to silencing

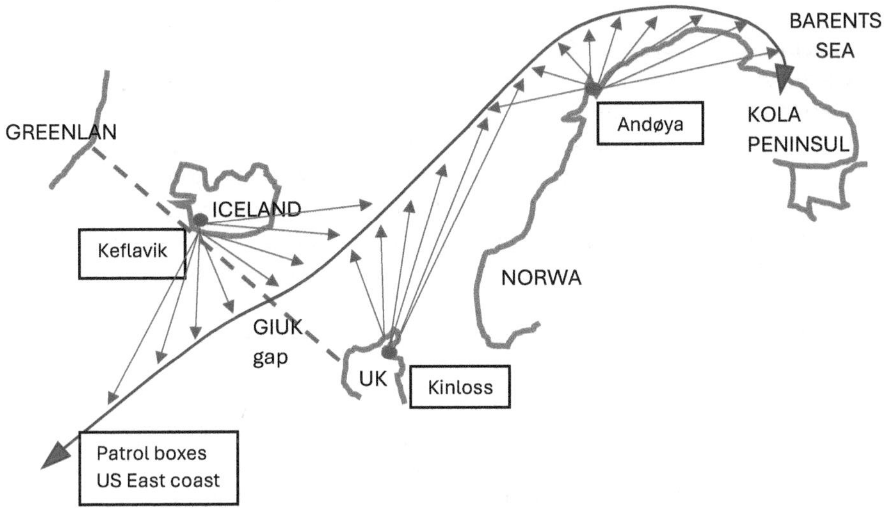

Figure 7.8 International cooperation to follow Soviet SSBNs on strategic patrol

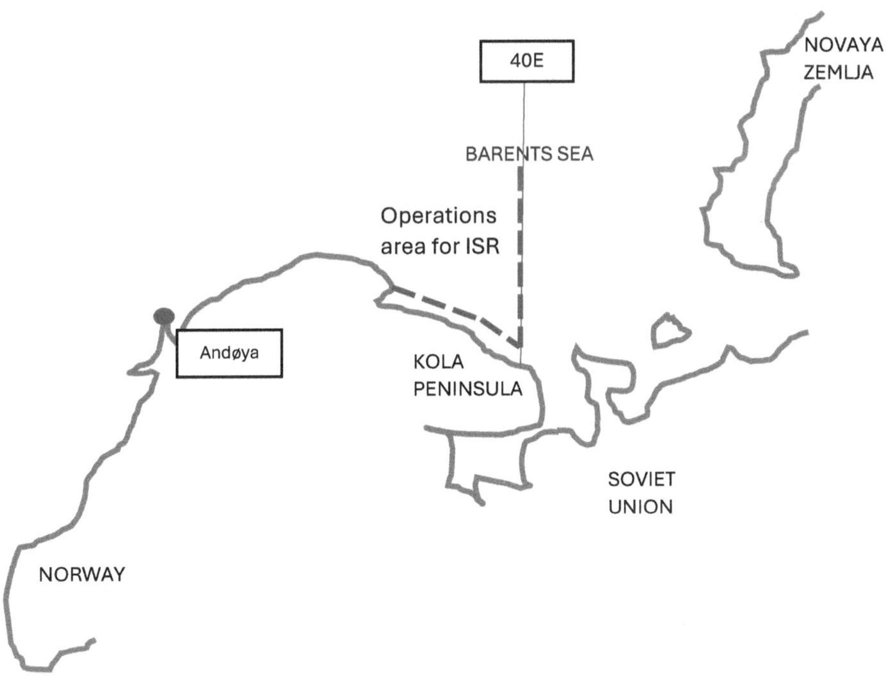

Figure 7.9 Eastern and southern limitation for airborne surveillance in the Barents Sea

efforts into the early 1980s.[14] The *Victor III* class was to surprise NATO navies with her quietness – she proved very hard to find. The *Victor III* class was armed with 65 wake-homing torpedoes, and the conventional successor to the SS-N-15, the SS-N-16 *Stallion* ASW missile.[15]

The nuclear-powered vessels were too large and too few in number to take over patrols close to shore, and a new conventionally powered submarine was required. The first unit of the *Kilo* class was completed in 1982. Only carrying torpedoes, the class was relatively simplistic and carried basic technology, and was soon marked for export. The *Kilo* class would become one of the most common diesel-electric submarines in the world.

As the last *Charlie II* was completed in 1982, the first *Oscar*-class submarine was set to sea. The *Oscar* was much larger than any of the previous anti-carrier submarines, and her displacement surpassed that of the *Delta*-class SSBNs. The *Charlie's* missiles had a relatively short range, necessitating the submarine to get close to her targets, which had a significant effect on American carrier battle group tactics and defences. This in turn led to the Soviets increasing the range of their anti-ship missiles in order to defeat the improved ASW defences of their adversary. The new main armament of the *Oscar* was the SS-N-19 *Shipwreck*, with a 300 NM range and a submerged launch capability. The range of the missiles of the *Oscar* integrated her into the surface and air forces in a way that her predecessors were not, and thus re-established the Soviet combined arms approach to anti-carrier operations.[16] Through the *Oscar*, it is clear that the Soviet Navy maintained capable anti-carrier elements as a part of its overall naval strategy.

The *Typhoon* class became operational in 1983, and it is the largest submarine ever constructed. The *Typhoon* carried the record large SS-N-20 *Sturgeon*, which was the first SLBM to be driven by solid-fuel propulsion (range 4,400 NM/8,300 km). The SS-N-20 carried six to nine MIRVs, with a significant increase in accuracy compared to its predecessors. The *Typhoon* was designed for calmly patrolling her Arctic home waters whilst being protected by other assets, thereby constituting an assured second-strike capability.[17] The Soviets also continued the production of the *Delta* series. The *Delta IV* was much less costly to produce than the *Typhoon*, it carried the SS-N-23 *Skiff* with a range of 4,400 NM/8,300 km and was equipped with four MIRV warheads.[18] She became operational in 1988.[19] The *Strategic Arms Limitation Treaty* (SALT I) Agreement of 1972 limited the number of missiles allowed in the inventory and thus indirectly the number of firing platforms permitted. This made further large-scale production of SSBNs unnecessary. It was thus a matter of replacing old SSBNs with newer and more capable ones. The 1970s trend of operationalizing cruise missiles was also a result of arms agreements, which gave incentives to turn long-range, land-based weapons into shorter-range sea-based ones. This was also the reason for several *Yankee*-class SSBNs being converted into SSGNs carrying SS-N-21 and SS-NX-24 cruise missiles, in order not to go past the SLBM ceiling of the *SALT I* Agreement.[20]

The bastion-operating concepts demanded a large number of capable attack boats for forward deployment for attacking lines of communication, in addition to providing a robust defence of the bastions. The mid-1980s saw the emergence of two such attack submarines, namely the *Sierra* and the *Akula* classes in 1984 and 1985, respectively. The *Akula* was markedly quieter than its predecessors, and the new submarine classes were a manifestation of the Soviet quieting efforts that had started to materialize in the early 1980s.[22] Internal improvements to submarine construction had taken the Soviets a long way by the late 1970s, but the silencing efforts were almost definitely taken even further by the acquisition of advanced milling machinery and software for very silent, multibladed propeller production in 1983/1984.[23] Owen Cote has described the entry of *Victor III*, *Sierra*, and *Akula* onto the operational scene as the end of the Western "ASW Happy Time."[24] The *Sierra* and the *Akula* were armed with the SS-N-21 *Sampson* submarine-launched cruise missile (SLCM), among other weapons. The *Sampson* LACM, the Soviet equivalent to the US Navy's *Tomahawk* land attack missile (T-LAM), has a range between 900 and 1,100 NM. When NATO in 1983 deployed various land-based cruise missiles and intermediate-range nuclear ballistic missiles at various European bases, the Soviet Union gave an "analogous response" by deploying submarines to the US coastline, including *Akula*-class attack submarines carrying the nuclear-capable SS-N-21.[25] Patrolling off the coast of the United States, a missile engagement would provide little to no warning of attack. The new submarines posed an entirely new threat to Western ASW forces that had been spoilt with noisy *Yankee*-class SSBNs patrolling in somewhat predictable patterns in the Atlantic. On the *Akula* deployment, a member of the ASW force said, "the entire Navy had to deploy in order to find and maintain contact on one submarine."[26] As shall be seen, this combination of quiet submarines with capable land-attack missiles marks the starting point of today's Russian submarine threat.

Evolution of Western naval strategy

Increased Soviet naval activity took place in the Norwegian Sea, and the Northern Fleet showed its prowess during large blue water exercises in 1968 and 1970 in the North Atlantic. The Northern Flank of NATO was increasingly being viewed as a crucial element of any overarching campaign by the Alliance. However, budgetary restrictions amongst all Allies resulted in specific warfare areas being downplayed in strategic plans, such as strengthening the flanks, maritime surveillance, and ASW.[27] An increasing awareness was thus not necessarily followed through by commitment in strategic plans. The 1970s did, however, see an increase in Allied exercises in the North Atlantic. Exercises included ASW and forced landings in Norway, marking a more forward-leaning approach to the North Atlantic at least during exercises, if not in explicit strategy. Although downplayed in strategic plans, new ASW systems and technology gave a sense of optimism about the Alliance's ability to tackle the Soviet submarine threat. In spite of this technological optimism, the 1970s, together with the previous decade, marked a low point in Allied presence in the North Atlantic. In the realm of ASW, the decade saw a number of changes. American ASW aircraft carriers were phased out during the Vietnam War, and focus was given to submarines, SOSUS, and MPA.[28] US strategic nuclear forces that, for the past decade, had been represented by aircraft carriers were now represented by submerged submarines, and the overall presence in the north eroded.

The emerging bastion approach of the Soviets occurred in the context of détente and negotiations between the United States and Soviet Union on reducing their respective nuclear stockpiles in 1969. In 1972, the SALT I Agreement was signed, in which a ceiling was set on a maximum number of launch platforms allowed for strategic ballistic missiles, to include strategic submarines.[29] Phasing out the *Yankees* gave way for more of the modern *Delta* submarines, thus staying below the maximum level of submarine missile launchers in accordance with SALT I.[30]

By the 1970s, the maritime surveillance of Soviet submarine movements and surface exercises carried on, but few initiatives were taken on a strategic level to ensure the defence of the Atlantic SLOC. The traditional approach of taking the initiative had to make way for a more defensive way of thinking.[31] With regard to technology, cruise missiles saw a renaissance in the 1970s on both sides of the Iron Curtain, and the American submarine-launched *Tomahawk* missiles are a prime example of this. Cruise missiles have shorter ranges than their ballistic brethren, and the submarines therefore had to get closer to their targets in order to fire their load. Thus, the seeds of a more forward-leaning maritime strategy could be seen in the West through cruise missile development.[32] And as Soviet capabilities and posture changed in the late 1970s, a change in NATO strategic reasoning occurred. Soviet submarines operated to an increasing degree in the North Atlantic and the Barents Sea, which lead to an increased focus on forward operations in NATO. NATO commenced a reconsideration of the Alliance's maritime concept, which culminated in the NATO Concept of Maritime Operations (CONMAROPS) in 1982.[33] Similar work was initiated in the United States, where forward operations and force projection were key elements. The Soviet bastion strategy in many ways invited a more forward-leaning West, and the revised Soviet posture of the mid-1970s was a key ingredient to the emerging strategic concepts of the West in the 1980s. With the emergence of the Soviet bastion approach came a renaissance in NATO interest in the High North, and the bastion concept became the basis for strategic naval thinking in the West. The US Navy conducted a comprehensive review of maritime strategy, eventually leading to the signing of the Maritime Strategy in 1984. The most important elements of this strategy were maritime superiority and forward power projection.[34] Part of the Soviet forward defence in the High North was their use of maritime bomber aircraft such as the Tu-16 *Badger* and the Tu-22 *Backfire*. With new and quiet submarines commissioned in the 1980s and a relative decline in the effectiveness of Western fixed undersea listening installations, the Russians had a significantly increased ability for sea denial in the North Atlantic. In addition, the barrier defence was no longer deemed adequate by SACLANT, who pointed out shortfalls in forces to tackle threats to the Northern Flank and simultaneously assure the safe passage of Western reinforcements across the Atlantic. The answer to the shortfalls would be, according to SACLANT, *forward* operations.[35] The concept of forwarding aircraft carriers into Norwegian fjords was revitalized from the 1950s.[36] Naval forward operations were first tested during exercise *Mainbrace* in 1952 and were again practiced during *Ocean Safari* in 1985 and *Team Work* in 1988.

The Americans had, in the early 1980s, initiated forward operations with submarines, which were highly classified.[37] Such operations were an integrated and important part of the Maritime Strategy and led to strategic, principled discussions on forward neutralization of the Soviet submarine force, so-called "strategic ASW." When the Norwegian Labour Party came back into power in 1986, Minister of Defence Holst stated that such strategic forward operations constituted a destabilizing element of the challenging balancing act to maintain low tensions in the north. He said that the operationalization of the US Maritime Strategy might escalate tensions and further lead to permanent presence of allied maritime offensive units.[38] At the same time, it was important for the Norwegian Government to have regular presence in Norway and surrounding areas, in order to demonstrate a credible reinforcement mechanism during times of war. The Maritime Strategy thus put further pressure on the demanding security policy balancing act in the High North.

The most formative agreement for surveillance in the North Atlantic for the remainder of the Cold War was the *GIN CLEAR* Agreement of 1969, the successor of the *BOX CAR* Agreement.[39] In essence, the focus of the arrangement was to maintain overview of Soviet naval movements in the Greenland, Iceland, Norway (GIN) gap. *GIN CLEAR* was based on bilateral agreements and was aimed at limiting foreign surveillance aircraft taking off from Norwegian soil in monitoring of Soviet naval deployments. From mid-1970, the agreement expanded to include Denmark, the United States, the United Kingdom, Canada, West Germany, and the Netherlands. However, the Norwegian Government showed a restrictive attitude towards increasing the number of countries involved in surveillance in the High North and maintained a restrictive policy on the foreign use of airfields. West Germany and the Netherlands, for example, were not approved to fly in the High North.[40] The reason for this was a politically conservative attitude towards operating German aircraft, in particular, in what was envisaged as a low-tension area by the Norwegians. It was also an ambition of Norway not to involve more parties than necessary in intelligence sharing and coordination. British squadrons, however, flew many missions under the *GIN CLEAR* Cooperation Agreement in 1970. A notably large number of *GIN CLEAR* missions were executed during the first 6 months of 1970, followed by more evenly spaced missions on the planning schedule. Having used Bodø in the 1960s as host base, *Nimrod* crews in the 1970s almost exclusively utilized Andøya as the Norwegian hosting base. The change likely took place because of the improvement in operational support at Andøya following the conversion from the Norwegian *Albatross* to the *Orion*. After the establishment of the *GIN CLEAR* Agreement, Allied interest in flying from Norwegian airfields went into a decline. After the geographical delineation of areas in the North Atlantic through the NORCANUSUK Agreement, both the British based in Scotland and the Americans based on Iceland were able to reach their dedicated patrol areas to the South and West, which were further west and south of the Norwegian areas. Also, the intelligence gathered in the Norwegian areas to the north and east of Norway were of such quality that additional surveillance operations were considered to be superfluous.[41] The *GIN CLEAR* arrangement remained in effect throughout the Cold War.

Norwegian procurement of the *Orion* included a formal agreement to share intelligence with the Americans. Further building on the general intelligence sharing agreement between Norway and the United States from 1954, the specific sharing of ELINT data took form through the NORUSA II Agreement of 1970.[42] Intelligence sharing was based on the increasingly close relationship between the two countries and was explicitly stated in the *Orion* procurement agreement. It was important for Norway to contribute to Western defence cooperation in general and NATO specifically. Also, Norwegian intelligence contributed to low tensions as the West would relate to facts as opposed to speculation. Norwegian surveillance precluded potentially provocative Allied operations in the same area. Also, intelligence from the North showed the exposed nature of the Northern Flank of NATO. And finally, technical information provided an improved basis for decision making by Western authorities.[43] Intelligence products that came out of *Orion* operations in the High North were initially of a *technical* character, but subsequently took on an increasingly important role in Norwegian security politics.[44]

The increased capabilities of the Norwegian *Orions* shifted the focus of Western surveillance in the North Atlantic from technological matters (areas apportioned based on sensor suite) to geographical questions (areas apportioned based on geographical proximity).[45] Even with lacking ASW capabilities, the Norwegian authorities had been restrictive with regard to the 24°E limitation, with the exception of specific periods when a new type of aircraft was introduced. The strict geographical delineation of responsibility after the introduction of the *Orion* for the Norwegians was facilitated by the fact that the three main nations conducting surveillance in the High North were equally ASW-capable. The aforementioned NORCANUSUK Agreement is a clear sign of a common understanding of common capabilities, particularly with respect to ASW.[46] In 1979, the Royal Norwegian Air Force formalized the requirements for maritime surveillance and reconnaissance. Norwegian MPA had for decades been operating based on intelligence requirements and not on a cognizant framework for operations. This led to requirements for improved capabilities on board the *Orions*, materializing in the *Chief of Defence Directive for Maritime Patrol Aircraft Operations* in 1982. The directive formalized surface and subsurface surveillance and reconnaissance as the primary task of Norwegian MPA, followed by support to civilian national entities, ensuring national jurisdiction and support to search and rescue as follow-on tasks.[47]

Although the 24°E restriction had been declared in early 1959, it had not been formalized in publicly available documentation. From having been a restriction enforced by the operational headquarters and written only in operational directives, the MoD finally stated the 24°E restriction in writing in 1980 through a review of the plans for Allied reinforcement.[48]

The evolution of Allied ASW and MPA capabilities

The 1970s saw a prevailing optimism regarding the American ability to maintain and even enhance the technological lead over the Russians.[49] Most importantly, for the MPA force, the improved computer processing enabled improved ability to distinguish between ambient noise and sounds emitted from a submarine in real time. The P-3B *Orion* was introduced into Norwegian service in 1969, and the internal ASW system now included the Delayed Time Compression (DELTIC) system that, together with the new JEZEBEL acoustic processor, allowed the operator to monitor 16 channels of acoustic data simultaneously. Overall, the new aircraft was able to fuse information from the radar and acoustic sensors into a single tactical picture. The new processors improved the accuracy of the information collected, and thus facilitated more time for *tactical* problems as opposed to *mathematical* issues.[50]

The SOSUS continued to provide essential cueing to the MPA community in the 1970s. Initial contact on approaching Soviet submarines was usually gained by fixed systems on the ocean seabed. For cases in which contact was lost on a submarine by "other sources," an extensive search and localization effort had to be initiated by the respective MPA fleets.[51] The Norwegians' P-3 *Orion* procurement included state-of-the-art analysis and processing tools for acoustic intelligence.[52] These systems were co-located with the analysis centre for the seabed systems at Andøya, similar to the American arrangement at Keflavik, Iceland. Such co-location of processing systems facilitated a close relationship between the fixed-array and MPA communities. Moreover, the previously mentioned *BRIDGE* system was expanded to include cables in the Barents Sea. The cables were laid in 1974 and were aimed at detecting Russian submarines at as early a stage into their deployment as possible; they functioned as an integrated part of the SOSUS.[53]

After the British conversion from *Shackleton* to the MR1 *Nimrod* in 1970–1971, the aircrews also enjoyed a significant leap forward in technology and capabilities. Combined with knowledge and experience, the new aircraft performed very well. The noise from the *Nimrod* jets was also harder to distinguish by a submerged submarine close to the surface than the noise from the turboprop *Orion* aircraft.[54]

In mid-1970, the Americans deployed their new P-3C *Orion* model to Keflavik. The P-3C brought with it the latest ASW technology available, including new computers, and improved navigation equipment and sensor functionalities. The capacity of the acoustic system's active elements was four times that of the previous aircraft.[55] The US Navy further transitioned to the P-3C *Orion* Update II in early 1979, which, among other things, for the first time, enabled it to carry the AGM-84A *Harpoon* missile aimed at Soviet surveillance trawlers in case of war.[56] US *Orions* also operated as airborne *Harpoon* missile platforms in support of the Iranian hostage crisis.[57] The British *Nimrods* initially had the AIM-9X *Sidewinder* air-to-air missile installed in the early 1980s and were carrying these missiles during the Falklands war. The *Nimrods* were thus prepared for an increased threat environment by the 1990s. In parallel with the P-3 *Orion* upgrades, the *Harpoon* ASuW missile was installed on the *Nimrods* as well.[58] The Norwegians did not have missiles installed on their aircraft.

The US Navy upgrade to the P-3C *Orion* Update II in 1979 came with further improved avionics and weapons systems. A new sensor consisted of an infrared detection system, which could identify targets regardless of weather and daylight, albeit only on the surface.[59] The new ASW-designated radar AN/APS-137 became operational and was first utilized in the North Atlantic in 1986.[60] The early 1980s saw an increase in the utilization and coordination of data link for ASW operations, enabling the passing of basic target information back and forth on a tactical network.[61] Further, the P-3C *Orion* Update III deployed for the first time to Keflavik in May 1990. The new on-board processor doubled once again the number of passive buoys an operator could monitor simultaneously to 32. The processor showed a four-fold improvement in isolating submarine sounds from ocean ambient noise.[62] Towards the end of the 1980s, there was a further increase in the exercising of *Harpoon* missile strike missions and the use of the APS-137 imaging radar.[63] The new radars also facilitated the use of P-3 *Orions* in providing targeting information for *Tomahawk* cruise missile strikes by submarines.[64] However, towards the end of the 1980s, there was a clear tendency of utilizing MPA to an increasing degree in joint operations, combining the MPA sensor and targeting suite with other assets' weapons capabilities.

By the 1980s, analyses by the Norwegian Air Force indicated that the aircraft would have to be replaced not long after 1986. Norwegian authorities were increasingly worried about the naval situation that was developing in the High North. Soviet exercises conducted just off the Norwegian coast in the early and mid-1980s demonstrated a capacity to project power in a manner reminiscent of a blue-water navy, and this was being noticed at the highest political levels on both sides of the Atlantic. Thus, in 1986, the Norwegian Government signed the *Maritime Surveillance Agreement* with the United States.[65] This was a similar arrangement to that of the procurement of the P-3B *Orion*, and this time the procurement consisted of four P-3C *Orion* Update III. What was new about this agreement was that the Norwegian Government had to bear a greater share of the cost than before. The degree of formalization of the intelligence sharing between the two nations was also new. The agreement from 1969 was founded more on a tacit understanding between the nations, but this time the requirement for sharing was explicit.[66] It is reasonable to believe that this was due to the increased value of naval intelligence from the Barents Sea, resulting from the fact that an increasing number of strategic missile submarine patrols were conducted east of the 24° delineation by the 1980s. The four P-3C *Orion* UIII were delivered to Norway between December 1989 and June 1990. This was the same type of aircraft that the US Navy deployed to Keflavik in 1990.

In the 1970s, there was a notable increase in the use of SURTASS. The SURTASS consisted of surface ships that towed a long line array (approximately 8,000 feet long) that functioned in a similar way as the SOSUS arrays. Through satellite communication and links, the SURTASS ships connected with the SOSUS system, and together with the shore processing systems comprised the *Integrated Undersea Surveillance System* (IUSS).[67] By the mid-1980s, the Western fixed arrays' ability to detect and track Soviet submarines had declined significantly as the Soviet submarines became quieter. This led to the deployment of the *fixed distributed system* (FDS) in 1985. The FDS is a large-area, distributed field of acoustic arrays. They are upward-looking and only capable of detecting submarines in their immediate vicinity, but this new method and system was necessitated by the low emission of sound from the new submarines.[68]

The use of satellites for surveillance and reconnaissance has by some been termed a "revolution in intelligence."[69] The first imaging and ELINT satellites of the 1960s were operationally proven by the 1970s. A satellite that could detect navigational radars and other emissions from ships was thought to complement the operational picture. The *White Cloud* passive ELINT satellite was launched in 1971.[70] *White Cloud* became the satellite workhorse for maritime surveillance for the US Navy during the Cold War and constituted the initial constellation of the *naval ocean surveillance system* (NOSS). The system was upgraded continuously during the Cold War and up until the present. In addition, a series of IR satellites were placed in geostationary orbit in order to detect the firing of Soviet Union ballistic missiles, as part of the *Defence Support Program* (DSP). In the early 1980s, the DSP was tested to support the general detection of missile and aircraft flight trajectories for early warning and for self-protection of naval vessels as well.[71]

Efforts further started in the mid-1970s in the United States to develop a synthetic aperture radar satellite for maritime surveillance; however, fully functional radar satellites for maritime surveillance did not become operational until after the Cold War.[72] The new *Charlie II* submarines were deployed along with the first Soviet ocean surveillance satellites, where the latter were meant to provide targeting to the former. However, the Russian surveillance satellites never achieved their promise, and the submarines had to rely on Soviet MPA for targeting.[73] In response to increasingly quiet Soviet submarines, the mid-1970s also saw significant interest in developing non-acoustic means of detecting submarines, and attempts were made to detect thermal wakes, internal waves, electric currents, bioluminescence, and surface activity from a passing submerged submarine.[74] None of the capabilities were integrated into operational satellites, however. Towards the end of the 1970s, imaging satellites were having a significant impact on naval operations, particularly in performing surveillance of adversary ports, naval bases, and shipyards. Satellites could fly with impunity directly over areas of vital naval interest, including inland targets, supporting strategic intelligence.[75] However, photographic satellites are incapable of looking through weather and clouds, and this led to efforts to develop high-resolution radar satellites such as the *Lacrosse*, which was launched in 1988. *Lacrosse* reportedly provided a resolution of approximately 1 foot, and worked in tandem with optical imaging satellites during the Gulf War in 1991.[76] However, satellites were still only able to perform surveillance on the surface.

Towards the end of the 1980s, American MPA had satellite communication (SATCOM) equipment installed. SATCOM enabled aircraft to perform immediate, reliable, over-the-horizon, two-way secure tactical communications. Among other tasks, SATCOM was utilized for *Harpoon* missile strikes and more detailed coordination of submarine prosecution with shore and surface facilities.[77] Through the SATCOM capability, tactical operations and on-station flexibility were greatly enhanced.

Satellite technology was starting to play an important role in maritime surveillance operations. These satellites, however, were geostationary, leading to poor and at times no coverage in the Arctic. This will be elaborated upon in the final chapter.

The deployment of allied MPA capabilities

When the *Yankee*-class SSBNs initiated their regular patrols off the coast of the United States, the allied airborne ASW community was ready for them. As the Soviet strategic submarines departed from the Kola Peninsula, they were followed first by the Norwegian *Orions* from Andøya, as they entered the North Atlantic. Further south, the Soviet submarine contact was handed over to American *Orions* from Keflavik and British *Nimrods* from Kinloss. Then the *Yankees* continued their course further south and passed through the GIUK gap. Here, the submarines in transit were tracked by the Americans, the British, the Dutch flying out of Iceland, and the Canadians, who did the same. As the Soviet submarines entered their patrol boxes south of Bermuda, they were tracked continuously by aircraft from Naval Air Station Bermuda and Naval Air Field Lajes, Azores.[78] Although not formalized, the handovers were facilitated by experience, NATO procedures, and close and daily interaction between the tasking agencies of the respective nations.

The Northern Fleet operations and exercises shaped the days of MPA crews during the Cold War.[79] British operational records show regular entries of "contact gained immediately," "Russian Type 2 Nuclear Possub," and "contact held for the remainder of the sortie and was handed to the relieving aircraft at Off-Task."[80] The allied aircraft were launched in carefully planned succession with a combined aim of tracking the target and gathering as much updated acoustic intelligence as possible. This intelligence was then processed at home base.[81] The squadron would send out aircrews to try to hold the target for as long as she was within reasonable reach of home base, and then hand it off to the next squadron or nation to pick up the tracking.[82] Close cooperation with Allied MPA crews continued into the 1970s, and were usually well-coordinated and problem-free. At times, however, miscommunication could lead to two nations quarrelling over who should track the Russian submarine.[83] Operational missions included both tracking of transiting strategic submarines, but also monitoring of general Soviet naval activities in order to chart tactics and manoeuvres. The largest Soviet naval exercises and routine activity took place in the north Norwegian Sea, and in the ocean areas north of Norway. Although also given other areas to cover, when it comes to documenting Soviet naval activity, the Allied aircrews benefitted the most by flying in the High North. At the turn of the decade, the high operational tempo of the Soviets and the corresponding surveillance efforts by NATO MPA had become routine. It was commonplace for American squadrons to report more than 40 different submarines during a deployment of 4–6 months to Keflavik.[84] MPA operations also included surface surveillance, which entailed following Soviet intelligence ships, monitoring the Soviet fishing fleet, and keeping track of Soviet Navy surface groups. Soviet surface forces were usually escorted by several submarines, at times more than eight submarines in the same formation.[85]

The *modus operandi* of the deployed American MPA squadrons stood out in comparison to other, allied nations' way of operating. The aircrews were away from their families for approximately 6 months at a time, remaining on standby for flying missions nearly around the clock, throughout the deployment. Their ability to surge squadron assets and fly non-stop over extended periods of time to track Soviet submarines meant that single *Yankee*-class submarines could be followed by the Keflavik squadrons for up to a week or more at a time.[86] The squadrons that regularly deployed to Keflavik amassed a significant amount of flight hours during the Cold War. They regularly accumulated more than 700 h a month throughout a deployment, which at times was quadruple the flight hours they put in whilst at home in the United States.[87] One squadron reported approximately 80,000 h over 10.5 years.[88] This averages to approximately 630 h a month for more than a decade consecutively, which is telling of the demands the Cold War put on the aircrews that worked on the front line. Numbers would routinely exceed 900–1,000 ha month during particularly busy periods.[89] Operational records from the deployments in the late 1970s show that the squadrons would encounter almost every type of Russian submarine, with aircrews prosecuting two and sometimes three submarines simultaneously.[90] This underlines the fact that the North Atlantic was not only a necessary transit area for their SSBNs, but also a crucial exercise area for the remainder of the fleet, not only in the 1960s–1970s, but also into the 1980s.

There was no decline in intensity of ASW operations towards the end of the Cold War.[91] This indicates that the bastion defence had not fully settled by the end of the 1980s, and that several patrols south of the bastion were still taking place. During an exercise in 1987, five *Victor III*-class submarines were prosecuted simultaneously.[92] By the Americans alone, 82 sorties and 800 flight hours were flown over 13 days of continuous prosecution against the five Soviet submarines, generating a total of 230 h of submarine acoustic contact. The Keflavik deployment between November 1988 and May 1989 flew 744 sorties locating 17 Soviet submarines in 6 months.[93] This equals to almost four sorties every day, which, in practical terms, means the aircraft belonging to the detachment were airborne almost continuously throughout the deployment.

The Americans conducted reconnaissance in the High North with aircraft types other than MPA as well. Between 1977 and 1989, American SR-71 *Blackbirds* flew regular missions into the Barents Sea, usually supported by KC-135Q air-to-air refuelling (AAR) aircraft based in the United Kingdom.[94] These aircraft flew close to both Russia and Norway, at one point violating Norwegian airspace.[95] However, the *Blackbirds* provided invaluable radar intelligence on military activity and bases on the Kola Peninsula, intelligence collection that would eventually be taken over by satellites.[96] In operationalizing the Maritime Strategy, the US Navy initiated regular clandestine submarine operations in the Barents Sea.[97] From the mid-1980s, the Royal Navy supported the Americans with submarine operations of their own in the High North. The British were integrated in the forward strategy, aiming at destroying the Soviet fleet early in a conflict.[98] As these were clandestine operations; they were not openly debated or confronted by the Soviet Union. The operations were opposed in policy terms by the Norwegian government, as discussed, but there was no mechanism for denying such operations, either by the Norwegians or by the Russians. They became an integrated part of the silent cold war in the abyss.

British *Nimrod* crews were sent to both Norway and Iceland to conduct surveillance and tracking of Russian submarines. Regardless of originating base and nationality of the MPA, the aircrews received orders from the nation they took off from and updates on targets and areas to cover by the headquarters with the most current information. The coordination for such a multinational effort was executed by the lead nation at any given time, which was the nation in whichever sector the target was located. That nation carried the weight of the operation and offered openings or gaps in coverage in terms of flying time to other nations.[99] For example, the Canadians sporadically deployed aircraft to Iceland to get an early feel for the target, before they picked up the submarine as it approached the Western Atlantic.[100]

The British MPA community also spent a significant amount of time working in the High North.[101] When Argentina invaded the Falkland Islands in April 1982, an urgent operational requirement for an AAR capability for the *Nimrod* arose.[102] The new AAR capability facilitated missions into areas without having to secure foreign landing approval for refuelling mid-operation.[103] A range of missions were thereafter conducted north of Norway. Their reports reveal that British crews were also exposed to a significantly higher level of Soviet activity in the High North as compared to further south in the Atlantic.[104] The AAR capability thus paid off in the North Atlantic as well as in support of the Falklands campaign. The *Orions* never had an AAR capability. The Norwegians were geographically close enough to Soviet operations for AAR requirement not to be a factor. US Navy squadrons, on their side, deployed to one location, such as Keflavik or Bermuda, and then sent detachments to other locations, such as the Azores, in order to fill gaps in surveillance.[105] The Americans thus met lack of range through deploying assets further out, or by operating collaborating assets in the areas of interest.

At least one Norwegian *Orion* surveillance mission was executed every single day for the remainder of the Cold War. The area covered was fairly constant throughout the time period, going as far as 40°E, and along the Soviet border on the Kola Peninsula.

From the mid-1970s until the end of the Cold War, the Norwegians ventured further east than 40° once a week. When the navigational equipment on board the P-3B *Orion* was upgraded in 1980, headquarters authorized extending the southern operational boundary further south, down to 15 nautical miles from land (3 miles outside the international 12-nautical-mile sovereign border).[106]

The *Orion*-era for Norwegian MPA was characterized by the mere localization of their operations area. The localization of the primary surveillance area for the Norwegians meant that the amount of Russian activity that the Norwegian operators witnessed during their daily operations was second to none in the Alliance. It was commonplace during the 1980s for a Norwegian MPA to track and follow several different submarines during a single sortie.[107] One Norwegian tactical coordinator operating at the Norwegian MPA squadron in the 1980s reports that he once recorded 14 different submarines in a single mission.[108] The combination of the level of equipment of the *Orion* aircraft, together with the proficiency that came with operating in such a "target rich environment," led to an unsurpassed overview of Soviet submarine activity in the High North. The NIS assessed in 1970 that it had awareness of the whereabouts and activities of 90% of the nuclear-propelled submarines at any given time, and 100% of the *Yankee* class in particular.[109] This number presumably declined as the new and silent submarines such as the *Akula* and the *Delta IV* came online. The intelligence collected was used by the NIS in support of providing a decision-making baseline for military and political leadership, and as a continuation of this, it was used for trading information, materiel, and access to partners' systems.[110]

Operational records show a different *modus operandi* by the MPA fleets when tracking attack submarines compared to tracking strategic missile submarines. For example, prosecution of *Charlie* and *Victor*-class submarines were "short, 'go for the throat' prosecutions," in which the MPA would spend as little time as possible attaining attack criteria and performing a simulated attack.[111] In following the nuclear strategic submarines, the MPA would take a more defensive, hold-at-risk approach in which the target would be tracked without revealing the aircraft's presence.

Widespread international cooperation took place within the MPA community. Interaction happened daily in operational flying, regularly during NATO exercises, and many times in the form of visits to each other's home bases. There were at times cases where coverage was not deemed adequate by other nations or there was some particular interest in a given target. In such cases, the Keflavik squadron would send a detachment to fly for a specified period of time from an allied country.[112] The US Navy at Keflavik regularly hosted aircrews from Canada, Norway, the Netherlands, and the United Kingdom in return.[113] In 1973, while tracking a *Hotel*-class SSBN, cooperation between US, Dutch, and Canadian aircrews flying out of Keflavik resulted in 90% contact time over 23 sorties, spanning 5 days and 1,300 NM of submarine transit.[114] Sorties were coordinated with those air operations centres that could have a mutual interest in a target or area.[115] As one former MPA air wing commander put it: "It was just friendly cooperation."[116] This multinational coordination, conducted by national C2 elements under national command, essentially functioned as the basis for NATO MPA C2 in crisis or times of war. These arrangements continued throughout the 1980s.

By the 1970s, the Russians continued their strategic patrols with the *Yankee* class that were initiated towards the end of the 1960s. The posture of the Soviet Navy was forward-leaning and assertive, but the *Yankee* class was somewhat of an interim solution to the challenge posed by Western ASW forces. When the *Delta*/SS-N-8 combination came online for the Northern Fleet in the mid-1970s, it facilitated a retreat to the bastion in the High North. Fewer and fewer strategic patrols took place south in the Atlantic, and the Russian fleet entered a period of a retracted posture that was to last for several decades.

Through all this, even as satellite technology facilitated increasing levels of intelligence gathering from space, MPA were the weapon of choice for sustaining knowledge on and pressure over Russian submarines as they exercised and patrolled. The Northern Fleet conducted several large-scale exercises in the North Atlantic, and continued several patrols with both surface vessels and submarines even after the bastion approach was introduced. The cooperation between American, British, and Norwegian MPA in the North Atlantic continued until the end of the Cold War. As we have seen, Russian engineers were able to quieten their submarines through research and Western technology. This put NATO ASW sensors, including those of MPA, significantly under pressure in the 1980s. The improvement to computer processing power kept modern MPA somewhat able to follow the emerging submarines. However, the significant quieting progress of Russian submarine technology gave a sense of NATO ASW force being "saved by the bell" with the collapse for the Soviet Union.

The initial decade covered in this chapter entailed a significant degree of predictability by the Russian submarines. Although far out to sea, the transit routes of the *Yankee* class were known to Western MPA. This meant that the range of the aircraft was used efficiently. It was the silencing of the new attack submarines in the 1980s and their non-predictable movements that really put the Western MPA force to the test.

Notes

1 Jacob Børresen, Gullow Gjeseth, og Rolf Tamnes, *Norsk forsvarshistorie – Allianseforsvar i endring, 1970–2000*, Norsk Forsvarshistorie 5 (Bergen: Eide Forlag, 2004), 36.

2 Børresen, Gjeseth, og Tamnes, 40–41.

3 Rolf Tamnes, *Oljealder, 1965–1995*, Bd. 6, Norsk Utenrikspolitikks historie (Oslo: Universitetsforlaget, 1997), 6:68.

4 Børresen, Gjeseth, og Tamnes, *Norsk forsvarshistorie – Allianseforsvar i endring, 1970–2000*, 67.

5 Tamnes, *Oljealder, 1965–1995*, 6:249–55.

6 Børresen, Gjeseth, og Tamnes, *Norsk forsvarshistorie – Allianseforsvar i endring, 1970–2000*, 114.

7 James Bamford, "The Walker Espionage Case," *United States Naval Institute Proceedings* 112, no. 5 (May 1986): 2; Cote, "The Third Battle – Innovation in the U.S. Navy's Silent War Struggle with Soviet Submarines," 73.

8 This becomes clear when comparing the Soviet project numeration of the two classes, where the *Yankee* class was given Project 667A, and the *Delta* class Project 667B. For more on the development of these classes, see for example Podvig, *Russian Strategic Nuclear Forces*, 240–43.

9 Podvig, 330–31.

10 Office of Naval Intelligence, *Overview of Soviet Union Ballistic Submarine Patrol Areas – 1960s–1980s* (Washington, DC: National Museum of American History, 2000), https://americanhistory.si.edu/subs/work/missions/warfare/index.html.

11 John B. Hattendorf og Peter M. Swartz, reds., "U.S. Naval Strategy in the 1980s – Selected Documents," *Naval War College Newport Papers* 33 (2008): 212–16.

12 Cote, "The Third Battle – Innovation in the U.S. Navy's Silent War Struggle with Soviet Submarines," 57–60; Klevberg, "Maritime Surveillance in the North," 394; Jordan, *Soviet Submarines – 1945 to the Present*, 126.

13 Jordan, *Soviet Submarines – 1945 to the Present*, 138.
14 Bamford, "The Walker Espionage Case"; Cote, "The Third Battle – Innovation in the U.S. Navy's Silent War Struggle with Soviet Submarines," 73.
15 Jordan, *Soviet Submarines – 1945 to the Present*, 146–48.
16 Jordan, 167–68.
17 Jordan, 157–60.
18 Podvig, *Russian Strategic Nuclear Forces*, 336; Jordan, *Soviet Submarines – 1945 to the Present*, 154–65.
19 Breemer, *Soviet Submarines – Design, Development and Tactics*, 144.
20 Tamnes, *The United States and the Cold War in the High North*, 274–78.
21 Overview by the Norwegian High Command. Depicted in Tamnes, *Oljealder, 1965–1995*, 6:34. Translation: *Ubåt-barriere*: Submarine barrier // *Hangarskipsgruppe*: Carrier strike group // *Forsterkninger til N. Norge*: Reinforcements to Northern Norway.
22 Cote, "The Third Battle – Innovation in the U.S. Navy's Silent War Struggle with Soviet Submarines," 69–71.
23 For an outstanding analysis of the Toshiba/Kongsberg scandal that erupted in 1987 based on the sale of Western high-tech machinery to the Soviet Union in 1983/1984, see Olav Wicken, *Stille propell i storpolitisk storm – KV/Toshiba-saken og dens bakgrunn* (Institutt for Forsvarsstudier, 1988); Tamnes, *The United States and the Cold War in the High North*, 293.
24 Cote, "The Third Battle – Innovation in the U.S. Navy's Silent War Struggle with Soviet Submarines," 66–67; Jordan, *Soviet Submarines – 1945 to the Present*, 184–86.
25 Rear Admiral Butts, Department of Defense Authorization for Appropriations for Fiscal Year 1986; Cote, "The Third Battle – Innovation in the U.S. Navy's Silent War Struggle with Soviet Submarines," 70; John Lehman, *Oceans Ventured: Winning the Cold War at Sea* (New York: W.W. Norton & Company Inc., 2018), 140–41.
26 Cote, "The Third Battle – Innovation in the U.S. Navy's Silent War Struggle with Soviet Submarines," 70.
27 Tamnes, *The United States and the Cold War in the High North*, 234–38.
28 Cote, "The Third Battle – Innovation in the U.S. Navy's Silent War Struggle with Soviet Submarines," 52–55; Tamnes, *The United States and the Cold War in the High North*, 256.
29 Podvig, *Russian Strategic Nuclear Forces*, 8–9.
30 Jordan, *Soviet Submarines – 1945 to the Present*, 103.
31 Tamnes, *The United States and the Cold War in the High North*, 225–26.
32 Tamnes, 228.
33 Tamnes, 261.
34 John B. Hattendorf, "The Evolution of the U.S. Navy's Maritime Strategy, 1977–1986," *Naval War College Newport Papers* 19 (2004): 54–55.
35 Hattendorf og Swartz, "U.S. Naval Strategy in the 1980s," 147.
36 Grove og Thomson, *Battle for the Fjords*, 9.
37 Lehman, *Oceans Ventured*, 100.
38 Børresen, Gjeseth, og Tamnes, *Norsk forsvarshistorie – Allianseforsvar i endring, 1970–2000*, 114.
39 Rolf Tamnes, "The Significance of the North Atlantic and the Norwegian Contribution," in *NATO and the North Atlantic – Revitalizing Collective Defense*, redigert av John A. Olsen, Whitehall Paper 87 (London: Royal United Services Institute (RUSI), 2017), 17.
40 Klevberg, "Maritime Surveillance in the North," 346–50.
41 Klevberg, 346.
42 Riste og Moland, *Strengt Hemmelig – Norsk etterretningstjeneste 1945–1975*, 214; Tamnes, *Oljealder, 1965–1995*, 6:69–74; Skogrand, *Alliert i krig og fred*, 216–18.
43 Børresen, Gjeseth, og Tamnes, *Norsk forsvarshistorie – Allianseforsvar i endring, 1970–2000*, 90.

44 Tamnes, *Oljealder, 1965–1995*, 6:69; Klevberg, "Maritime Surveillance in the North," 342–43.
45 Klevberg, "Maritime Surveillance in the North," 345.
46 In addition to the operational agreements, a logistics agreement was signed in 1960 between the US Navy and the Royal Norwegian Air Force. The agreement facilitated the operational agreements through logistical arrangements such as the storage of mines, ammunition, fuel, lubricants and spare parts. This was the forerunner of the 1971 *Invictus* agreement, which had a particular focus on MPA equipment. In 1980, this was broadened to include aircraft carrier equipment. See Tamnes, *The United States and the Cold War in the High North*, 118, 251, 260.
47 Klevberg, "Maritime Surveillance in the North," 462–68.
48 Klevberg, 376.
49 Simon, "The 'Third' US Offset Strategy and Europe's 'Antiaccess' Challenge," 425; Tamnes, *The United States and the Cold War in the High North*, 254.
50 Reade, *The Age of Orion – The Lockheed P-3 Story*, 20–24.
51 "Command History Report (CHR) for VP-10 (1970)."
52 Riste og Moland, *Strengt Hemmelig – Norsk etterretningstjeneste 1945–1975*, 231–32.
53 Skogrand, Njølstad, og Tamnes, "Hot Spot – Cold War, the High North, and Grand Strategy," 24. Initial cables were laid in 1972, but these were severed and had to be replaced.
54 Air Commodore (r) Garfield "Garry" Porter, *British Maritime Patrol Aircraft Operations During and After the Cold War* (London: Royal Air Force Club, May 31, 2017); AIR 27/3161, "Operational Records – 201 squadron – 1966–1968."
55 "Command History Report (CHR) for VP-49 (1970)."
56 Roberts, *Dictionary of American Naval Aviation Squadrons*, 2:173.
57 "Command History Report (CHR) for VP-23 (1979)"; "Command History Report (CHR) for VP-23 (1980)."
58 Kenneth B. Sherman, "Feet Dry: Maritime Patrol Goes Ashore – New Roles and Missions Mean No Safe Harbor for Land Forces," *The Journal of Electronic Defense (JED)* (March 2001): 40–41.
59 Roberts, *Dictionary of American Naval Aviation Squadrons*, 2:310.
60 Roberts, 2:249.
61 "Command History Report (CHR) for VP-56 (1984)."
62 Roberts, *Dictionary of American Naval Aviation Squadrons*, 2:293–94.
63 "Command History Report (CHR) for VP-44 (1986)."
64 "Command History Report (CHR) for VP-44 (1987)."
65 Klevberg, "Maritime Surveillance in the North," 473–75.
66 Klevberg, 474.
67 Tamnes, *The United States and the Cold War in the High North*, 256; Edward Whitman, "SOSUS – The 'Secret Weapon' of Undersea Surveillance," *Undersea Warfare* 7, no. 2 (Winter 2015).
68 Whitman, "SOSUS – The 'Secret Weapon' of Undersea Surveillance"; Jonathan White og Sean Filipowski, "Know the Environment, Know the Enemy, Know the Target," *United States Naval Institute Proceedings* 140, no. 7 (July 2014).
69 Tamnes, *The United States and the Cold War in the High North*, 211.
70 Friedman, *Seapower and Space*, 177.
71 Friedman, 242–45.
72 Friedman, 205–8.
73 Cote, "The Third Battle – Innovation in the U.S. Navy's Silent War Struggle with Soviet Submarines," 61.
74 Friedman, *Seapower and Space*, 201–5.
75 Friedman, 99–100.
76 Friedman, 98.

77 "Command History Report (CHR) for VP-44 (1987)."
78 Robideau, "Third Battle for the North Atlantic 1962–1991," 14–19; Perkins, *Alliance Airborne Anti-Submarine Warfare*, 12–14.
79 Porter, British Maritime Patrol Aircraft Operations during and after the Cold War.
80 AIR 27/3357, "Operational Records – 201 Squadron – 1972–1975," October 13, 1972.
81 "Command History Report (CHR) for VP-56 (1971)"; "Command History Report (CHR) for VP-24 (1974)."
82 "Command History Report (CHR) for VP-24 (1974)," January 22, 1974.
83 AIR 27/3162, "Operational Records – 201 Squadron – 1969–1970."
84 "Command History Report (CHR) for VP-56 (1973)."
85 "Command History Report (CHR) for VP-56 (1971)"; "Command History Report (CHR) for VP-56 (1973)"; "Command History Report (CHR) for VP-56 (1978)."
86 "Command History Report (CHR) for VP-10 (1970)."
87 "Command History Report (CHR) for VP-56 (1970)."
88 "Command History Report (CHR) for VP-49 (1972)."
89 "Command History Report (CHR) for VP-56 (1973)."
90 "Command History Report (CHR) for VP-44 (1979)"; "Command History Report (CHR) for VP-56 (1978)"; "Command History Report (CHR) for VP-56 (1984)."
91 "Command History Report (CHR) for VP-44 (1986)," 3.
92 "Command History Report (CHR) for VP-44 (1987)." This has among ASW veterans colloquially been termed "the Victor surge".
93 Roberts, *Dictionary of American Naval Aviation Squadrons*, 2:87–90.
94 Crickmore, *Lockheed Blackbird*, 305–27.
95 Børresen, Gjeseth, og Tamnes, *Norsk forsvarshistorie – Allianseforsvar i endring, 1970–2000*, 116.
96 Crickmore, *Lockheed Blackbird*, 349, 398.
97 Lehman, *Oceans Ventured*, 100.
98 Peter Hennessy og James Jinks, *The Silent Deep – The Royal Navy Submarine Service Since 1945* (Penguin Random House, 2015), 546, 556.
99 AIR 27/3357, "Operational Records – 201 Squadron – 1972–1975."
100 Porter, *British Maritime Patrol Aircraft Operations During and After the Cold War.*
101 Porter.
102 AIR 27/3697, "Operational Records – 201 Squadron – 1981–1982."
103 Porter, *British Maritime Patrol Aircraft Operations During and After the Cold War.*
104 AIR 27/3694, "Operational Records – 201 Squadron – 1983–1984."
105 Roberts, *Dictionary of American Naval Aviation Squadrons*, 2:46.
106 Klevberg, "Maritime Surveillance in the North," 378–80.
107 Reade, *The Age of Orion – The Lockheed P-3 Story*, 68.
108 Harald Håvoll, "Norwegian Maritime Patrol Aircraft Operations During and After the Cold War," Nesodden, Norway, October 22, 2018.
109 Riste og Moland, *Strengt Hemmelig – Norsk etterretningstjeneste 1945–1975*, 232.
110 Klevberg, "Maritime Surveillance in the North," 496–97.
111 "Command History Report (CHR) for VP-44 (1987)."
112 "Command History Report (CHR) for VP-56 (1973)"; "Command History Report (CHR) for VP-56 (1975)"; "Command History Report (CHR) for VP-56 (1978)"; Roberts, *Dictionary of American Naval Aviation Squadrons*, 2:87, 293.
113 "Command History Report (CHR) for VP-56 (1971)"; "Command History Report (CHR) for VP-56 (1978)."
114 "Command History Report (CHR) for VP-56 (1973)."
115 Angus, "Command and Control of MPA Assets on the Northern Flank."
116 Porter, *British Maritime Patrol Aircraft Operations During and After the Cold War.*

Bibliography

AIR 27/3161. "Operational Records – 201 Squadron – 1966–1968." Royal Air Force, 1968–1966. UK National Archives, Kew.

AIR 27/3162. "Operational Records – 201 Squadron – 1969–1970." Royal Air Force, 1970. UK National Archives, Kew.

AIR 27/3357. "Operational Records – 201 Squadron – 1972–1975." Royal Air Force, 1972–1975. UK National Archives, Kew.

AIR 27/3694. "Operational Records – 201 Squadron – 1983–1984." Royal Air Force, 1984–1983. UK National Archives, Kew.

AIR 27/3697. "Operational Records – 201 Squadron – 1981–1982." Royal Air Force, 1982–1981. UK National Archives, Kew.

Angus, Dave. "Command and Control of MPA Assets on the Northern Flank." *Maritime Patrol Aviation* 2, no. 1 (September 1991).

Bamford, James. "The Walker Espionage Case." *United States Naval Institute Proceedings* 112, no. 5 (May 1986).

Børresen, Jacob, Gullow Gjeseth, og Rolf Tamnes. *Norsk forsvarshistorie – Allianseforsvar i endring, 1970–2000*. Norsk Forsvarshistorie 5. Bergen: Eide Forlag, 2004.

Breemer, Jan S. *Soviet Submarines – Design, Development and Tactics*. Surrey, UK: Jane's Information Group Ltd., 1989.

"Command History Report (CHR) for VP-10 (1970)." US Navy, 1970. FOIA case number: DON-NAVY-2018-001782. Naval History and Heritage Command.

"Command History Report (CHR) for VP-23 (1979)." US Navy, 1979. FOIA case number: DON-NAVY-2018-001782. Naval History and Heritage Command.

"Command History Report (CHR) for VP-23 (1980)." US Navy, 1980. FOIA case number: DON-NAVY-2018-001782'. Naval History and Heritage Command.

"Command History Report (CHR) for VP-24 (1974)." US Navy, 1974. FOIA case number: DON-NAVY-2018-001782. Naval History and Heritage Command.

"Command History Report (CHR) for VP-44 (1979)." US Navy, 1979. FOIA case number: DON-NAVY-2018-001782. Naval History and Heritage Command.

"Command History Report (CHR) for VP-44 (1986)." US Navy, 1986. FOIA case number: DON-NAVY-2018-001782. Naval History and Heritage Command.

"Command History Report (CHR) for VP-44 (1987)." US Navy, 1987. FOIA case number: DON-NAVY-2018-001782. Naval History and Heritage Command.

"Command History Report (CHR) for VP-49 (1970)." US Navy, 1970. FOIA case number: DON-NAVY-2018-001782. Naval History and Heritage Command.

"Command History Report (CHR) for VP-49 (1972)." US Navy, 1972. FOIA case number: DON-NAVY-2018-00574. Naval History and Heritage Command.

"Command History Report (CHR) for VP-56 (1970)." US Navy, 1970. FOIA case number: DON-NAVY-2018-001782. Naval History and Heritage Command.

"Command History Report (CHR) for VP-56 (1971)." US Navy, 1971. FOIA case number: DON-NAVY-2018-001782. Naval History and Heritage Command.

"Command History Report (CHR) for VP-56 (1973)." US Navy, 1973. FOIA case number: DON-NAVY-2018-001782. Naval History and Heritage Command.

"Command History Report (CHR) for VP-56 (1975)." US Navy, 1975. FOIA case number: DON-NAVY-2018-001782. Naval History and Heritage Command.

"Command History Report (CHR) for VP-56 (1978)." US Navy, 1978. FOIA case number: DON-NAVY-2018-001782. Naval History and Heritage Command.

"Command History Report (CHR) for VP-56 (1984)." US Navy, 1984. FOIA case number: DON-NAVY-2018-00574. Naval History and Heritage Command.

Cote, Owen R. Jr. "The Third Battle – Innovation in the U.S. Navy's Silent War Struggle with Soviet Submarines." *Naval War College Newport Papers* 16 (2003).

Crickmore, Paul F. *Lockheed Blackbird – Beyond the Secret Missions*, revised ed. Oxford: Osprey Publishing, 2016.
Friedman, Norman. *Seapower and Space*. Annapolis, MD: Naval Institute Press, 2000.
Grove, Eric, og Graham Thomson. *Battle for the Fjords*. Annapolis, MD: Naval Institute Press, 1991.
Hattendorf, John B. "The Evolution of the U.S. Navy's Maritime Strategy, 1977–1986." *Naval War College Newport Papers* 19 (2004).
Hattendorf, John B., og Peter M. Swartz, reds. "U.S. Naval Strategy in the 1980s – Selected Documents." *Naval War College Newport Papers* 33 (2008).
Håvoll, Harald. "Norwegian Maritime Patrol Aircraft Operations During and After the Cold War." Nesodden, Norway, October 22, 2018.
Hennessy, Peter, og James Jinks. *The Silent Deep – The Royal Navy Submarine Service Since 1945*. Penguin Random House, 2015.
Jordan, John. *Soviet Submarines – 1945 to the Present*. London: Arms & Armour Press, 1989.
Klevberg, Håvard. "Maritime Surveillance in the North: 333 Squadron in Norwegian Security Politics." PhD, History, University of Oslo, 2011.
Lehman, John. *Oceans Ventured: Winning the Cold War at Sea*. New York: W.W. Norton & Company Inc., 2018.
Office of Naval Intelligence. *Overview of Soviet Union Ballistic Submarine Patrol Areas – 1960s–1980s*. Washington, DC: National Museum of American History, 2000. https://americanhistory.si.edu/subs/work/missions/warfare/index.html.
Perkins, William. *Alliance Airborne Anti-Submarine Warfare – A Forecast for Maritime Air ASW in the Future Operational Environment*. Joint Air Power Competence Centre (JAPCC), June 2016.
Podvig, Pavel, red. *Russian Strategic Nuclear Forces*, 2004 paperback ed. Cambridge, MA: MIT Press, 2001.
Porter, Air Commodore (r) Garfield "Garry". *British Maritime Patrol Aircraft Operations During and After the Cold War*. London: Royal Air Force Club, May 31, 2017.
Reade, David. *The Age of Orion – The Lockheed P-3 Story*. Atglen, PA: Schiffer Publishing Ltd., 1998.
Rear Admiral Butts, John. "Department of Defense Authorization for Appropriations for Fiscal Year 1986." Pub. L. No. Part 8, § Committee on Armed Services, U.S. Senate, 4364, 1985.
Riste, Olav, og Arnfinn Moland. *Strengt Hemmelig – Norsk etterretningstjeneste 1945–1975*. Oslo: Universitetsforlaget, 1997.
Roberts, Michael D. *Dictionary of American Naval Aviation Squadrons*, vol. 2. Washington, DC: Naval Historical Center, 2000.
Robideau, Larry. "Third Battle for the North Atlantic 1962–1991." *Cold War Times* 6, no. 1 (February 2006). www.coldwar.org/text_files/coldwartimesfeb2006.pdf.
Sherman, Kenneth B. "Feet Dry: Maritime Patrol Goes Ashore – New Roles and Missions Mean No Safe Harbor for Land Forces." *The Journal of Electronic Defense (JED)* (March 2001): 37–43.
Simon, Luis. "The 'Third' US Offset Strategy and Europe's 'Antiaccess' Challenge." *Journal of Strategic Studies* 39, no. 3 (2016): 417–45.
Skogrand, Kjetil. *Alliert i krig og fred*. Norsk Forsvarshistorie 4. Bergen: Eide Forlag, 2004.
Skogrand, Kjetil, Olav Njølstad, og Rolf Tamnes. "Hot Spot – Cold War, the High North, and Grand Strategy." *Institutt for Forsvarsstudier (IFS) Info* 5 (1998).
Tamnes, Rolf. *Oljealder, 1965–1995*, Bd. 6. Norsk Utenrikspolitikks historie. Oslo: Universitetsforlaget, 1997.
Tamnes, Rolf. "The Significance of the North Atlantic and the Norwegian Contribution." In *NATO and the North Atlantic – Revitalizing Collective Defense*, redigert av John A. Olsen. Whitehall Paper 87. London: Royal United Services Institute (RUSI), 2017.

Tamnes, Rolf. *The United States and the Cold War in the High North*. Oslo: Ad Notam, 1991.

White, Jonathan, og Sean Filipowski. "Know the Environment, Know the Enemy, Know the Target." *United States Naval Institute Proceedings* 140, no. 7 (July 2014).

Whitman, Edward. "SOSUS – The 'Secret Weapon' of Undersea Surveillance." *Undersea Warfare* 7, no. 2 (Winter 2015).

Wicken, Olav. *Stille propell i storpolitisk storm – KV/Toshiba-saken og dens bakgrunn*. Institutt for Forsvarsstudier, 1988.

8 A change in focus for intelligence gathering, 1991–2015

The collapse of the Soviet Union halted much of the submarine production and threw Russia into political and economic turmoil. The US ONI has estimated that fully three-quarters to five-sixths of the Soviet-era naval inventory was scrapped, including vessels that were fully operational but deemed too costly to maintain.[1] The steady decline of the Russian fleet in the 1990s fuelled speculations that the Russian navy of the future would be capable of no more than basic coastal defence.[2] The Presidential Nuclear Initiatives (PNIs) of the early 1990s between Russia and the United States further resulted in a series of reciprocal pledges to substantially reduce both the American and the Russian nuclear arsenals.[3]

The collapse of the Soviet Union effectively removed the one, obvious security threat to NATO. The 1990s was thus characterized by a fledging understanding of threats, which led to several warfare specialties losing their explicit mission. A prime example of this was NATO's MPA fleet. With Russian submarine patrols in the 1990s at times ebbing to just a handful per year, many nations started using submarine hunting aircraft in support of the land battle. MPA carry several sensors that can be put to extensive use outside the realm of hunting submarines. This, however, further devolved ASW competency in the Alliance.

In political terms for Norway, the 1990s in the High North were characterized by significant efforts to mould the political dynamics in the Barents region into formal, practical arenas for international cooperation. Early on, this manifested itself into bilateral science programmes, arrangements for maritime search-and-rescue, and cooperation on safely dismantling the many nuclear reactors on the Kola Peninsula. Bilateral fishery surveillance and resource management were further developed. The requirement to establish formal arrangements for political cooperation was based on a desire to strengthen the framework within which both low tensions could be maintained, and natural resources could be explored and managed.[4] In many ways, these soft power moves coincided neatly with the decreasing military strategic focus on the High North due to the diminishing activity levels. Within NATO, however, Norway remained one of the most active advocates for a continued focus on collective defence as the overarching strategic concept increasingly turned towards a reduced command structure and expeditionary warfare capabilities.[5]

DOI: 10.4324/9781003626015-8

By the beginning of the new century, Norway finally restructured its armed forces from a counter-invasion force to a structure tuned for expeditionary warfare, like most other allies. The terrorist attack on the United States in 2001 further drove organizational changes within the Alliance. Norway took active part in these changes for two main reasons: First, because the threat from Russia had decreased and therefore took up less space in Norwegian strategic thinking, and second, because there was an imminent requirement to secure a continued American engagement in the defence of Europe.[6]

Strategic thinking in the first decade of the new millennium was by shaped by the conflicts in the Middle East. By 2006 and 2007, however, there had been a resurgence in Russian activities in the North Atlantic and very harsh rhetoric from the Russian political leadership, exemplified by Vladimir Putin's speech at the Munich Security Conference in 2007. Norway then sought to refocus the Alliance's attention towards challenges closer to home. There was a growing concern regarding NATO's actual capability to defend its own territory, after years of focussing on expeditionary warfare. This led to the Norwegian initiative in 2008 to "raise NATO's profile." This initiative would be the starting point for the Alliance's return to its roots with regard to defending its members. In the strategic concept released 2 years later, several elements from the initiative were incorporated, as we shall see.

By the 2010s, however, the Russians had begun to acquire new submarine capabilities, resulting in increased levels of naval activity and a resumption of inter-state rivalry. This forced the strategic focus back towards prospects of peer-to-peer conflicts, leading to a renewed focus on MPA and ASW as a whole.

This chapter portrays the final historical element of the intersection between the development of Russian submarines and the utilization of the Western MPA fleet in meeting that threat. We shall see that in the mid-2010s, the United States, the United Kingdom, and Norway had decided once again to move ahead with a capable MPA force.

The 1990s – Russian decline and Western strategic ambivalences

In general, the economic situation simply demanded evolving the Russian Navy into a leaner force. At the same time, the military and political leadership did not disregard the possibility of war, even after the collapse of the Soviet Union. These overarching strategic trends seem to have provided the foundation for renewed work on the bastion concept in the mid-1990s.[7] The Northern Fleet had, throughout the Cold War, established itself as the main strategic hub for Russian naval operations. The Russian leadership also explicitly stated that there was a requirement to "concentrate Russia's sparse forces and means in a limited number of directions," thus providing safer grounds for President Yeltsin to announce the establishment of the Northern Strategic Bastion in August 1998.[8] The main function of the bastion would be, first, to provide a credible nuclear deterrent in the North and, second, to constitute an "independent naval force able to secure Russia's naval interests on the world oceans."[9] This was likely meant as much for an internal Russian audience as for an external international one. The Northern Strategic Bastion was a political instrument: The mission of the Northern Fleet was to secure Russian military and economic interests in the Arctic and on the world oceans.[10] The concept only really gained traction and credibility in the 2000s, however, after the initiation of an expensive Russian weapons programme under President Putin.

In the cutbacks and triage that took place, an implicit policy of emphasizing quality over quantity and multirole over single-mission vessels permeated the Russian Navy. It was in this context that the keel of a new, nuclear multirole nuclear submarine was laid in 1993, that of the *Severodvinsk* class.[11] The new multirole submarine was meant to take over from the *Sierra* and *Akula* attack submarines and the *Oscar* cruise missile submarines. Three years later, in 1996, the keel was laid for a new SSBN to replace the *Deltas* and *Typhoon*, namely the *Dolgorukiy* class. But successive failures in the development of the associated SS-N-32 *Bulava* SLBM caused the development of the submarine to take more than a decade and a half.[12] The *Typhoons* were decommissioned in the late 1990s due to cost of production, maintenance, and the absence of a mid-life update regime.[13]

The operational tempo of the Russian SSBN fleet declined significantly in the 1990s. There were 37 patrols in 1991, but there were only seven in 1999.[14] The plight of the Russian economy, as well as the complete breakdown of established logistics, led to fuel shortages, forcing the cancellation of high-profile foreign deployments for the Russian Navy mid-decade, with only few exceptions.[15] The 1990s would thus be characterized by a significant drop in operational activity, vast fleet decommissioning, and a stumbling and agonizingly slow construction of only two, distinct, new submarines.

Within NATO, the years following the downfall of the Soviet Union were characterized by a desire to realize the peace dividend. Significant cutbacks in force numbers were enacted by most nations. As early as 1991, NATO issued a new strategic concept. The concept emphasized that the "threat of a simultaneous, full-scale attack" had effectively been removed, and that in contrast to the predominant threat of the past, the Alliance now had to prepare to meet threats that were "multifaceted in nature and multidirectional."[16] In naval terms, the focus on Russian submarine deployments more or less evaporated in parallel with the decline in Russian submarine activity.[17] NATO's new naval mission was "to ensure sea control in order to safeguard the Allies' sea lines of communication, to support land and amphibious operations, and to protect the deployment of the Alliance's sea-based deterrent."[18] While still maintaining a role in deterrence operations and the mission to protect SLOCs, we see an early shift towards support to joint operations, with collaboration with air and land forces.

Conceptual developments in the 1990s emphasized naval support to land operations and joint operations in the littorals, and downgraded ASW and anti-air warfare at sea.[19] Stress was further placed on power projection, strategic nuclear deterrence, and a credible forward presence. With an increasing requirement for close joint coordination, ideas for interconnectedness provided the foundation for "network-centric warfare."[20] In NATO, the focus turned to understanding the new threat environment and manoeuvring towards Alliance expansion without increasing tensions with Russia. Russia had joined the North Atlantic Cooperation Council in 1991, and NATO's Partnership for Peace (PfP) programme 3 years later. As such, developments in the maritime domain were thus not a high priority within the Alliance.[21]

NATO's large C2 structure of the Cold War was slowly but surely downsized to a fraction of its former size.[22] As the Russian Navy's operational activity in the Atlantic diminished, more focus was placed on the Mediterranean, where Russian activity continued into the 1990s. In 1994, a common interest in following Russian vessels in the Mediterranean led to the establishment of a memorandum of understanding (MoU) organization based in Naples, called *Commander Maritime Air Naples* (CMAN). CMAN performed the overall coordination of national MPA missions in the Mediterranean. The national Maritime Air Control Authorities (MACAs) would coordinate their missions through CMAN but were ultimately in charge of their own assets during the respective missions. When NATO commenced its lone Article 5 operation, *Operation Active Endeavour* (OAE) in 2001 in the Mediterranean following the 9/11 attacks on the United States, the CMAN construct constituted the backbone for coordination of MPA missions in that operation. Nations that deployed their MPA to Sicily, for example, and nations that flew in so-called *direct support* of the operation, were under C2 of CMAN for the (OAE) mission. The MACA construct did not apply outside the Mediterranean, however. In the North Atlantic, C2 coordination between the Americans at Keflavik, the British at Kinloss, and the Norwegians at Andøya continued to be characterized by "friendly cooperation," and was based on the *modus operandi* for surveillance coordination established during the Cold War.[23]

In the mid-1990s, concerns were raised in Norwegian political circles regarding the validity of the self-imposed restrictions in Norwegian security policy. Among other activities in 1988, the substantial NATO exercise *Teamwork 88* had raised concerns surrounding the possibility of Allied vessels capable of carrying nuclear weapons making port calls in Norway without explicitly confirming the absence of such weapons to the Norwegian Government.[24] The long-standing Norwegian basing policy forbade any storage of nuclear weapons in Norway during peacetime, and large exercises created uncertainties in this regard. There was a recognized requirement to review the restrictions in the context of the new security environment.[25] The Norwegian Government conducted an internal review of the self-imposed restrictions and national basing policy in 1995. This led to a few changes, such as the term "self-imposed restrictions" in several instances being replaced by the more flexible term "political guidelines."[26] The new description was more in line with both the dialogue between Norwegian and Russian authorities in the North that had been initiated after the Cold War, as well as the overarching thaw in the relationship between NATO and Russia. Instead of mentioning only "allied units," the guidelines would henceforth cite "foreign units," in order also to take into account countries operating under the new PfP construct.[27] It was also important for Norwegian authorities not to *restrict* Allied activities, but rather stimulate allied activity within the framework of low tensions in the North. Before

1995, very few dispensations were given from the 24° restriction. After 1995, the Norwegian MoD could authorize Allied flights east of 24°E. The objective of the guidelines was always to facilitate Allied activity and presence *in combination with* maintaining low tensions in the North.[28] Norway had always been dependent on allied activity as a part of her deterrence posture, but it was essential to put restrictions on this activity.[29] In the new security environment, emphasis was even given to conducting joint search-and-rescue exercises with Russian units, to facilitate coordination and low tension in the High North.

With the collapse of the Soviet Union, high-end Russian technology became available to whichever nation could afford it. Western strategists raised concerns over nations such as Libya, China, Iran, and Algeria acquiring relatively capable, small, conventional diesel-electric submarines, which would pose a credible threat in the littorals.[30] The regular Keflavik Tactical Exercise (KEFTACEX) in the 1990s became an important event to practice this scenario. NATO MPA were tested against NATO conventional submarines. One important lesson that came out of the exercises of the 1990s was that upon detection of a small, conventional submarine, the aircrew had to prepare for attack *immediately*. The open-ocean approach established during the Cold War of detection, localization, and tracking before settling into attack mode was no longer applicable when prosecuting a submarine in the littorals.[31] The demanding acoustic environment with much ambient noise against the very quiet electric propulsion of the submarines required a quick attack while the chance was there – it might not come back.

With the diminishing operational tempo of Russian submarines, there was a fear in the MPA community that lower priority would be given to ASW training and maintaining ASW capabilities and knowledge within the Alliance.[32] This concern was justified. Missions increasingly entailed surveillance of smuggling, exercising joint and combined operations, littoral surface surveillance, and research and development. NATO *tactics* and *procedures* for ASW were still adjusted for Cold War strategic patrols of Russian SSBNs, as opposed to smaller submarines posing a threat closer to shore.[33] Regular NATO exercises compensated for some of the reduction in adversary activities, although the exercises, to an increasing degree, focused on anti-surface engagements as opposed to anti-subsurface operations.[34] A lack of training opportunities was less prominent for the Norwegian aircrews that continued to cover the Barents Sea on a regular basis. After all, even with a severely reduced operational and exercise tempo, the Russians continued to execute missions, albeit of a local geographical nature. The downturn in operational activity on the part of the Russians had somewhat less of an impact on submarine exposure for the average Norwegian operator.[35]

Over the years, the improvements made to the capabilities of MPA mostly involved existing sensors, such as improved capacity and memory of the central computers, more processing capacity for acoustic processors, and better navigational systems. Some improvement programmes initiated in the late 1980s met with fiscal constraints in the 1990s, and together with schedule delays and cost overruns, several upgrade programmes were cancelled.[36] It seemed difficult to justify spending money on airborne ASW when the perception was that the threat had evaporated. In line with the new security environment and strategic development, the US Navy instead initiated the Anti-Surface Warfare Improvement Program in 1994. The AIP upgrade included radar upgrades to both inverse synthetic aperture radar (ISAR) and synthetic aperture radar capabilities, improved electronic support measures (ESM), and the advanced imaging multispectral sensor (AIMS) with a capable EO/IR camera. The AIP aircraft were also equipped with the AGM-85E *stand-off land attack missile* (SLAM) and the AGM-65D/F *Maverick* missile. The upgraded aircraft were given a SATCOM capability and a communications suite that facilitated closer integration with the joint force. A limited number of AIP aircraft also had the tactical common data link (TCDL), which enabled real-time streaming of sensors from the aircraft to other units.[37] In the mid-1990s, a few selected US Navy P-3 *Orions* were also updated through the new command, control, computers, and communications for ASW (C4 for ASW) programme. This included tactical data link through the Link-16, and satellite connectivity through an encrypted INMARSAT connection.[38] These innovations reflect efforts to achieve greater integration of MPA with other warfare assets and domains. The US Navy initiated studies in the 1990s to find a replacement for the ageing P-3C *Orion*, and the decision was made in 2004 to replace the latter with the P-8 *Poseidon*, a modified and militarized Boeing 737 jet aircraft.[39] A UK MoD study in the early 1990s similarly evaluated options to replace the *Nimrod* MR2 that essentially mirrored the alternatives of the Americans, including a P-3 option.[40] The decision was made in 1996 to build the *Nimrod* MR4 as a continuation of the MR2, with significant upgrades to the entire aircraft, including the airframe and the sensor suite.[41]

The Norwegian P-3C *Orion* fleet was also upgraded at the end of the 1990s through the upgrade improvement programme (UIP), largely mirroring the American AIP suite. The mission for the Norwegian *Orions* was thus still to monitor the Russian Northern Fleet and its submarines in the High North. The Norwegian *Orions* were therefore not upgraded with the aforementioned TCDL system. Nor were the Norwegian aircraft upgraded to carry missiles. The Norwegian P-3 *Orion* UIP included several Norwegian-specific solutions and sensors, such as a dedicated ESM/ELINT station.[42] The US Navy has flown the P-3 *Orion* for ASW and acoustic intelligence (ACINT) missions, and the EP-3 *Aries* for SIGINT collection. The Norwegians only had one type of operational MPA, and the Norwegian UIPs could thus be seen as a hybrid between the American P-3 *Orion* and the EP-3 *Aries*. The UIP model included a significant upgrade to the radar, the ESM suite, the computer processing capability, GPS installation, improved self-protection systems, and SATCOM.[43] The tailored Norwegian stations and the consequent low volume of spare parts for the sensors led to costly and time-consuming upgrades of a very customized aircraft in the 2000s. Throughout the entire period since the UIP upgrade, the Norwegian MPA community has struggled to get a hold of spare parts. The discussion of replacing the *Orions* began in earnest in the late 2000s, but would take 6 years to come to a conclusion.[44]

It was not until the 1990s that unmanned aircraft technology matured to the point that capable systems could be controlled from the other side of the world and be integrated with conventional land or maritime forces. And *Operation Desert Storm* in 1991 raised new requirements for ISR. Several targets were attacked simultaneously in a systematic fashion, with targets "linked" through their relative importance to the Iraqi regime and infrastructure.[45] This demanded almost instant information on target updates and attack results, so-called BDA. Satellite imagery was neither responsive nor flexible enough. This opened the door for what is termed "organic assets" to conduct intelligence gathering, such as the *Pioneer* unmanned aircraft.[46] Based on these operations, combined with historical strategic airborne intelligence gathering, the concept of a long-endurance unmanned aircraft was born in the early 1990s. This idea would eventually lead to the *Global Hawk* drone flying with the US Air Force today and the *Alliance Ground Surveillance* (AGS) system about to become operational for NATO.[47]

Satellite operations had been prohibitive for most nations in at least two aspects: Cost and the level of technology required. The 1990s were characterized by the proliferation and further miniaturization of satellite technology, and an abrupt decrease in cost associated with developmental breakthroughs. Globalization created a global market for satellite services, which in turn facilitated a market for launch services. A combination of these factors has led to what is today termed "New Space."[48] "Old Space" was the realm of the few nations that could afford the immense investments required to send big, heavy satellites into orbit. "New Space" is for all nations, commercial companies, universities, and anyone that sees benefit or profit in satellite services. New Space has facilitated new possibilities in the realm of maritime surveillance, as we shall see.

Fewer Russian submarine deployments led to fewer intelligence collection opportunities for NATO surveillance aircraft. It did not take long until MPA squadrons were explicitly tasked to perform duties that diverged from wartime MPA roles, such as postal drops in remote locations and ice edge reconnaissance.[49] By the mid-1990s, mission areas shifted more towards NATO ASW exercises with an emphasis on quiet diesel-electric submarines and surface surveillance, training with national forces such as transiting aircraft carrier battle groups, long-range navigation training, and ice reconnaissance.[50]

US squadrons that had deployed to Iceland increasingly sent detachments to the Mediterranean.[51] The British had always paid attention to the Baltic Sea with regard to maritime surveillance, and this effort continued into the 1990s with operational missions flown out of Nordholz, Germany.[52] MPA that flew daily to maintain awareness of the whereabouts of the Soviet Navy during the Cold War were utilized in operations that did not entail ASW at all.[53] American *Orions* and British *Nimrods* took part in operations *Desert Shield* and *Desert Storm* during the first Gulf War in 1991, as well as in the NATO embargo operation that was part of the Adriatic *Operation Sharp Guard* during the war on Serbia in 1992.[54] Balkan engagements culminated in the controversial *Operation Allied Force* in the spring of 1999, which saw an MPA fire land-attack missiles in a live operation for the first time.[55] The Balkan and Adriatic operations served as catalysts for increased utilization of MPA in an over-land role.[56] Norwegian MPA did not take part in these operations, but remained focussed on the High North.

The 2000s – Russian ambitions and Western expeditionary warfare

When Vladimir Putin came to power in 2000, he halted the decline of the Navy and later reversed it.[57] He continued cooperation with NATO at the same time as he channelled more funds towards restructuring of and investments in the military. The new president saw the need for an explicit naval strategy. That same year he promulgated a new national security concept and a new naval doctrine. The naval doctrine emphasized the "qualitative renewal" of the forces, and old systems would be replaced by new ones in the coming decades.[58] With the Russian economy taking a slight upturn at the time of issue, the naval doctrine demanded the renewal of outdated submarines and surface vessels, development of multirole, ship-based as well as land-based aircraft, increased readiness, improved C2 systems, the reduction of diversity in systems, and improved standardization of weapons system production.[59] The doctrine also emphasized the requirement for information systems and smart weapons. Critics underlined, however, that the doctrine attempted to "preserve at least the appearance of a navy worthy of a superpower, even at the cost of sacrificing quality and safety."[60] The doctrine was seemingly aimed at pleasing hawks in the Kremlin, without having any obvious impact on NATO strategy.

In 2007–2008, there was a pivot towards what can be termed a Cold War mentality in the military and political leadership in Russia. This became evident through both rhetoric and military posture. One example is Vladimir Putin's speech at the Munich Security Conference in 2007, where he attacked the "architecture of global security," condemned the United States' "disdain for basic principles of international law," and accused NATO of moving in on Russian borders from the West.[61] The new military posture was manifested not only through possessing the largest stockpile of nuclear weapons in the world and showing an increased interest in them, but also through an increase in exercises and test missile launches.[62] In late 2007/early 2008, the *Admiral Kuznetsov* aircraft carrier led a 2-month task force deployment from the Barents Sea to the Mediterranean. The goal of the deployment was "to ensure naval presence in tactically important regions of the world oceans,"[63] and to emphasize the renewed importance of strategic assets. Oil revenues had facilitated a 150% increase in spending for the strategic nuclear forces since the early 2000s.[64] But with the immense backlog in maintenance of ageing vessels, what can be termed overambitious political goals at the beginning of the century were still not met by matching levels of funding.[65] This negatively impacted modernization efforts.

The Georgian war in 2008 would show Russian leadership the state-of-affairs in the Russian military: Poor ability to mobilize, inflexible leadership, poorly trained conscripts, and lack of synchronization of units.[66] The deficiencies led to expedited modernization programmes for new equipment, readiness, and C2.[67] Russian land forces began executing snap exercises regularly, in order to improve readiness.[68]

Russia had also initiated work to modernize existing strategic nuclear weapons, the finalization of which would become permissible only after the *START Treaty* expired in 2009. As a consequence, the ageing *Delta IV* class was fitted with the new SS-N-23 *Sineva* strategic missile. In September 2008, the Russian Navy finally conducted a successful test-firing of a SS-N-32 *Bulava* missile, indicating that the new *Dolgorukiy*-class SSBN was approaching operational status. This test-firing was, however, followed by several failed firings, further delaying the programme. In mid-2009, a *Delta IV*-class SSBN test-launched two SS-N-23 *Sineva* missiles from a position close to the North Pole. According to Russian sources, the *Delta IV* was escorted and protected by a task force of several nuclear-powered attack submarines.[69]

In 2009, Russia also published her new national Arctic strategy. The most important aspects of the strategy were exploitation of natural resources, securing Russia's borders, and strengthening international cooperation in the region. Russia has maintained significant cooperation with Norway with regard to common regulations of fish stocks in the Norwegian and the Barents Sea, which extends back to the height of the Cold War. The financial setback of 2008 did not affect the implementation of the strategy to any mentionable degree.[70] There seemed, however, to be a lack of a distinguishable mission for the Russian Navy.[71]

The financial crisis of 2008 put a brake on the force modernization programme.[72] However, facilitated by an improved fiscal foundation by 2010, the Russians published a new military strategy that year. Significant weapons programmes were initiated, aiming for an overhaul of 70% of the prioritized units of the armed forces by 2020. The new National Security Strategy aimed to clarify Russia's nuclear weapons employment policy, and how Russian forces would be utilized in a modern, complex, and nuanced threat environment.[73]

The *Dolgorukiy*-class SSBN was finally accepted by the navy in January of 2013 after more than a decade and a half in development, production, and testing.[74] The issues delaying particularly the SS-N-32 *Bulava* v (range 4,600 NM/8,600 km) missiles' acceptance onto the vessel were said to be associated with technical malfunctions and quality control issues in final assembly.[75] The first *Dolgorukiy* class commenced its first operational patrol in the fall of 2014.[76]

The modernization of the non-strategic nuclear capability seems, however, to be centred around the introduction of the *Severodvinsk*-class SSGN. The *Severodvinsk*-class multi-purpose submarine also took 17 years to complete.[77] The mission of the *Severodvinsk* is ASW, ASuW, and land-attack missions. The first *Severodvinsk*-class submarine was finally declared combat-ready in early 2016. Although not as quiet as her American counterparts, the new Russian SSGN is considered to be the least detectable of the vessels in the Russian Navy order of battle.[78] An important feature with non-strategic nuclear forces, however, is that they are able to carry both nuclear and conventional missiles. Important aspects of this are both intentional and unintentional signalling and posture.[79] This ambiguity can lead to strategic uncertainty on the part of the adversary. A significant area of emphasis for Russia is to make herself less unilaterally dependent on nuclear weapons for regional and global deterrence. Focus is therefore given to developing conventional precision-guided, long-range weapons, in addition to the ability to utilize and exploit all the state's elements of power in a crisis situation.[80] The development of the nuclear-capable *Kalibr* family of missiles that can be placed on nuclear as well as conventional attack submarines stands out as a centre-point for Russian doctrinal development.

By the year 2000, nearly two-thirds of the Russian SSBN fleet had been withdrawn from service. The Russian Navy was able to maintain one SSBN on strategic patrol in the Arctic and one in the Pacific, although there were periods lasting up to 3 months at a time where there were no patrols at all due to safety concerns.[81] These were well justified. In August 2000, the *Oscar*-class SSGN *Kursk* sank in the Barents Sea, prompting a further decrease in operational tempo on Russian submarines.[82] In 2001, the Russian Navy conducted one strategic submarine patrol; and in 2002 it conducted none.[83] Russian submarine activity would eventually start to recover, and in 2005 the Russian Navy conducted five strategic patrols. Several of those were concurrent, meaning that Russia in the 2000s did not have a continuous operational sea-based strategic deterrent. However, by 2009, the patrol tempo of Russian strategic submarines had risen to nine patrols per year. But the number of patrols would once again drop in the following years.[84] In 2007, Russia's attack and cruise missile submarines increased their number of patrols to seven, an increase from four patrols the previous year. These submarines had averaged roughly 12 patrols a year during the 1990s.[85]

NATO issued a revised strategic concept concurrent with the war against Serbia in 1999.[86] This revision emphasized mobile and flexible forces with the endurance to conduct expeditionary operations. Specific tasks for the different services were downplayed, and emphasis was given to joint operations.[87] Cooperation between NATO and Russia continued through the NATO-Russian Founding Act of 1997, and deepened further in 2002 with the establishment of the NATO-Russia Council. By then, Russian submarine patrols in the Atlantic had been more or less discontinued.[88] NATO's relationship to Russia had morphed into an explicit political framework for security cooperation. It seems evident that changes to threats and threat perceptions induce changes to strategic thinking. The revised Russian strategy of the early 2000s was not backed by new equipment or expanded operations. The 9/11 attacks in 2001 on the United States were therefore the primary factor shaping the focus of the strategic discussions of NATO.[89] The Alliance continued to focus on expeditionary warfare while making changes to its command structure, which, among other things, led to the disestablishment in 2003 of SACLANT in Norfolk, Virginia, United States.[90] This removed the sole headquarters that maintained a primary focus on and responsibility of the Atlantic, with an emphasis on the Russian Northern Fleet. SACLANT had the peacetime responsibility for contingency plans, force requirements, exercises, and strategic and doctrinal development in the Atlantic area. The removal of SACLANT from the NATO Command Structure seems to have played a part in the atrophy of ASW skills within the Alliance in the decade that followed. In what was likely a combination of faith in emerging technology, a focus on the Middle East, and the lack of Russian operational submarine deployments, the Americans closed down Keflavik as a naval air base in

2006. American ASW C2 was pulled back to the East Coast for the Atlantic. The British continued to fly out of Kinloss in Scotland, with C2 situated in Northwood, London.

The 9/11 attacks on the United States in 2001 for the first time led to a member nation calling for Alliance members' support through NATO's Article 5.[91] The ensuing operation, *Operation Active Endeavour (OAE)*, was launched in late 2001, and was a maritime anti-terror operation tasked with tracking and controlling ships in the Mediterranean.[92] MPA were used extensively in OAE. In addition to US squadrons from Keflavik, the Mediterranean NATO nations routinely supported the operation. So did, for the first time, the Norwegian P-3 *Orion* squadron, as it deployed to Sicily to support as well.

An overwhelming focus on the Middle East in the mid-2000s had led to a sentiment in Norwegian security policy circles of reduced focus on the maritime domain, particularly in the High North. Norway did not deploy *Orions* to the Middle East, but sent them to the Mediterranean in support of OAE in 2005 and 2006. In September 2008, the Norwegian minister of defence presented at an informal ministerial meeting in London a paper on raising NATO's profile in the member states. The initiative sought a better balance between out-of-area operations and in-area activities and emphasized the importance of revitalizing collective defence and deterrence. In Norwegian terms, this meant an increased NATO focus on the North Atlantic.[93] After the paper had been presented, NATO initiated work to strengthen the links between the NATO Command Structure and the military establishments in the member nations. Progress, however, was slow, as was the implementation of changes in NATO activities. A key difference from the Cold War was the wide array of threat perceptions within the Alliance, as opposed to the one common threat in the past that provided a focus for all Alliance members.

However, even with a more forward-leaning Russia (the Munich speech, the Georgia war), there were other pressing challenges elsewhere. By 2009, NATO had initiated *Operation Allied Protector* off the Horn of Africa at the request of the United Nations, in order to counter piracy and armed robbery at sea.[94] MPA, including Norwegian units, were extensively utilized in order to build a pattern-of-life both ashore and at sea.[95] Later in 2009, however, NATO declared that "developments in the High North have generated increased international attention."[96] Concurrently, sporadic reports emerged of Russian submarines patrolling the East coast of the United States.[97] Starting in 2010, the need for a renewed focus on ASW began gaining traction within the Alliance.[98] This, however, was never reflected through changes in long-term investment among the Alliance members.[99]

In 2010, NATO also revised its strategic concept.[100] The concept built further on force flexibility and international stabilization efforts but incorporated more aspects of defending Alliance territory than the previous one.[101] Important to Norwegian security politics, the strategic concepts incorporated several elements from the Norwegian initiative from 2008. Among these were an increased focus on situational awareness and early warning, a closer link between national headquarters and the NATO Command Structure, improved readiness times for NATO forces, and the execution of more regular and relevant exercises.[102] All of these elements were emphasized to an increasing degree in the Alliance in the coming decade. The Norwegian initiative and active engagement with other Alliance members on such issues were not the only factors that played into the restructuring of NATO in the 2010s, but they certainly seem to have played a significant role. In 2011, the *Alliance Maritime Strategy* was issued, with a combined emphasis on expeditionary warfare at distance, and security operations closer to home.[103] Further changes to the NATO Command Structure were initiated in 2010–2012.[104] The two maritime commands in Naples, Italy, and Northwood, UK, were consolidated into a single maritime command at Northwood, namely the Allied Maritime Command (MARCOM). Up until then, Naples had been in charge of the Mediterranean and surrounding areas to the south and east, such as the Indian Ocean, and counter-piracy operations. Northwood were in charge of the English Channel, the Baltic Sea, and the Atlantic, to a larger extent focussing on exercises. Now, Northwood was in charge of all things maritime, with only one office in charge of coordinating maritime air affairs, that of *Commander Maritime Air NATO* (COMMARAIRNATO). Building on the MACA construct from the Mediterranean, COMMARAIRNATO had to adjust to a smaller command structure in addition to an increased geographical area of responsibility (AOR). COMMARAIRNATO's focus areas stretched from north of Norway, to the Baltic Sea, the Mediterranean, and all the way down to the Indian Ocean. Described in detail in MARCOM's classified document *Alliance Maritime Governance* from 2016, COMMARAIRNATO sought to exercise his C2 through a distributed network of national MACAs and other national command and control constructs, named the "*Maritime Air Network.*"[105] The NATO command structure did not have any dedicated ASW forces nor a mandate to coordinate them in peacetime outside of exercises. ASW in peacetime had always been coordinated by the respective *national* headquarters, although interaction and coordination between these headquarters certainly continued.

The PNI from 1991, together with the *START I Treaty*, had prevented Russia and the United States from placing tactical nuclear SLCM on attack submarines.[106] *START I* expired in December 2009, and in early 2011, the two nations ratified the *New START Treaty*, in which both countries committed to further reduction in deployed strategic nuclear weapons, as well as an improved inspection regime.[107] The latter was also specifically emphasized in NATO's strategic concept the previous year.[108] NATO also pointed to the disparity between the Russian and Western non-strategic nuclear weapon stockpile and their deployment close to NATO countries' borders. More importantly for the discussion in this book, however, the *New START Treaty* of 2010 did not put any limitations on placing SLCMs on attack submarines. The treaty solely focusses on capping the deployed *strategic* warheads allowed on long-range delivery vehicles.[109] As a result, the Russian SLCM capability has now risen to be one of the main naval threats for US and European targets.

After 9/11, NATO's engagement in Afghanistan drove force requirements, budgets, and consumed the general attention of the Alliance.[110] A wide array of troops and weapons systems were deployed to the Middle East, and MPA were no exception. The counter-insurgency operations were intelligence-demanding and intelligence-driven, and the requirement for airborne intelligence assets was more or less insatiable throughout. This high demand for ISR services was the root cause of the explosion in utilization of unmanned ISR platforms in the 2000s.[111] As the MPA were increasingly being deployed for over-land missions, they were tasked to operate in a fundamentally different surface-to-air threat environment. This demanded a new approach to missile defence and self-protection. The new threat environment led to the installation of improved self-protection systems.[112] The migration to over-land missions in the 1990s and the 2000s seems to have been facilitated by a symbiosis between operational demands and the MPA community adapting to a new threat environment. The aircraft was increasingly being tasked to fly over-land missions because they already carried the sensors that could support those missions. The dangerous element of such flexibility towards the mission portfolio is that knowledge and competence for niche missions and capabilities are perishable and will atrophy. This, critics claim, is what has happened to the NATO ASW capability in general, and to the NATO airborne long-range ASW capability, specifically.[113]

In spite of a decreasing strategic focus on adversary submarines at the turn of the century, the United States ONR initiated work on developing *multi-static active coherent* (MAC) technology for airborne assets in the early 2000s. In order to meet the challenge of increasingly quiet adversary submarines, the so-called *multi-static* systems have been developed that enable the search for a submerged target under difficult acoustic conditions. Monostatic acoustic systems consist of one active buoy that sends out a sound and searches for a return. Bistatic systems consist of two platforms, for example, two ships or two separate buoys, where one platform is the active one and the other is passive.[114]

Multi-static systems have several passive platforms or buoys and often more than one active or explosive buoy. The system will note the timing of the transmitted sounds, and then record the returns on all the passive sensors. This will give the computer the opportunity to triangulate the position of the enemy submarine. These are capable systems, but they have their limitations. The multi-static systems are active and reveal the presence of the ASW force. Multi-static systems are also

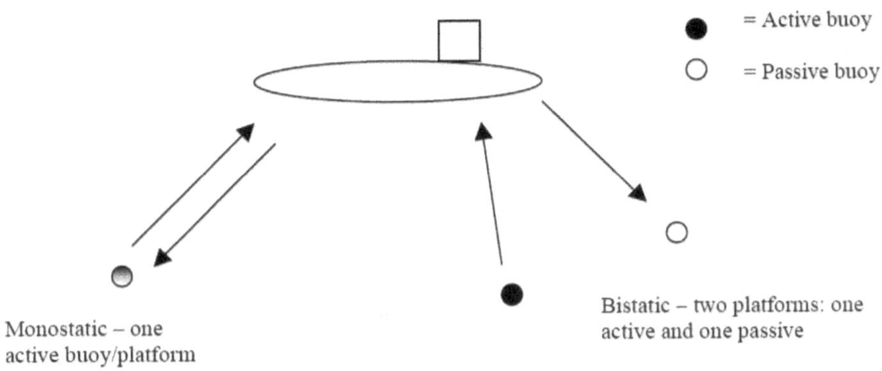

Figure 8.1 Monostatic and bistatic systems[115]

not as efficient in littoral waters, due to false and confusing returns from elements other than submarines. Also, fresh water, differences in temperature, turbulence, and currents in the littorals will create layers in the water, which complicate all types of acoustic searches, no matter how efficient they are.[116] But with further development of both buoys and processing systems, MAC systems are expected to have a significant impact on modern ASW in the near future.

The technology will be a welcome addition to any ASW asset whenever the use of active buoys is required. MAC technology has been further refined and developed, and will be installed on the P-8 *Poseidon*.[118] The US Navy's decision to procure the P-8 *Poseidon* MPA is a significant leap forward in terms of sensor capacities, such as the radar, the EO/IR, and the ESM system.[119] The acoustic suite brings a significant improvement to the processing capacity, and the tactical computer provides a full tactical picture of the battlespace the aircraft is operating in for every crew station. This is crucial for, among other things, the MAC concept just discussed. With regard to communications, the P-8 is set up with SATCOM, UHF, VHF, HF, and Link-16.[120] The Link-16 provides a near-instantaneous sharing of data with other elements for the forces that are connected to the network.

In the mid-2000s, the US Navy underlined the requirement for airborne wide-area, ocean surveillance, in order to monitor ship traffic and military naval movements worldwide. The concept was termed *broad area maritime surveillance* (BAMS) and set the requirement for a capability that could provide persistent coverage in any weather, and be able to investigate targets in a manner that other strategic assets, such as satellites, could not. As a supplement to MPA, unmanned technology was pursued. In 2008, the US Navy moved ahead with a derivative of the *Global Hawk* drone. The aircraft was modified to carry a maritime radar with 360° coverage, electro-optic camera providing full-motion video, and an *automatic identification system* (AIS) receiver. It was constructed for long-range surface surveillance, but was not capable of conducting underwater search. Additionally, the drone had its wings and fuselage strengthened in order to enable descent to lower altitudes, which would entail both more turbulence and icing conditions compared to the high-altitude operations of the *Global Hawk*.[121] The drone was given the

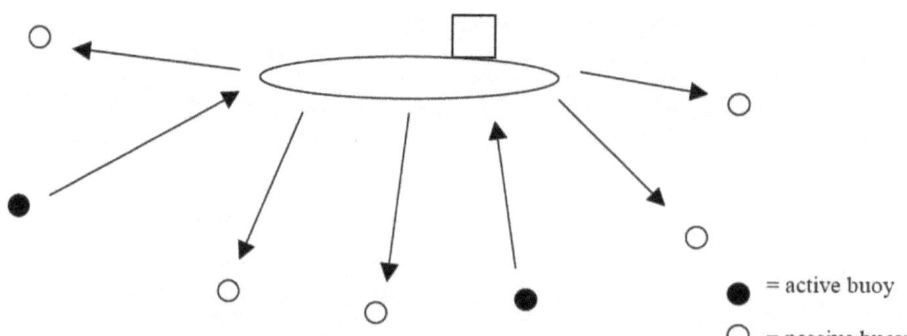

Figure 8.2 Multi-static systems[117]

interim name *BAMS-Demonstrator* (BAMS-D). One BAMS-D aircraft deployed to the Arabian Gulf in support of naval operations there, and proved its relevance to naval commanders through updating emerging threats and targets over the horizon.[122] Satellites and organic assets to the naval force were unable to provide similar services with corresponding speeds and update intervals.

As a continuation of the New Space movement of the 1990s, space-based maritime surveillance also continued to grow in the 2000s. Although not able to penetrate below water, space-based sensors provide a crucial element of the total picture when compiling maritime situational awareness. The NOSS constitutes the US Navy's main space-based ELINT capability and provides an emission-based, global maritime surface picture.[123] The DSP from the 1970s was replaced with the *space-based IR system* (SBIRS), which is still operating in support of air and missile defence.[124] Other systems for SIGINT have been derived from the satellites launched in the 1970s, and as naval vessels for the most part are underway, the most prominent surveillance tool from space is based on SIGINT.[125] The most obvious downgrading factor of EO systems is that of cloud coverage disturbing the line-of-sight between the satellite and the target. Radar imaging systems penetrate clouds and do not face this problem. A prominent radar imaging satellite system launched by the United States in 2010 is the *TOPAZ*, which will likely be replaced by a new generation of surveillance satellites in the near future.[126] The targets of surveillance, however, have, for the most part, been static infrastructure and sites of activity on land and close to the shore. Ports, shipyards, and docks are obviously important elements of maritime surveillance, but they are not the primary subject of surveillance when fleets or submarines are at sea. The more adaptive and flexible asset for surveillance and reconnaissance of submarines is still, however, the MPA in its primary role.

The 2010s – annexation of Crimea and renewed tensions

In March 2014, Russia illegally annexed the Crimean peninsula of Ukraine, which changed NATO-Russia relations for the foreseeable future.[127] The Russians revised their national strategy in December 2014. It is defensive in nature; NATO is still portrayed as the most pressing threat to Russia's existence, and calculated ambiguity is a prominent feature of the document.[128] The strategy did, however, take into account new fiscal constraints that arose both from external sanctions due to the situation in Ukraine and from internal economic turmoil and slow growth.[129] Even before the sanctions were imposed in response to the Crimean annexation, the decline in oil prices had complicated the Russian military's ambitious modernization plans.[130] And even in times of economic growth under Putin, the significant funds allocated to the Navy were simply not enough to recover from the massive naval decline of the 1990s. There remained significant backlogs of maintenance, training, and dismantlement tasks.[131] In 2015, a new Russian maritime doctrine was released, to be in effect through 2020. The document focussed on a variety of issues, of which naval activities were but one fraction of the national portfolio of tasks and interests, such as exploitation of natural resources, climatological issues, and fisheries science. The national maritime doctrine in fact tells us little about the role of the Navy.[132] In order to gain an understanding of that role, particularly in relation to the Northern Fleet, one instead needs to keep track of Russian exercises and weapons testing in the High North. This is where the requirement for a renewed MPA fleet comes into play.

After the Russian annexation of Crimea, the Alliance initiated the *NATO Readiness Action Plan*, which included increased readiness of the NATO Response Force and the initiation of a range of *Immediate Assurance Measures*. These consisted of an increased presence and forward posture, including "maritime patrol aircraft flights along our eastern borders."[133] Through the Alliance Maritime Strategy, the importance of the maritime domain was referenced as growing in relevance both geopolitically and economically.[134]

In 2016, the NATO Alliance reiterated its dismay with Russian aggression and posture, with particular emphasis on Ukraine and Crimea.[135] NATO enhanced its posture and presence along the Eastern periphery and cancelled all civilian and military cooperation with Russia, whilst remaining open for dialogue. The maritime domain was given enhanced attention, and NATO recognized both the Baltic and the Black Sea as challenging regions requiring close cooperation and support, and increased situational awareness. The North Atlantic was described in terms recognizable from the Cold War years:

> The Alliance will be ready to deter and defend against any potential threats, including against sea lines of communication and maritime approaches of NATO territory. In this context, we will further strengthen our maritime posture and comprehensive situational awareness.[136]

But in introducing the term "strategic anticipation" (as opposed to "indications and warnings"), the importance of overall enhanced situational awareness is underlined, "particularly in the east, the south, and in the North Atlantic. Our ability to understand, track, and ultimately, anticipate, the actions of potential adversaries through Intelligence, Surveillance and Reconnaissance (ISR) capabilities and comprehensive intelligence arrangements is increasingly important."[137]

A set of questions from the Foreign Affairs and Defence Committee forwarded to the Norwegian MoD in 2016 included two on the military guidelines on foreign military activity in Norway. The MoD emphasized in its response that the restrictions have played a fundamental part in Norwegian security policy since the 1950s, manifesting a key tool in the balancing act between deterrence and assurance towards Russia. The explicit response was:

> All Allied flying missions over Finnmark with foreign military aircraft further East than 24 degrees East, with the exception of transport and passenger aircraft, shall be approved by the Ministry of Defence. There is an absolute limitation in place for foreign military aircraft, with the exception of transport and passenger aircraft, to fly east of 28 degrees East.[138]

This was a reaffirmation of past restrictions. In its effort to stimulate Allied activities in the North Atlantic, it will be interesting to follow the Norwegian Government's approach to the 24° restriction in the future. Possibly, a more important question is, however, how the Norwegian government will handle foreign assets operating in the Barents Sea after launching from bases outside of Norwegian territory. Unmanned, long-endurance aircraft, for example, can easily launch from abroad and operate in the Barents Sea for extended periods of time.[139]

The Commander of MARCOM pointed in 2017 to increased Russian activities; he emphasized that NATO was attempting to reverse the general atrophy in ASW competency, among other things by prioritizing ASW exercises.[140] By the following year, it was recognized at the strategic level of NATO that there was a requirement for an increased focus on ASW, a notion gaining traction with many nations, not least the United States, the United Kingdom, and Norway.[141] The "collective maritime warfighting skills in key areas" were to be reinvigorated, with emphasis on "anti-submarine warfare, amphibious operations, and protection of sea lines of communications."[142] In the summer of 2018, NATO announced several changes to its command structure. The most noteworthy of which in the context of the North Atlantic was the establishment of a Joint Force Command in Norfolk, United States, to "focus on protecting the transatlantic lines of communication."[143] The requirement for a refocussing on ASW included an acknowledgement that ASW is a "combined arms" endeavour, requiring multiple forces working closely in an integrated manner. The MPA cannot solve the problem alone, neither can the ASW frigate, nor the lone submarine. It will thus not be sufficient to procure one or two new ASW assets – a systemic approach is required to handle modern ASW challenges.[144]

In the United Kingdom, the MR4 development programme was costly and fell significantly behind schedule. In addition, a *Nimrod* MR2 deployed to Afghanistan crashed in 2009, claiming the lives of 11 crewmembers. The ensuing accident investigation pointed to significant systemic faults in many areas from procurement to aircraft maintenance. Combined with the significant cost-overruns of the MR4 programme, the review seems to have played a prominent role in the cancelling of the *Nimrod* programme in the *Strategic Defense and Security Review* (SDSR) of 2010. The primary reason for the MR4 cancellation was economic. However, the MoD officially regretted the decision. The very fast decision to terminate the *Nimrod* on economic grounds was quickly undermined by the clear strategic requirement for maritime surveillance a long-range ASW capability. The MoD therefore established the "Seedcorn Initiative" in 2011, which consisted of Royal Air Force personnel being sent to allied air forces to operate maritime surveillance aircraft in order to maintain and develop surveillance and ASW skills. It would also shorten the time required to regenerate an airborne ASW organization in spite of several years of lacking this capability.[145] In what is likely the most humiliating occurrence in the history of British maritime air operations, the United Kingdom had to call on France and Canada for support in 2015 in the search for a Russian submarine operating off the Scottish coast, close to Faslane, the home port of the British strategic nuclear submarines.[146] The new SDSR in 2015 assessed that Great Britain could no longer live without an airborne maritime ISR capability. The decision was presented to procure nine P-8 *Poseidon*, with the goal of flying them operationally by 2019.[147]

In the 2010s, uncertainties arose in Norway as to whether the *Orions* should continue to go through costly upgrades, or be replaced by a new aircraft such as the P-8 *Poseidon*. The comprehensive review of the force structure in 2014 and 2015 culminated in the CHOD White Paper in 2015. In it, the Norwegian CHOD recommended the discontinuation of the P-3 *Orion* fleet after its operational lifetime, and proposed its replacement by medium-altitude/long-endurance (MALE) unmanned aircraft.[148] By design, the CHOD white paper was a recommendation to the Norwegian MoD. However, because of the processes of the long-term plan development at the strategic level in the Norwegian Defence Staff, political considerations concerning security policy were not adequately incorporated in the CHOD's formal advice to the minister of defence. The main emphasis was rather put on long-term economic sustainability. Thus, when the proposal to phase out the P-3 *Orions* was made in 2015, it was made without adequate security policy considerations.[149]

One of the Norwegian Armed Forces' primary tasks is to conduct intelligence operations, conduct resource and traffic management in areas under Norwegian jurisdiction, and claim sovereignty by showing presence in areas of national interest.[150] The fact that it might take some time before Allied support can arrive in Norway only emphasizes the importance of early warning, or strategic anticipation.[151] When the Norwegian Government discusses national contributions to the Alliance, the surveillance effort in the High North is regularly brought forward as one of the most important NATO contributions.[152] In addition, the notion that intelligence represents political capital has increasingly gained traction also in the civilian community in Norway.[153] With the government's explicit ambition to prioritize the High North and the nation's maritime areas and resources, political traction was gained to maintain the P-3s and to consider a replacement, as opposed to the military advice from the CHOD to replace them with drones. Airborne drones are not capable of conducting underwater search for submarines, although they are expected to achieve this in the near future. This means that replacing the Norwegian MPA with drones would remove the national long-range airborne ASW capability, a capability recognized as being of strategic importance to Norway. The ensuing long-term planning (LTP) process at the Norwegian MoD in the spring of 2016 thus overturned the P-3 proposal: Norway would instead replace the P-3 *Orions* with a new fleet of P-8 *Poseidons*.[154] The official explanation underlined that the maritime domain and the High North have always been fundamental to Norwegian security politics.[155] Norwegian and adjacent maritime areas are seven times the size of the Norwegian land mass, and the distances are vast. "These areas have regained their military-strategic importance, and there are no indications on this changing in the near future. The new security policy situation requires improved situational awareness in our areas of interest."[156] The Norwegian Government also pointed to the international dimension in their justification for procuring the P-8. As NATO is increasingly focussing on the maritime domain, MPA have become a sought-after resource. As a nation with significant maritime interests, it is of fundamental importance that Norway contributes to "ensuring freedom of navigation on the high seas, particularly for the transatlantic sea lines of communication, which are fundamental to allied safety and security."[157] Finally, it is emphasized that the "combination of a new security policy realm and new generations of submarines entail that modern sensors and systems for surveillance are paramount."[158] The Norwegian MoD deemed the deliveries from the MPA fleet too important to cut. It was thus a political decision to override the advice of the CHOD.[159] The LTP for the Norwegian Armed Forces was published in the spring of 2016, and additionally included the decision to shut down the Andøya airbase and move the MPA fleet to the Evenes airbase.[160]

Norway depends on her MPA for airborne monitoring of Russian testing and development activities. However, according to reports in the media, the Norwegian MPA capability is somewhat limping into the future as of 2020. Sensor upgrades and airframe improvements have been discontinued due to the new incoming aircraft, creating challenges in retaining experienced personnel.[161] The latter seems to be partly due to the decision in the current LTP to close down the Andøya airbase and establish the new P-8 capability at the Evenes airbase.[162] These challenges have been acknowledged by the MoD, and the focus is to get the P-8 operational as soon as possible after 2023.[163]

After a few years of operational experimentation, testing, and evaluation in the US Navy, the large unmanned aircraft *BAMS-D* was given the name *Triton*, and the technical designation MQ-4. The US Navy decided that the *Triton* would complement the *Poseidon* as a part of the overarching BAMS approach mentioned earlier. From focussing on expeditionary warfare on land in the 2000s, the US Navy was, in the 2010s, returning to a holistic approach to maritime airborne surveillance. In addition to manned and unmanned systems, space-based surveillance has increasingly become an integrated part of maritime surveillance. Through New Space, there are now several capable commercial surveillance services available, such as *Airbus*, *Planet*, or *Digital Globe*. These provide imagery with resolution down to a few centimetres, with obvious intelligence value.[164] They are, for the most, however, focussed over land and near shore, and have limited value in the search for submarines. The New Space environment facilitated Norway in launching her own satellite for maritime surveillance in 2012, a micro-satellite that carries an AIS receiver.[165] New Space, as we shall see, has also facilitated the broadband communications satellite that Norway launched and put into orbit in the fall of 2024.

In the years following the Cold War, Russia initially attempted to establish herself in a new geopolitical setting, before pursuing renewed ambitions both militarily and geopolitically, massing the economic means to support those ambitions. After this, Russia initiated several military campaigns to underscore the Russian comeback on the geopolitical scene. NATO, meanwhile, struggled to find a common area of focus, and the fluctuating strategic focus in the 1990s did not settle until the campaigns in the Middle East began after 9/11 in 2001. Those land-centric campaigns were executed to the detriment of the maritime theatre of operations.

MPA were increasingly used to support the over-land battle in the late 1990s, and even more so in the 2000s. In the cases where MPA were utilized in the maritime domain, it was to conduct surface surveillance, mostly of civilian movements. This led to a general atrophy in the perishable ASW competency within NATO and marginalized ASW in the MPA mission portfolio. The significant increase in the use of unmanned aircraft and satellite services has fed into the insatiable requirement for ISR services in counter-insurgency warfare.

This seems to have masked some of the new requirements for modern airborne ASW. New technology has been operationalized that might challenge the assumption of MPA being the sole answer to handling the new submarines in the north. Satellites and unmanned aircraft seem to have an important role to play in the future. For the time being, however, these new services remain inadequate in the search for submerged submarines.

However, the modern Russian submarine threat poses challenges to the traditional MPA approach. The adversary submarines are now much quieter, they move unpredictably, and they require intra-theatre connectivity and understanding on the part of the hunter. The question remains whether the MPA is the sole answer to the modern Russian submarine threat.

Notes

1 Office of Naval Intelligence, "The Russian Navy – A Historic Transition," Office of Naval Intelligence (ONI), December 2015, vi, www.oni.navy.mil/Portals/12/Intel%20agencies/russia/Russia%202015print.pdf?ver=2015-12-14-082038-923; Robert S. Norris og William N. Arkin, "Russian (C.I.S.) Strategic Nuclear Forces End of 1995," *Bulletin of the Atomic Scientists* 52, no. 2 (April 1996): 63.

2 Richard Staar, "Russia's Navy Remains in Decline," *United States Naval Institute Proceedings* 124, no. 8 (August 1998); Kristian Atland, "The Introduction, Adaptation and Implementation of Russia's 'Northern Strategic Bastion' Concept, 1992–1999," *Journal of Slavic Military Studies* 20, no. 4 (2007): 501.

3 Daryl Kimball, "The Presidential Nuclear Initiatives (PNIs) on Tactical Nuclear Weapons at a Glance," Arms Control Association, July 2017, 1, www.armscontrol.org/printpdf/111.

4 Tamnes, *Oljealder, 1965–1995*, 6:332–35.

5 Børresen, Gjeseth, og Tamnes, *Norsk forsvarshistorie – Allianseforsvar i endring, 1970–2000*, 119.

6 Børresen, Gjeseth, og Tamnes, 126–27.

7 Atland, "The Introduction, Adaptation and Implementation of Russia's 'Northern Strategic Bastion' Concept, 1992–1999," 505–7.

8 Atland, 508.

9 First Deputy Minister Andrei A. Kokoshin, as quoted in Atland, 509.

10 Atland, 513.

11 Office of Naval Intelligence, "The Russian Navy – A Historic Transition," 18.

12 Nuclear Threat Initiative (NTI), "Russia Submarine Capabilities," NTI, June 10, 2014, www.nti.org/analysis/articles/russia-submarine-capabilities/.

13 Victor Mizin og Michael Jasinski, "The Future of the Russian Sea-Based Deterrent," *The Journal of Slavic Military Studies* 16, no. 1 (2003); Nuclear Threat Initiative (NTI), "Russia Submarine Capabilities"; Norman Friedman, "World Naval Developments: The Typhoon Saga Ends," *United States Naval Institute Proceedings* 125, no.

2 (February 1999); Robert S. Norris og Hans M. Kristensen, "Russian Nuclear Forces, 2006," *Bulletin of the Atomic Scientists* 62, no. 2 (April 2006). The Russian Navy kept one *Typhoon* class submarine as a test platform for new strategic missiles. The first such test was performed through the firing of an SS-N-32 *Bulava* SLBM in September 2005 from the modified *Typhoon*

14 Robert S. Norris, William N. Arkin, Hans M. Kristensen, og Joshua Handler, "Russian Nuclear Forces, 2001," *Bulletin of the Atomic Scientists* 57, no. 3 (June 2001): 78.

15 Norris og Arkin, "Russian (C.I.S.) Strategic Nuclear Forces End of 1995," 63; Staar, "Russia's Navy Remains in Decline."

16 NATO, "The Alliance's New Strategic Concept," NATO, November 8, 1991, NATO, Online Archive.

17 Birkeland, "The Potential of LIDAR as an Anti-Submarine Warfare Sensor," 54–56; Perkins, *Alliance Airborne Anti-Submarine Warfare*, 45.

18 NATO, "The Alliance's New Strategic Concept."

19 Hattendorf, "U.S. Naval Strategy in the 1990s – Selected Documents," 9.

20 Hattendorf, 17–20.

21 NATO, "Declaration of the Heads of State and Government – Brussels Summit," NATO, January 11, 1994, NATO, Online Archive; NATO, "Final Communique – Berlin," NATO, June 3, 1996, NATO, Online Archive.

22 See for example NATO, *NATO Handbook 1995* (Brussels: NATO Office of Information and Press, 1995), 164.

23 Porter, *British Maritime Patrol Aircraft Operations During and After the Cold War*; Håvoll, "Norwegian Maritime Patrol Aircraft Operations During and After the Cold War."

24 Innst. S no. 151 (1995–1996), "Recommendation to Parliament by the Defense Committee – Confirmation of Norway's Self-Imposed Restrictions in Security Policy," Stortinget, March 13, 1996, Stortinget, Online Archive.

25 Norwegian Minister of Defence Jørgen Kosmo, "Long Term Challenges for the Armed Forces," Speech, Oslo Militære Samfund, Oslo, January 8, 1996, Stortinget, Online Archive.

26 Norwegian Minister of Defence Jørgen Kosmo, "Letter from the Minister of Defence to the Defense Committee," Official Correspondence, February 26, 1996, Stortinget, Online Archive.

27 The PfP program was initiated in 1994, and is a framework for cooperation, coordination and for creating trust between NATO and former members of the Soviet Union.

28 Svein Efjestad, "Norwegian Self-Imposed Restrictions in the High North," Interview, September 2, 2019, Norwegian Ministry of Defense.

29 Norwegian Minister of Defence Jørgen Kosmo, "Letter from the Minister of Defence to the Defense Committee."

30 Perkins, *Alliance Airborne Anti-Submarine Warfare*, 31.

31 David Reade, "NAS Keflavik, Iceland – The ASW Training Capital of the World," *Maritime Patrol Aviation* 3, no. 2 (March 1995): 36.

32 Reade, 41.

33 "Command History Report (CHR) for VP-16 (1995)," Encl. (3).

34 Roberts, *Dictionary of American Naval Aviation Squadrons 2*, 2:48, 256–57.

35 Håvoll, "Norwegian Maritime Patrol Aircraft Operations During and After the Cold War."

36 Gary Ambrose, "Transforming Maritime Patrol Aviation" (Master's thesis, USMC Command & Staff College, 2003), 6.

37 Ambrose, 7–8.

38 Geoff Fein and Grace Jean, "Changing the Game: U Navy Applies New Approaches to Submainres Threats," *IHS Jane's Navy Internatioinal*, 2014, www.janes.com/images/assets/836/46836/US_Navy_applies_new_approaches_to_submarine_threats.pdf.

39 GE Aviation, "Boeing Team Wins $3 Billion Multi-Mission Maritime Aircraft Program," *GE Aviation Online*, June 14, 2004, www.geaviation.com/press-release/jv-archive/boeing-team-wins-3-billion-multi-mission-maritime-aircraft-program.

40 Flight International, "RMPA Contenders Step Up Campaign," *Flight International* 1 (May 7, 1996).
41 Ministry of Defence, *Major Projects Report 2005 – Project Summary Sheets* (London: National Audit Office, November 25, 2005), 67.
42 Marco Borst, "Under Arctic Conditions in Europe: Norwegian Orions," *AIR International*, January 1998, 58.
43 Borst, 58.
44 This author worked as the primary staff officer for maritime patrol aircraft at the Air Operations Inspectorate of the Royal Norwegian Air Force between 2009 and 2011, and took part in those initial discussions.
45 For an introduction to the thinking and core ideas behind the air campaign in *Operation Desert Storm* in 1991, in particular the concept of "parallel attack", see John A. III Warden, "The Enemy as a System," *Airpower Journal* 9, no. 1 (1995).
46 In brief, *organic assets* are those belonging to the commander that tasks them. For *Pioneer* operations, see Thomas, *Global Hawk*, 51–52.
47 NATO, "First NATO AGS Remotely Piloted Aircraft Ferries to Main Operating Base in Italy," NATO, November 21, 2019, www.nato.int/cps/en/natohq/news_171171.htm. The momentum for unmanned aircraft led to the development of other capable aircraft as well, such as the *Predator*, *Reaper, Heron, Hermes 900,* to name a few. These will not be discussed in this thesis.
48 For articles on "New Space", see for example Jeff Foust, "Current Issues in New Space," *The Space Review*, March 5, 2007, www.thespacereview.com/article/823/1; Stephen Cass, "We May Be Heading for a Space Bubble," *MIT Technology Review Online*, September 14, 2010, www.technologyreview.com/s/420776/we-may-be-heading-for-a-space-bubble; Jeff Foust, "The Evolving Ecosystem of New Space," *The Space Review*, August 15, 2011, www.thespacereview.com/article/1906/1; Sheraz Sadiq, "Silicon Valley Goes to Space," *kqed.org*, November 18, 2013, www.kqed.org/science/11135/silicon-valley-goes-to-space-2; Eric Berger, "Blue Origin Just Validated the New Space Movement," *arstechnica.com*, June 10, 2016, https://arstechnica.com/science/2016/10/blue-origin-just-validated-the-new-space-movement/.
49 "Command History Report (CHR) for VP-5 (1992)," September 1992.
50 Håvoll, "Norwegian Maritime Patrol Aircraft Operations During and After the Cold War"; "Command History Report (CHR) for VP-49 (1993)"; "Command History Report (CHR) for VP-5 (1993)."
51 "Command History Report (CHR) for VP-5 (1992)," Encl. (4).
52 Porter, *British Maritime Patrol Aircraft Operations During and After the Cold War*.
53 Borgeld, "Keflavik – from the Past to the Present," 41.
54 "Command History Report (CHR) for VP-5 (1992)," Encl. (4); Charles Haddon-Cave, *The Nimrod Review – An Independent Review into the Broader Issues Surrounding the Loss of the RAF Nimrod MR2 Aircraft XV230 in Afghanistan in 2006*, Report to the House of Commons (London: House of Commons, October 2009), 18; "Command History Report (CHR) for VP-5 (1993)," Encl. (1).
55 Sherman, "Feet Dry," 39; Naval Air Museum Barbers Point, "Lockheed P-3C Orion – BuNo 160770"; Perkins, *Alliance Airborne Anti-Submarine Warfare*, 45.
56 Roberts, *Dictionary of American Naval Aviation Squadrons*, 2:88, 146, 355, 357, 360.
57 Rolf-Inge V. Andresen og Tor Bukkvoll, *Russian Weapons Development Towards 2020*, Forsvarets Forskningsinstitutt (FFI) FFI-rapport 2008/01957, January 12, 2009, 4, 9–10.
58 International Institute for Strategic Studies, "Reviving Russia's Navy," *Strategic Comments* 6, no. 6 (July 2000).
59 International Institute for Strategic Studies.
60 Mizin og Jasinski, "The Future of the Russian Sea-Based Deterrent," 78–79.
61 Vladimir Putin, "Speech and the Following Discussion at the Munich Conference on Security Policy," *Kremlin*, February 19, 2007, http://en.kremlin.ru/events/president/transcripts/24034.

62 Robert S. Norris og Hans M. Kristensen, "Russian Nuclear Forces, 2008," *Bulletin of the Atomic Scientists* 64, no. 2 (June 2008): 54.
63 Guy Faulconbridge, "Russian Navy to Start Sorties in Mediterranean," *Reuters*, December 5, 2007, www.reuters.com/article/us-russia-navy/russian-navy-to-start-sorties-in-atlantic-tass-idUSL0518563620071205.
64 Nuclear Threat Initiative (NTI), "Russia Submarine Capabilities."
65 Nuclear Threat Initiative (NTI).
66 Gustav Gressel, *Russia's Quiet Military Revolution, and What It Means for Europe*, Policy Brief (European Council on Foreign Relations, October 2015), 2; Robert S. Norris og Hans M. Kristensen, "Russian Nuclear Forces, 2009," *Bulletin of the Atomic Scientists* 65, no. 3 (June 2009): 55. A reaction to Western posturing materialized a little more than 20 years earlier as well, through the "Analogous Response" that we saw earlier in the mid-1980s with the deployment of the *Akulas* carrying nuclear capable SS-N-21 SLCMs.
67 Svein Efjestad, "Norway and the North Atlantic: Defence of the Northern Flank," in *NATO and the North Atlantic – Revitalizing Collective Defence*, ed. John A. Olsen, Whitehall Paper 87 (London: Royal United Services Institute (RUSI), 2017), 62–64.
68 John A. Olsen, red., *NATO and the North Atlantic – Revitalizing Collective Defence*, Whitehall Paper 87 (London: Royal United Services Institute (RUSI), 2017), 68.
69 Robert S. Norris og Hans M. Kristensen, "Russian Nuclear Forces, 2010," *Bulletin of the Atomic Scientists* 66, no. 1 (February 2010): 77.
70 Norwegian Intelligence Service, "FOKUS 2011 – The Assessment of the Intelligence Service," Annual Unclassified Assessment, Oslo, 2011, 7, https://forsvaret.no/fakta_/ForsvaretDocuments/FOKUS-2011.pdf.
71 Andrei Martyanov, "Russia's Navy in Search of a Mission," *United States Naval Institute Proceedings* 140, no. 12 (December 2014).
72 Katarzyna Zysk, "Mot et moderne russisk forsvar? Utviklingstrender i militær modernisering og strategisk tenkning," *Norsk Militært Tidsskrift* 183, no. 3 (2013): 10.
73 Hans M. Kristensen og Robert S. Norris, "Russian Nuclear Forces, 2011," *Bulletin of the Atomic Scientists* 67, no. 3 (2011): 69; Zysk, "Mot et moderne russisk forsvar?" 10–12.
74 RT, "Finaly Flying Colors: Yury Dolgoruky Nuclear Sub Joins Russian Navy," *RT.com*, January 10, 2013, www.rt.com/news/yury-dolgoruky-submarine-ceremony-678/; Steve Gutterman, "New Russian Nuclear Submarine Goes into Service," *Reuters*, January 10, 2013, www.reuters.com/article/us-russia-submarine/new-russian-nuclear-submarine-goes-into-service-idUSBRE9090YR20130110; Hans M. Kristensen og Robert S. Norris, "Russian Nuclear Forces, 2013," *Bulletin of the Atomic Scientists* 69, no. 3 (2013): 75.
75 Office of Naval Intelligence, "The Russian Navy – A Historic Transition," 17.
76 Hans M. Kristensen and Robert S. Norris. "Russian Nuclear Forces, 2015," *Bulletin of the Atomic Scientists* 71, no. 3 (2015): 91.
77 Norman Polmar, "The Russian Navy – Beneath the Waves," *United States Naval Institute Proceedings* 138, no. 6 (June 2012); Korolkov, "Russia's Top-Secret Nuclear Submarine Comes into Service."
78 Tamnes, "The Significance of the North Atlantic and the Norwegian Contribution," 24–25.
79 Kristensen og Norris, "Russian Nuclear Forces, 2015," 95.
80 Norwegian Intelligence Service, "FOKUS 2013 – The Assessment of the Intelligence Service," Annual Unclassified Assessment, Oslo, 2013, 18.
81 In comparison, the US Navy SSBN patrol rate in the 2000s was steadily numbering more than 60 patrols a year. See Robert S. Norris og William N. Arkin, "Russian Nuclear Forces, 2000," *Bulletin of the Atomic Scientists* 56, no. 4 (August 2000): 70–71; Robert S. Norris og Hans M. Kristensen, "Russian Nuclear Forces, 2004," *Bulletin of the Atomic Scientists* 60, no. 4 (August 2004): 74.

82 Robert S. Norris, William N. Arkin, Hans M. Kristensen, og Joshua Handler, "Russian Nuclear Forces, 2002," *Bulletin of the Atomic Scientists* 58, no. 4 (August 2002): 73.
83 Norris et al., 73; Robert S. Norris og Hans M. Kristensen, "Russian Nuclear Forces, 2007," *Bulletin of the Atomic Scientists* 63, no. 2 (April 2007): 64.
84 Jeffrey Lewis, "Led Zeppelin Comes to Washington," *Foreign Policy*, January 5, 2015, https://foreignpolicy.com/2015/01/05/led-zeppelin-comes-to-washington-russia-nukes-putin-arms-control/; Hans M. Kristensen og Robert S. Norris, "Russian Nuclear Forces, 2014," *Bulletin of the Atomic Scientists* 70, no. 2 (2014): 80; Norris og Kristensen, "Russian Nuclear Forces, 2007," 64.
85 Norris og Kristensen, "Russian Nuclear Forces, 2008," 57.
86 Øyvind Bergstrøm, "NATOs maritime strategi etter 1990 – Fra sjømakt til ordensmakt?" (Master's thesis, Norwegian Armed Forces Staff College, 2009), 24.
87 Bergstrøm, 30.
88 Reference previous section of this chapter, and also Olsen, *NATO and the North Atlantic – Revitalizing Collective Defence*, 4.
89 NATO, "Prague Summit Declaration," NATO, November 21, 2002, NATO, Online Archive.
90 NATO, para 4.b.
91 In summary, Article 5 of the North Atlantic Treaty states that an attack on an Alliance member shall be considered an attack on the Alliance as a whole.
92 NATO, "Operation Active Endeavour," NATO, October 27, 2016, NATO, Online Archive, www.nato.int/cps/en/natolive/topics_7932.htm.
93 The Norwegian Ministry of Foreign Affairs, "St.meld.nr.38 (2008–2009) – Om samarbeidet i NATO i 2008," The Norwegian Government, 2008, 7.
94 NATO, "Counter-Piracy Operations," NATO, December 19, 2016, NATO, Online Archive, www.nato.int/cps/en/natolive/topics_48815.htm#Protector.
95 The operation later turned into *Operation Ocean Shield* in 2009 and ran until the end of 2016.
96 NATO, "Strasbourg/Kehl Summit Declaration," NATO, April 4, 2009, NATO, Online Archive.
97 Mark Mazzetti og Thom Shanker, "Russian Subs Patrolling Off East Coast of U.S," *New York Times*, August 4, 2009.
98 Perkins, *Alliance Airborne Anti-Submarine Warfare*, 31.
99 Peter Hudson og Peter Roberts, "The UK and the North Atlantic: A British Military Perspective," in *NATO and the North Atlantic – Revitalizing Collective Defence*, redigert av John A. Olsen, Whitehall Paper 87 (London: Royal United Services Institute (RUSI), u.å), 80–83.
100 NATO, "Active Engagement, Modern Defense – Strategic Concept for the Defense and Security for the Members of NATO," NATO, November 20, 2010, NATO, Online Archive.
101 NATO, "Lisbon Summit Declaration," NATO, November 20, 2010, NATO, Online Archive.
102 Efjestad, *Raising NATO's Profile – the Norwegian NATO Initiative*.
103 Smith-Windsor, "NATO's Maritime Strategy and the Libya Crisis as Seen from the Sea," 2–3.
104 NATO, "Lisbon Summit Declaration."
105 Rear Admiral Thomas Ernst, "Agile command and control in a degraded environment."
106 Nuclear Threat Initiative (NTI), "Russia Submarine Capabilities."
107 BBC, "US Senate votes to begin debate on new Start treaty"; nti.org, "U.S. Senate Ratifies New START in 71–26 Vote, Despite Top GOP Opposition"; Kristensen og Norris, "Russian Nuclear Forces, 2011," 67.
108 NATO, "Active Engagement, Modern Defense," 24.
109 Kristensen og Norris, "Russian Nuclear Forces, 2011," 67.

110 NATO, "Istanbul Summit Communique," NATO, June 28, 2004, NATO, Online Archive.
111 Birkeland og Dyndal, "Fremtidig autonom droneteknologi og -konsepter."
112 Sherman, "Feet Dry," 39.
113 Håvoll, "Norwegian Maritime Patrol Aircraft Operations During and After the Cold War"; Perkins, *Alliance Airborne Anti-Submarine Warfare*; Birkeland, "The Potential of LIDAR as an Anti-Submarine Warfare Sensor."
114 Birkeland, "The Potential of LIDAR as an Anti-Submarine Warfare Sensor," 24.
115 Birkeland, 24.
116 Birkeland, 25.
117 Birkeland, 24.
118 The Office of the Director, Operational Test and Evaluation, "FY2014 DOT&E Annual Report," Annual Report, 2014, 227–28.
119 Fein og Jean, "Changing the game: U Navy applies new approaches to submainres threats."
120 Fein og Jean.
121 Thomas, *Global Hawk*, 241–42.
122 Thomas, 232–38.
123 Forden, "Viewpoint: China and Space War," 148; Friedman, *Seapower and Space*, 177–79; Dwayne A. Day, "Above the Clouds: The White Cloud Ocean Surveillance Satellites," *The Space Review*, April 13, 2009, Online utgave, www.thespacereview.com/article/1351/1; Richelson, *The US Intelligence Community*, 230–31.
124 Kevin McLaughlin, "Would Space-Based Systems Defenses Improve Security?" *Washington Quarterly* 25, no. 3 (2002): 179–81.
125 Richelson, *The US Intelligence Community*, 232–33.
126 Berkowitz, *The National Reconnaissance Office at 50 years: A Brief History*, 26; Richelson, *The US Intelligence Community*, 189–94.
127 NATO, "Wales Summit Declaration," para 1, para 18.
128 Kristensen og Norris, "Russian Nuclear Forces, 2015," 86–87.
129 Oliker, "Unpacking Russia's New National Security Strategy."
130 Norwegian Intelligence Service, "FOKUS 2015 – The Assessment of the Intelligence Service," 14; Kristensen og Norris, "Russian Nuclear Forces, 2015," 86.
131 Nuclear Threat Initiative (NTI), "Russia Submarine Capabilities."
132 Kremlin, "Maritime Doctrine of the Russian Federation," 18.
133 NATO, "NATO's Readiness Action Plan (RAP) – Fact Sheet."
134 NATO, "Wales Summit Declaration," para 71.
135 NATO, "Warsaw Summit Communique," para 5.
136 NATO, para 23.
137 NATO, para 47.
138 Norwegian Ministry of Defense, "Response from the Ministry of Defense to the Foreign Affairs and Defense Committee on questions related to the Long Term Plan – Prop. 151 S (2015–2016)."
139 Keflavik is not permanently operational, but is sporadically used by the US Navy for P-8 operations, as well as for NATO flights such as Icelandic Air Policing (IAP).
140 Johnstone, "NATO's Maritime Moment: A Watershed Year in Alliance Sea Power."
141 Hanley, "Creating the 1980s maritime strategy and implications for today," 5–6.
142 NATO, "Brussels Summit Declaration," para 19.
143 NATO, para 29.
144 Cote, "The Third Battle – Innovation in the U.S.Navy's Silent War Struggle with Soviet Submarines," 83–84; Hanley, "Creating the 1980s maritime strategy and implications for today," 5–6.
145 UK Defense Committee, ""Future Maritime Surveillance," 41; Allison, "First British P-8A Poseidon will be 'ready to fly with a UK crew' on day one of arriving in the UK."

146 Gibbons-Neff, "With no sub-chasing aircraft of its own, UK calls on allies to help find Russian submarine"; Avionews, "The United Kingdom has no maritime patrol aircraft, asks NATO for help."
147 UK MoD, "National Security Strategy and Strategic Defense and Security Review (SDSR) 2015," 37; Defense Contracts Online, "P-8A programme progresses"; Allison, "First British P-8A Poseidon will be 'ready to fly with a UK crew' on day one of arriving in the UK."
148 Norwegian Chief of Defense (CHOD), "A Defense Undergoing Change – Fagmilitært Råd (FMR) – Defense White Paper," 85.
149 Efjestad, "Norwegian Self-Imposed Restrictions in the High North."
150 Barth Eide, "Forsvarets oppdrag i nordområdene ('The mission of the Armed Forces in the High North')."
151 NTB, "Det vil ta én til to måneder for NATO å hjelpe Norge."
152 Forsvarsdepartementet, "NATO-toppmøtet: Viktige avklaringer om fremtidens NATO."
153 Madsen, "Etterretning er politisk kapital."
154 St.prop. 151 S (2015–2016), "Long-term Plan for the Norwegian Armed Forces," 76.
155 The Norwegian Government, "Prop. 27 S (2016–2017) – Endringer i statsbudsjettet 2016 under Forsvarsdepartementet."
156 Forsvarsdepartementet, "Norge har inngått kontrakt om kjøp av fem nye P-8A Poseidon."
157 Forsvarsdepartementet.
158 Forsvarsdepartementet.
159 Håvard Klevberg, *Discussion – Norwegian MPA Capability* (Oslo: Norwegian Ministry of Defense, May 24, 2018); Johnsen, "Solberg: Vil overvåke mer med fly som forsvarssjefen vil skrote."
160 Among other things, the closure of Andøya Air Base and establishment of Evenes as the sole air base in the north, the closure of the coastal national guard and the eventual phasing out of the Navy's corvettes of the Skjold class.
161 See, for example, Stormark, "Bare ETT operativt overvåkningsfly."
162 Budalen og Pettersen, "Orion-flyene er på vingene igjen." Although also located in the High North, Evenes is too far from Andøya for anyone to engage in a daily commute. The move therefore means that the personnel has to move or lose their workplace
163 Stortinget, Skriftlig spørsmål fra Willfred Nordlund (Sp) til forsvarsministeren; Sivertsen, "Må forvente noe lavere flyaktivitet."
164 Richelson, *The US Intelligence Community*, 195. See digitalglobe.com, planet.com and intelligence-airbusds.com for examples.
165 In accordance with International Maritime Organization (IMO) regulations, it is mandatory for all civilian ships with a tonnage surpassing 300 tons (and all passenger ships irrespective of size) to emit AIS signals that state the ship's name, course, speed, destination, and other details. A satellite is able to pick up these signals and create an overview of maritime traffic.

Bibliography

Allison, George. "First British P-8A Poseidon Will Be 'Ready to Fly with a UK Crew' on Day One of Arriving in the UK." *UK Defence Journal*, July 25, 2018. https://ukdefencejournal.org.uk/first-british-p-8-poseidon-will-be-ready-to-fly-with-a-uk-crew-on-day-one-of-arriving-in-the-uk/.

Ambrose, Gary. "Transforming Maritime Patrol Aviation." Master's thesis, USMC Command & Staff College, 2003.

Andresen, Rolf-Inge V., og Tor Bukkvoll. *Russian Weapons Development Towards 2020*. Forsvarets Forskningsinstitutt (FFI) FFI-rapport 2008/01957, January 12, 2009.

Atland, Kristian. "The Introduction, Adaptation and Implementation of Russia's 'Northern Strategic Bastion' Concept, 1992–1999." *Journal of Slavic Military Studies* 20, no. 4 (2007): 499–528.

Avionews. "The United Kingdom Has No Maritime Patrol Aircraft, Asks NATO for Help." *Avionews*, October 12, 2014. www.avionews.com/item/1164814-the-united-kingdom-has-no-maritime-patrol-aircraft-asks-for-nato-help.html.

Barth Eide, Espen. "Forsvarets oppdrag i nordområdene ('The Mission of the Armed Forces in the High North')." Speech, Bodø, March 30, 2006.

BBC. "US Senate Votes to Begin Debate on New Start Treaty." *BBC.com*, Åpnet October 4, 2018. www.bbc.com/news/world-us-canada-12004945.

Berger, Eric. "Blue Origin Just Validated the New Space Movement." *arstechnica.com*, June 10, 2016. https://arstechnica.com/science/2016/10/blue-origin-just-validated-the-new-space-movement/.

Bergstrøm, Øyvind. "NATOs maritime strategi etter 1990 – Fra sjømakt til ordensmakt?" Master's thesis, Norwegian Armed Forces Staff College, 2009.

Berkowitz, Bruce. *The National Reconnaissance Office at 50 Years: A Brief History*. Chantilly, VA: Center for the Study of National Reconnaissance, 2011.

Birkeland, John O. "The Potential of LIDAR as an Anti-Submarine Warfare Sensor." M. Phil (R), War Studies, University of Glasgow, 2009. http://theses.gla.ac.uk/1252/1/2009birkelandmphil.pdf.

Birkeland, John O., og Gjert Lage Dyndal. "Fremtidig autonom droneteknologi og -konsepter." In *Når dronene våkner*, redigert av Tor Arne Berntsen, Gjert Lage Dyndal, og Sigrid Redse Johansen. Oslo: Cappelen Damm Akademiske, 2016.

Borgeld, Warner. "Keflavik – from the Past to the Present." *Maritime Patrol Aviation*, 2004.

Børresen, Jacob, Gullow Gjeseth, og Rolf Tamnes. *Norsk forsvarshistorie – Allianseforsvar i endring, 1970–2000*. Norsk Forsvarshistorie 5. Bergen: Eide Forlag, 2004.

Borst, Marco. "Under Arctic Conditions in Europe: Norwegian Orions." *AIR International*, January 1998, 57–58.

Budalen, Andreas, og Beth Pettersen. "Orion-flyene er på vingene igjen." *NRK Online*, November 8, 2016. www.nrk.no/nordland/orion-flyene-er-pa-vingene-igjen-1.13216832.

Cass, Stephen. "We May Be Heading for a Space Bubble." *MIT Technology Review Online*, September 14, 2010. www.technologyreview.com/s/420776/we-may-be-heading-for-a-space-bubble.

"Command History Report (CHR) for VP-5 (1992)." US Navy, 1992. FOIA case number: DON-NAVY-2018-00574. Naval History and Heritage Command.

"Command History Report (CHR) for VP-5 (1993)." US Navy, 1993. FOIA case number: DON-NAVY-2018-00574. Naval History and Heritage Command.

"Command History Report (CHR) for VP-16 (1995)." US Navy, 1995. FOIA case number: DON-NAVY-2018-001782. Naval History and Heritage Command.

"Command History Report (CHR) for VP-49 (1993)." US Navy, 1993. FOIA case number: DON-NAVY-2018-00574. Naval History and Heritage Command.

Cote, Owen R. Jr. "The Third Battle – Innovation in the U.S. Navy's Silent War Struggle with Soviet Submarines." *Naval War College Newport Papers* 16 (2003).

Day, Dwayne A. "Above the Clouds: The White Cloud Ocean Surveillance Satellites." *The Space Review*, April 13, 2009, Online utgave. www.thespacereview.com/article/1351/1.

Defense Contracts Online. "P-8A Programme Progresses." *Defense Contracts Online*, date unknown. www.contracts.mod.uk/do-features-and-articles/p-8a-programme-progresses/.

Efjestad, Svein. "Norway and the North Atlantic: Defence of the Northern Flank." In *NATO and the North Atlantic – Revitalizing Collective Defence*, edited by John A. Olsen. Whitehall Paper 87. London: Royal United Services Institute (RUSI), 2017.

Efjestad, Svein. "Norwegian Self-Imposed Restrictions in the High North." Interview, September 2, 2019. Norwegian Ministry of Defense.

Efjestad, Svein. *Raising NATO's Profile – The Norwegian NATO Initiative*. NUPI, December 9, 2020.

Faulconbridge, Guy. "Russian Navy to Start Sorties in Mediterranean." *Reuters*, December 5, 2007. www.reuters.com/article/us-russia-navy/russian-navy-to-start-sorties-in-atlantic-tass-idUSL0518563620071205.

Fein, Geoff, og Grace Jean. "Changing the Game: U Navy Applies New Approaches to Submarines Threats." *IHS Jane's Navy International*, 2014. www.janes.com/images/assets/836/46836/US_Navy_applies_new_approaches_to_submarine_threats.pdf.
Flight International. "RMPA Contenders Step Up Campaign." *Flight International* 1 (May 7, 1996): 16.
Forden, Geoffrey. "Viewpoint: China and Space War." *Astropolitics* 6, no. 2 (2008): 138–53.
Forsvarsdepartementet. "NATO-toppmøtet: Viktige avklaringer om fremtidens NATO." The Norwegian Government, September 3, 2014. www.regjeringen.no/no/aktuelt/NATO-toppmotet-Viktige-avklaringer-om-fremtidens-NATO/id766528/.
Forsvarsdepartementet. "Norge har inngått kontrakt om kjøp av fem nye P-8A Poseidon." The Norwegian Government, March 29, 2017. www.regjeringen.no/no/aktuelt/norge-har-inngatt-kontrakt-om-kjop-av-fem-nye-p-8a-poseidon-maritime-patruljefly/id2546045/.
Foust, Jeff. "Current Issues in New Space." *The Space Review*, March 5, 2007. www.thespacereview.com/article/823/1.
Foust, Jeff. "The Evolving Ecosystem of New Space." *The Space Review*, August 15, 2011. www.thespacereview.com/article/1906/1.
Friedman, Norman. *Seapower and Space*. Annapolis, MD: Naval Institute Press, 2000.
Friedman, Norman. "World Naval Developments: The Typhoon Saga Ends." *United States Naval Institute Proceedings* 125, no. 2 (February 1999).
GE Aviation. "Boeing Team Wins $3 Billion Multi-Mission Maritime Aircraft Program." *GE Aviation Online*, June 14, 2004. www.geaviation.com/press-release/jv-archive/boeing-team-wins-3-billion-multi-mission-maritime-aircraft-program.
Gibbons-Neff, Thomas. "With No Sub-Chasing Aircraft of Its Own, UK Calls on Allies to Help Find Russian Submarine." *Washington Post*, November 23, 2015. www.washingtonpost.com/news/checkpoint/wp/2015/11/23/with-no-sub-chasing-aircraft-of-its-own-uk-calls-on-allies-to-help-find-russian-submarine/?noredirect=on&utm_term=.fd82147b45bf.
Gressel, Gustav. *Russia's Quiet Military Revolution, and What It Means for Europe*. Policy Brief. European Council on Foreign Relations, October 2015.
Gutterman, Steve. "New Russian Nuclear Submarine Goes into Service." *Reuters*, January 10, 2013. www.reuters.com/article/us-russia-submarine/new-russian-nuclear-submarine-goes-into-service-idUSBRE9090YR20130110.
Haddon-Cave, Charles. *The Nimrod Review – An Independent Review into the Broader Issues Surrounding the Loss of the RAF Nimrod MR2 Aircraft XV230 in Afghanistan in 2006*. Report to the House of Commons. London: House of Commons, October 2009.
Hanley, John T. Jr. "Creating the 1980s Maritime Strategy and Implications for Today." *Naval War College Review* 67, no. 2 (Spring 2014): 11.
Hattendorf, John B., red. "U.S. Naval Strategy in the 1990s – Selected Documents." *Naval War College Newport Papers* 27 (2006).
Hudson, Peter, og Peter Roberts. "The UK and the North Atlantic: A British Military Perspective." In *NATO and the North Atlantic – Revitalizing Collective Defence*, redigert av John A. Olsen. Whitehall Paper 87. London: Royal United Services Institute (RUSI), u.å.
Håvoll, Harald. "Norwegian Maritime Patrol Aircraft Operations During and After the Cold War." Nesodden, Norway, October 22, 2018.
Innst. S no. 151 (1995–1996). "Recommendation to Parliament by the Defense Committee – Confirmation of Norway's Self-Imposed Restrictions in Security Policy." Stortinget, March 13, 1996. Stortinget, Online Archive.
International Institute for Strategic Studies. "Reviving Russia's Navy." *Strategic Comments* 6, no. 6 (July 2000).
Johnsen, Alf Bjarne. "Solberg: Vil overvåke mer med fly som forsvarssjefen vil skrote." *VG*, October 12, 2015. www.vg.no/nyheter/innenriks/i/22RKr/solberg-vil-overvaake-mer-med-fly-som-forsvarssjefen-vil-skrote.

Johnstone, Clive. "NATO's Maritime Moment: A Watershed Year in Alliance Sea Power." Speech, Residence of Belgian Ambassador to the United Kingdom, London, January 17, 2017. https://mc.nato.int/media-centre/news/2017/nato-maritime-moment-a-watershed-year-in-alliance-sea-power.aspx.

Kimball, Daryl. "The Presidential Nuclear Initiatives (PNIs) on Tactical Nuclear Weapons at a Glance." Arms Control Association, July 2017. www.armscontrol.org/printpdf/111.

Klevberg, Håvard. *Discussion – Norwegian MPA Capability*. Oslo: Norwegian Ministry of Defense, May 24, 2018.

Korolkov, Alexander. "Russia's Top-Secret Nuclear Submarine Comes into Service." *Russia Beyond*, June 17, 2014. www.rbth.com/defence/2014/06/17/russias_top-secret_nuclear_submarine_comes_into_service_37483.html.

Kremlin. "Maritime Doctrine of the Russian Federation." Oversatt av Anna Davis. U.S. Naval War College, 2015.

Kristensen, Hans M., og Robert S. Norris. "Russian Nuclear Forces, 2011." *Bulletin of the Atomic Scientists* 67, no. 3 (2011): 67–74.

Kristensen, Hans M., og Robert S. Norris. "Russian Nuclear Forces, 2013." *Bulletin of the Atomic Scientists* 69, no. 3 (2013): 71–81.

Kristensen, Hans M., og Robert S. Norris. "Russian Nuclear Forces, 2014." *Bulletin of the Atomic Scientists* 70, no. 2 (2014): 75–85.

Kristensen, Hans M., og Robert S. Norris. "Russian Nuclear Forces, 2015." *Bulletin of the Atomic Scientists* 71, no. 3 (2015).

Lewis, Jeffrey. "Led Zeppelin Comes to Washington." *Foreign Policy*, January 5, 2015. https://foreignpolicy.com/2015/01/05/led-zeppelin-comes-to-washington-russia-nukes-putin-arms-control/.

Madsen, Anders. "Etterretning er politisk kapital." *Aftenposten*, February 24, 2016. www.aftenposten.no/meninger/kommentar/i/vAkl/etterretning-er-politisk-kapital-per-anders-madsen.

Martyanov, Andrei. "Russia's Navy in Search of a Mission." *United States Naval Institute Proceedings* 140, no. 12 (December 2014).

Mazzetti, Mark, og Thom Shanker. "Russian Subs Patrolling Off East Coast of U.S." *New York Times*, August 4, 2009.

McLaughlin, Kevin. "Would Space-Based Systems Defenses Improve Security?" *Washington Quarterly* 25, no. 3 (2002): 177–91.

Ministry of Defence. *Major Projects Report 2005 – Project Summary Sheets*. London: National Audit Office, November 25, 2005.

Mizin, Victor, og Michael Jasinski. "The Future of the Russian Sea-Based Deterrent." *The Journal of Slavic Military Studies* 16, no. 1 (2003): 69–83.

NATO. "Active Engagement, Modern Defense – Strategic Concept for the Defense and Security for the Members of NATO." NATO, November 20, 2010. NATO, Online Archive.

NATO. "The Alliance's New Strategic Concept." NATO, November 8, 1991. NATO, Online Archive.

NATO. "Brussels Summit Declaration." NATO, July 11, 2018. NATO, Online Archive.

NATO. "Counter-Piracy Operations." NATO, December 19, 2016. NATO, Online Archive. www.nato.int/cps/en/natolive/topics_48815.htm#Protector.

NATO. "Declaration of the Heads of State and Government – Brussels Summit." NATO, January 11, 1994. NATO, Online Archive.

NATO. "Final Communique – Berlin." NATO, June 3, 1996. NATO, Online Archive.

NATO. "First NATO AGS Remotely Piloted Aircraft Ferries to Main Operating Base in Italy." NATO, November 21, 2019. www.nato.int/cps/en/natohq/news_171171.htm.

NATO. "Istanbul Summit Communique." NATO, June 28, 2004. NATO, Online Archive.

NATO. "Lisbon Summit Declaration." NATO, November 20, 2010. NATO, Online Archive.

NATO. *NATO Handbook 1995*. Brussels: NATO Office of Information and Press, 1995.

NATO. "NATO's Readiness Action Plan (RAP) – Fact Sheet." NATO, July 2016. NATO, Online Archive.

NATO. "Operation Active Endeavour." NATO, October 27, 2016. NATO, Online Archive. www.nato.int/cps/en/natolive/topics_7932.htm.

NATO. "Prague Summit Declaration." NATO, November 21, 2002. NATO, Online Archive.

NATO. "Strasbourg/Kehl Summit Declaration." NATO, April 4, 2009. NATO, Online Archive.

NATO. "Wales Summit Declaration." NATO, September 5, 2014. NATO, Online Archive.

NATO. "Warsaw Summit Communique." NATO, July 9, 2016. NATO, Online Archive.

NavalAir Museum Barbers Point. "Lockheed P-3C Orion – BuNo 160770." Åpnet September 6, 2018. http://nambp.org/lockheed-p-3c-orion.

Norris, Robert S., og William N. Arkin. "Russian (C.I.S.) Strategic Nuclear Forces End of 1995." *Bulletin of the Atomic Scientists* 52, no. 2 (April 1996): 62–63.

Norris, Robert S., og William N. Arkin. "Russian Nuclear Forces, 2000." *Bulletin of the Atomic Scientists* 56, no. 4 (August 2000): 70–71.

Norris, Robert S., William N. Arkin, Hans M. Kristensen, og Joshua Handler. "Russian Nuclear Forces, 2001." *Bulletin of the Atomic Scientists* 57, no. 3 (June 2001): 78–79.

Norris, Robert S., William N. Arkin, Hans M. Kristensen, og Joshua Handler. "Russian Nuclear Forces, 2002." *Bulletin of the Atomic Scientists* 58, no. 4 (August 2002): 71–73.

Norris, Robert S., og Hans M. Kristensen. "Russian Nuclear Forces, 2004." *Bulletin of the Atomic Scientists* 60, no. 4 (August 2004): 72–74.

Norris, Robert S., og Hans M. Kristensen. "Russian Nuclear Forces, 2006." *Bulletin of the Atomic Scientists* 62, no. 2 (April 2006): 64–67.

Norris, Robert S., og Hans M. Kristensen. "Russian Nuclear Forces, 2007." *Bulletin of the Atomic Scientists* 63, no. 2 (April 2007): 61–67.

Norris, Robert S., og Hans M. Kristensen. "Russian Nuclear Forces, 2008." *Bulletin of the Atomic Scientists* 64, no. 2 (June 2008): 54–62.

Norris, Robert S., og Hans M. Kristensen. "Russian Nuclear Forces, 2009." *Bulletin of the Atomic Scientists* 65, no. 3 (June 2009).

Norris, Robert S., og Hans M. Kristensen. "Russian Nuclear Forces, 2010." *Bulletin of the Atomic Scientists* 66, no. 1 (February 2010).

Norwegian Chief of Defense (CHOD). *A Defense Undergoing Change – Fagmilitært Råd (FMR) – Defense White Paper*. Norwegian Defense Staff (FST), u.å.

Norwegian Intelligence Service. "FOKUS 2011 – The Assessment of the Intelligence Service." Annual Unclassified Assessment. Oslo, 2011. https://forsvaret.no/fakta_/ForsvaretDocuments/FOKUS-2011.pdf.

Norwegian Intelligence Service. "FOKUS 2013 – The Assessment of the Intelligence Service." Annual Unclassified Assessment. Oslo, 2013. https://forsvaret.no/fakta_/ForsvaretDocuments/FOKUS-2013.pdf.

Norwegian Intelligence Service. "FOKUS 2015 – The Assessment of the Intelligence Service." Annual Unclassified Assessment. Oslo, 2015. https://forsvaret.no/ForsvaretDocuments/FOKUS2015-endelig.pdf.

The Norwegian Government. "Prop. 27 S (2016–2017) – Endringer i statsbudsjettet 2016 under Forsvarsdepartementet." The Norwegian Government, November 25, 2016. www.regjeringen.no/contentassets/a38d11296b8f4d2a8df432ac430335c4/nn-no/pdfs/prp201620170027000dddpdfs.pdf.

Norwegian Ministry of Defense. "Response from the Ministry of Defense to the Foreign Affairs and Defense Committee on Questions Related to the Long Term Plan – Prop. 151 S (2015–2016)." Stortinget, August 17, 2016. Stortinget, Online Archive.

Norwegian Minister of Defence Jørgen Kosmo. "Letter from the Minister of Defence to the Defense Committee." Official Correspondence, February 26, 1996. Stortinget, Online Archive.

Norwegian Minister of Defence Jørgen Kosmo. "Long Term Challenges for the Armed Forces." Speech. Oslo Militære Samfund, Oslo, January 8, 1996. Stortinget, Online Archive.

The Norwegian Ministry of Foreign Affairs. "St.meld.nr.38 (2008–2009) – Om samarbeidet i NATO i 2008." The Norwegian Government, 2008.

NTB. "Det vil ta én til to måneder for NATO å hjelpe Norge." *NRK*, April 5, 2016.

nti.org. "U.S. Senate Ratifies New START in 71–26 Vote, Despite Top GOP Opposition." *NTI Online*, December 22, 2010. www.nti.org/gsn/article/us-senate-ratifies-new-start-in-71-26-vote-despite-top-gop-opposition/.

NuclearThreat Initiative (NTI). "Russia Submarine Capabilities." NTI, June 10, 2014. www.nti.org/analysis/articles/russia-submarine-capabilities/.

The Office of the Director, Operational Test and Evaluation. "FY2014 DOT&E Annual Report." Annual Report, 2014.

Office of Naval Intelligence. "The Russian Navy – A Historic Transition." Office of Naval Intelligence (ONI), December 2015. www.oni.navy.mil/Portals/12/Intel%20agencies/russia/Russia%202015print.pdf?ver=2015-12-14-082038-923.

Oliker, Olga. "Unpacking Russia's New National Security Strategy." *csis.org*, January 7, 2016. www.csis.org/analysis/unpacking-russias-new-national-security-strategy.

Olsen, John A., red. *NATO and the North Atlantic – Revitalizing Collective Defence*. Whitehall Paper 87. London: Royal United Services Institute (RUSI), 2017.

Perkins, William. *Alliance Airborne Anti-Submarine Warfare – A Forecast for Maritime Air ASW in the Future Operational Environment*. Joint Air Power Competence Centre (JAPCC), June 2016.

Polmar, Norman. "The Russian Navy – Beneath the Waves." *United States Naval Institute Proceedings* 138, no. 6 (June 2012).

Porter, Air Commodore (r) Garfield "Garry". *British Maritime Patrol Aircraft Operations During and After the Cold War*. London: Royal Air Force Club, May 31, 2017.

Putin, Vladimir. "Speech and the Following Discussion at the Munich Conference on Security Policy." *Kremlin*, February 19, 2007. http://en.kremlin.ru/events/president/transcripts/24034.

Reade, David. "NAS Keflavik, Iceland – The ASW Training Capital of the World." *Maritime Patrol Aviation* 3, no. 2 (March 1995).

Rear Admiral Thomas Ernst. "Agile Command and Control in a Degraded Environment." In *Preparing NATO for Joint Air Operations in a Degraded Environment*. Joint Air & Space Power Conference 2016. Kalkar, Germany: Joint Air Power Competence Centre, 2016.

Richelson, Jeffrey T. *The US Intelligence Community*, 7th ed. Boulder, CO: Westview Press, 2016.

Roberts, Michael D. *Dictionary of American Naval Aviation Squadrons*, vol. 2. Washington, DC: Naval Historical Center, 2000.

RT. "Finaly Flying Colors: Yury Dolgoruky Nuclear Sub Joins Russian Navy." *RT.com*, January 10, 2013. www.rt.com/news/yury-dolgoruky-submarine-ceremony-678/.

Sadiq, Sheraz. "Silicon Valley Goes to Space." *kqed.org*, November 18, 2013. www.kqed.org/science/11135/silicon-valley-goes-to-space-2.

Sherman, Kenneth B. "Feet Dry: Maritime Patrol Goes Ashore – New Roles and Missions Mean No Safe Harbor for Land Forces." *The Journal of Electronic Defense (JED)* (March 2001): 37–43.

Sivertsen, Tony. "Må forvente noe lavere flyaktivitet." *Vestrålen Online (VOL)*, January 5, 2019. www.vol.no/pluss/2019/01/05/%E2%80%93-M%C3%A5-forvente-noe-lavere-fly-aktivitet-18189636.ece.

Smith-Windsor, Brooke A. *NATO's Maritime Strategy and the Libya Crisis as Seen from the Sea*. Research Paper. Rome: NATO Defense College, March 2013.

Staar, Richard. "Russia's Navy Remains in Decline." *United States Naval Institute Proceedings* 124, no. 8 (August 1998).

Stormark, Kjetil. "Bare ETT operativt overvåkningsfly." *aldrimer.no*, October 29, 2016. www.aldrimer.no/norge-har-bare-ett-operativt-overvakingsfly/.

Stortinget. "Skriftlig spørsmål fra Willfred Nordlund (Sp) til forsvarsministeren." Pub. L. No. 15:607 (2018–2019), 2019.

St.prop. 151 S (2015–2016). *Long-Term Plan for the Norwegian Armed Forces*. Norwegian Ministry of Defense, June 16, 2016.
Tamnes, Rolf. *Oljealder, 1965–1995*, Bd. 6. Norsk Utenrikspolitikks historie. Oslo: Universitetsforlaget, 1997.
Tamnes, Rolf. "The Significance of the North Atlantic and the Norwegian Contribution." In *NATO and the North Atlantic – Revitalizing Collective Defense*, redigert av John A. Olsen. Whitehall Paper 87. London: Royal United Services Institute (RUSI), 2017.
Thomas, Rick. *Global Hawk – The Story and Shadow of America's Controversial Drone*. Great Britain: Amazon, 2015.
UK Defense Committee. *Future Maritime Surveillance*, vol. I. London: House of Commons, September 5, 2012.
UK MoD. *National Security Strategy and Strategic Defense and Security Review (SDSR) 2015*. HM Government, November 2015.
Warden, John A. III. "The Enemy as a System." *Airpower Journal* 9, no. 1 (1995): 40–55.
Zysk, Katarzyna. "Mot et moderne russisk forsvar? Utviklingstrender i militær modernisering og strategisk tenkning." *Norsk Militært Tidsskrift* 183, no. 3 (2013).

9 The Russian submarine threat of the future

Russian maritime doctrine aims at countering Western ASW technologies and challenging NATO and the US presence in the Atlantic, the Baltic Sea, the Mediterranean, and the Black Sea through increasing Russian presence in those areas, including the Arctic. The Russians are open about the fact that they are conducting testing and exercises in the Barents Sea, and have historically and will regularly publish the coordinates for such activity in order to ensure that no ship will sail or aircraft fly into the danger areas.[1] Russian ships and coastal defences will fire air defence missiles and strategic missiles; Russian maritime infantry units will exercise coastal landings and assaults on Russian islands in the Arctic, and the Northern Fleet will conduct advanced exercises with submarines. According to the NIS, these activities are important for technical intelligence collection, but they also play a crucial role in demonstrating Russian capabilities and ambitions in the Arctic.[2] Vessels under development in the Russian navy will typically undergo sea trials in the Barents Sea before being commissioned in their respective fleets.[3] The Northern Fleet also regularly announces test firings of strategic weapons concurrently with Western naval exercises, such as the announcement of the firing of a salvo of *Bulava* missiles during the NATO exercise *Cold Response* in Norway in 2016.[4]

In addition to actual operations and strategic patrols in the High North with SSBNs, Russian units have also conducted aggressive signalling through exercises and training in the North. According to the NIS, the Russians have flown attack profiles towards the intelligence service's radar in Vardø, against NATO units exercising off the coast of Norway, and against the National Joint Headquarters in Bodø.[5] The increase in Russian activity is also evident through the higher intensity and regularity in Russian military aircraft flying out from the Kola Peninsula and down the Norwegian coast. Although not at the level of the Cold War, the numbers are much higher now than at the turn of the millennium.[6] In addition, the Russian Armed Forces have deployed cruise missile batteries to the Kola Peninsula and towards the Norwegian border during the Russian exercise *Zapad 2017*. In the fall of 2023, the Northern Fleet conducted exercise Ocean Shield. This included the forward placement of submarines along the line between Bear Island and the Norwegian mainland, underlining the Russians' emphasis on denying any foreign powers access to the Barents Sea.[7]

DOI: 10.4324/9781003626015-9

However, NATO units operating in other theatres have experienced significantly more tension than those operating in the Barents Sea: Attack profiles were flown against the *USS Donald Cook* in the Baltic Sea in 2016; a US surveillance aircraft was intercepted by a Russian fighter aircraft within 15 feet that same year; and an EP-3 *Aries* was intercepted and harassed in the Black Sea in early 2018.[8] In June of 2019, a RQ-4 *Global Hawk* was shot down by Iranian air defence systems when flying over the Strait of Hormuz.[9] In March of 2023, a US MQ-9 *Reaper* drone flying over the Black Sea was intercepted several times by Russian fighter aircraft, likely bringing it to malfunction and crash into the ocean.[10]

The primary actors in the Barents Sea, the Russians and the Norwegians, have managed to maintain relatively low tensions in the High North and aim to continue to do so in the foreseeable future. It seems fair to assume that the Norwegian policy of integration/deterrence and screening/reassurance has contributed to this. Comparing the High North with other areas of fairly high activity for the Russian Navy (Black Sea, Sea of Okhotsk, Baltic Sea), seen through the lens of the inherent potential for confrontation in such a militarily sensitive area, it seems that the Norwegian policy of integration/deterrence and screening/reassurance has contributed towards maintaining low tensions.

Even so, there has been a marked increase of Russian exercises and operations when compared to the previous two decades. The Northern Fleet exercised in mid-June 2018, with the largest deployment of ships seen in 10 years.[11] The Russian Navy followed up with a major naval exercise in the Western Barents Sea in March 2019.[12] In mid-August 2019, the Russian Navy conducted another major naval exercise off the western and northern Norwegian coast. The Russian activities were closely monitored by Norwegian, American, and Canadian MPA flying out of Andøya.[13] The Russian exercise, named *Ocean Shield*, included coordinated cross-theatre manoeuvres between the Baltic, Barents, and Black Sea fleets. It demonstrated the bastion defence strategy, and clearly positioned the northern part of Norway as encircled by the Russian bastion.[14]

Then, in October 2019, the Russian Navy initiated the largest submarine exercise conducted in the North Atlantic since the Cold War.[15] The exercise had similarities to the *Ocean Shield* surface combatant exercise conducted the previous month, but this time only submarines took part. NIS reported the submarines operated west, south, and east of Bear Island in order to monitor the entry points for the Barents Sea, and further south into the Norwegian Sea. The latter was executed in order to exercise deep diving in the significant depths there, compared to the Barents Sea. The objective of the exercise, according to the NIS, was to show that the Russian Navy can deploy submerged into the North Atlantic, and be able to reach North America with cruise missiles from forward firing positions. A further objective was to test the West's ability to respond to such a substantial deployment of strategically important assets.[16] The United States and Canada sent MPA to Andøya in Norway to cover the Russian activities together with the Norwegians.[17] Large Russian naval exercises just to the west of the Norwegian coast seem to exercise fundamental elements of the bastion concept of sea control and sea denial. The mission of the Northern Fleet is still to protect the strategic nuclear deterrent in the High North.[18] Although the role of disrupting Western SLOCs across the Atlantic seems to be somewhat downplayed, the threat against the SLOCs in times of crisis and war should not be discarded.

The Russian invasion into Ukraine in February 2022 tipped the worsening tensions between Russia and the West over the edge. The years of full-scale war in Ukraine have shaken the security infrastructure of Europe to its core and have likely changed it for the foreseeable future. Due to the high conventional losses on the battlefield in Ukraine, the importance of nuclear weapons has increased. This in turn means that the defence of the Kola Peninsula and exercising deterrence with strategic submarines have become even more important.[19]

In addition to the expanded conventional military activities on display from the Russians and the increased importance of their nuclear capabilities, there is another aspect of Russian warfare that has cemented itself in strategic discourse in the past years: Hybrid warfare. The publishing in February 2013 of a military article by the Chief of the Russian General Staff, General Valery Gerasimov, did not gain much attention at first.[20] But when the Russians annexed Crimea mostly using tactics associated with what we term hybrid warfare 1 month later, the article stood out as a blueprint for methods used during that annexation. Hybrid warfare can be described as "the synchronized use of multiple instruments of power tailored to specific vulnerabilities across the full spectrum of societal functions to achieve synergistic effects."[21] This is neither new nor revolutionary. However, the systematic use and scale of hybrid warfare tools as means to achieve ends outside of kinetic conflict is something to which the West must adapt quickly. There are discussions ongoing as to the validity of the Gerasimov doctrine as a doctrine, *per se*.[22] But regardless of these discussions, one fundamental factor on which most analysts concur is that Russia will work to achieve her security policy objectives while remaining below the threshold of war. The Gerasimov doctrine provides a depiction of the vast range of tools available before the threshold of conventional military power is reached.

Hybrid warfare in the maritime domain has not been discussed as much as in the land and informational domains. However, it seems increasingly clear that the maritime will be equally exploited as other domains in an effort to achieve policy goals without entering a conventional military conflict. As James Stavridis explains,

> The fundamental idea of hybrid warfare is to find the space short of clear-cut military action with direct and recognizable tactical, operational, and strategic impact and compress it into a zone wherein sufficient ambiguity is created to allow an offensive actor a better chance of accomplishing an objective without full-blown, overt offensive action.[23]

The toolbox in maritime hybrid warfare includes the use of civilian and fishery vessels; larger ships that function as mother ships for unmanned underwater vessels or smaller surface vessels; fishermen that deploy fixed seabed listening devices on the seabed; manipulation or destruction of underwater infrastructure such as pipelines for fossil fuels, communication cables and such; C2 over civilian, commercial-off-the-shelf (COTS) open networks; laser dazzlers and small jamming devices; and more.[24] These tools will be supported by conventional means such as submarines and special forces, in a mix that will constitute maritime hybrid warfare. Russian captains on civilian vessels calling on Norwegian ports have reportedly been instructed to familiarize themselves as much as possible with the Norwegian coastline and ports, in case of a conflict.[25] This mixing of operating platforms, personnel, and domains is and will be a fundamental challenge for determining the "normal situation," for example, in the High North. In order to properly be able to handle this challenge, it seems inevitable that the surveillance platforms of the future must be able to operate cross-domain, persistently, and with the ability to communicate with other sensors as well as their home base and other elements of the intelligence cycle, in real time.

One example of such unconventional military activities is the Russian Navy's intelligence collection on undersea cables laying on the seabed in international waters.[26] The Russian navy deploys special-purpose submarines that can carry other, smaller submarines and autonomous or remotely piloted vessels. These are utilized for both research purposes and intelligence gathering. Rebuilt strategic submarines from the *Oscar* and *Delta* classes are being used for gathering intelligence on Western seabed communication systems.[27] NATO reports that Russian special-purpose submarines have over time conducted operations around civilian, underwater communication cables between Europe and the United States.[28]

Figure 9.1 Overview over transatlantic communications cables[29]

Reports suggest that the special-purpose submarine AS-31 *Losharik*, which experienced a fatal fire on board whilst operating in the Barents Sea on 1 July 2019, is a part of an underwater intelligence unit that specializes in tapping into and preparing for the destruction of undersea communication cables.[30] It seems that with special purpose-built submarines, such as the *Losharik*, *Paltus*, and *Uniform*, which able to conduct advanced seabed operations, the submarine branches of the Russian armed forces are also an integrated part of hybrid warfare.[31] This type of activity has historically also been carried out by the US Navy.[32]

Submarines

Submarines remain the capital ships of the Russian Navy.[33] Open sources reveal that the Northern Fleet has somewhere between 20 and 30 operational submarines in its inventory.[34] Legacy SSN and SSGN submarines from the Cold War that are still operational are the *Victor III*, the *Akula*, the *Sierra II*-class SSNs, the *Oscar II*-class SSGN, and the *Delta IV*-class SSBN. In addition, several new classes are inbound. The *Severodvinsk* is meant to replace the SSNs and SSGNs mentioned above. It is a multipurpose, nuclear attack submarine which is incredibly capable, exceptionally quiet, and vastly expensive.[35] The overall plan is to build eight to ten Severodvinsk-class submarines.[36] However, the production tempo of the *Severodvinsk* class has slowed due to budgetary constraints. The submarine class is likely one of the more expensive pieces of military hardware in the Russian inventory, writ large.[37] The SSGN carries a wide array of missiles, namely the SS-N-26 *Strobile* supersonic anti-ship missile (range 190 NM/350 km), the SS-N-21 *Sampson* missile, similar to the US *Tomahawk* missile with the option to carry a nuclear warhead (land-attack, range 1,600 NM/3,000 km), the SS-N-27 *Sizzler* (anti-ship, range over 270 NM/500 km) and 32 units of the SS-N-30 *Kalibr* missile (land-attack, range approximately 1,300 NM/2,500 km).[38] Work is reportedly ongoing to find a more affordable follow-on for the *Severodvinsk* class.[39]

The *Husky* class may take on two separate designs in order to fulfil two particular roles: First, to protect surface strike groups from hostile submarines; and second, to function as a platform for cruise missiles.[40] The *Husky* will, in addition to *Kalibr* missiles and torpedoes, likely carry the hypersonic SS-N-33 *Zircon* anti-ship cruise missile, capable of flying more than eight times the speed of sound. The submarine will cost far less than the *Severodvinsk* class, but is expected to continue the trend of ever more quiet acoustic characteristics for Russian submarines.[41] Between 15 and 20 *Husky* submarines are expected to be operational by 2040, although it is fair to assume that the war in might affect this timeline.[42]

The *Dolgorukiy* seems to be the future workhorse of the Russian SSBN fleet. Three have been commissioned, and plans are to build eight to ten *Dolgorukiy*-class submarines in total. The follow-on vessels of the initial boat are expected to have redesigned hulls, leading to improved acoustics and lower sound levels.[43] These vessels carry 16 missiles of the type SS-N-32 *Bulava*, with six to ten nuclear warheads in each missile and a range of 4,600 NM/8,500 km.[44] The *Dolgorukiy*-class SSBN has been operationalized, and is patrolling at fairly regular intervals in tandem with remaining *Delta IV* SSBNs.

The diesel-electric *Kilo* class has been a successful design for both domestic use and export. It has proven its worth in modern conflict through the firing of *Kalibr* missiles into Syria.[45] The replacement submarine for the *Kilo* was planned to be the *Petersburg* class, but this project has been wrought with failures during development. So much so that it is reported that the *Petersburg* class may be terminated altogether, in favour of a new design, the *Kalina*-class SSK.[46] The *Kalina* will be an *air-independent propulsion* (AIP) submarine and will take over the duties of the *Kilo* class as part of the layered defence of the bastion and conduct patrols in the littorals.[47]

Cruise missiles

The development of capable cruise missiles has been a priority for the Russian armed forces, to the point where the weapons being fielded are considered to be among the most capable in the world.[48] Cruise missiles provide a combination of tactical flexibility and strategic leverage.[49]

They are a tactical and operational threat to any adversary unit or force, and they can simultaneously hold strategic targets at risk from great distances. Russian cruise missiles are an increasing concern for strategic force planning in NATO.[51]

Likely the most prominent and versatile Russian cruise missile is the *Kalibr* family of missiles. There are more than ten variants of the *Kalibr* missile depending

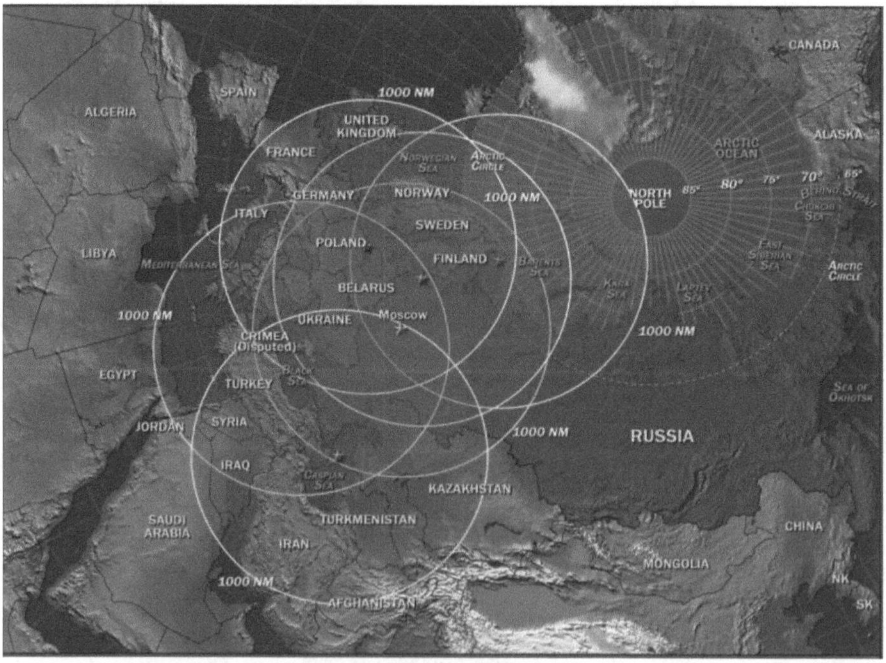

Figure 9.2 Nominal Kalibr LACM ranges from fleet areas[50]

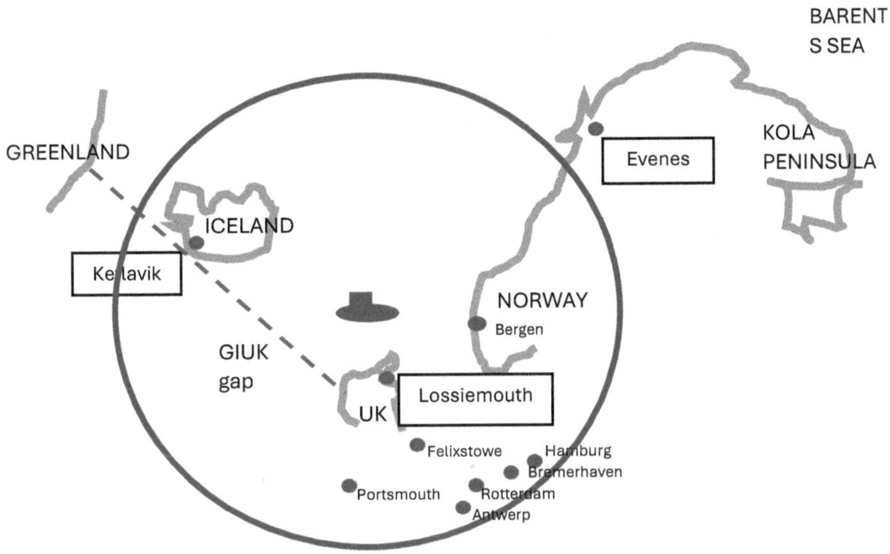

Figure 9.3 Russian submarine-launched Kalibr missile holding Northern European ports at risk from a position well north of the GIUK line

on launch platform, range, target profile, and speed. The most common categories are the anti-ship and land-attack versions. Both carry high-precision conventional munitions weighing between 220 and 450 kg but are also capable of carrying nuclear payloads. The land-attack version is designated SS-N-30A *Kalibr* by NATO and has a maximum range of 1,350 NM/2,500 km at subsonic speeds.[52] Figure 9.2 depicts the firing range for missiles launched approximately from the home ports of the respective Western Russian fleets (nominal range depicted is 1,000 NM). Figure 9.3 shows the nominal range of a *Kalibr* missile launched by a submarine situated north of the GIUK line. Note that Northern Europe's main ports are all within range.[53]

LACMs are complex weapons, more so than their anti-ship brethren. LACMs have to navigate terrain, change communications environments, navigate through air defence layers, and differentiate their target from the surrounding landscapes. One example of such a complex route to the missiles' target is the firing of *Kalibr* cruise missiles into Syria by two surface vessels in the Caspian Sea in autumn 2015. The missiles had to negotiate Iranian, Iraqi, and Syrian territory before hitting their targets.[54]

The anti-ship version has a shorter range of approximately 160 NM/300 km and skims low over the sea in order to avoid detection. The missile uses active radar-homing and performs evasive manoeuvres as it approaches its target. In doing so, it increases from Mach 0.8 in cruising speed to Mach 3, and lowers its altitude to less than 5 m over the water, making it very hard to detect and engage.[55] The NATO designation of the anti-ship version is SS-N-27 *Sizzler*. The weapon can be launched against a single target or a group of targets, and the target(s) can be approached from different directions, significantly complicating the defensive task.[56] An anti-submarine variant is reportedly under development, a tactical ballistic missile with the range of 27 NM/50 km for the submarine-launched version and 22 NM/40 km for the ship-launched version.[57] Compared to a torpedo, this missile will have a much longer range and a significantly shorter delivery time.

The range and flexibility that are inherent in these weapons will be a significant factor for several theatres of operations, simultaneously. In short, a submarine from the Northern Fleet, operating in the North Atlantic, will potentially have enormous impact in the Baltic Sea and on the European continent. This intra-theatre approach and connectivity has, among other things, been exemplified by the aforementioned Caspian Sea flotilla firing *Kalibr* missiles into Syria. Commentators also noted that the way the relatively small flotilla carried out the attacks was a showcase for export of *Kalibr* weapons: One does not require a great missile cruiser or destroyer to deliver formidable naval fire power.[58] It is also a sign of a distributed force structure, with numerous smaller vessels distributed across the theatre being able to provide formidable firing power both in a tactical as well as a strategic setting.[59]

In addition to the Caspian Sea firings of missiles, on 8 December 2015, Russia fired four *Kalibr* cruise missiles into Syria for the first time from a *Kilo*-class submarine in the eastern Mediterranean Sea.[60] The updated *Kilo* class is slightly longer and is fitted with improved engines and an improved combat system, in addition to new noise reduction technology. The submarine is reportedly able to operate on patrols for 45 days at a time.[61] It can fire both torpedoes and cruise missiles, the latter proven several times through the aforementioned firing into Syria.[62] This was a demonstration of the new, credible Russian naval threat: The combination of quiet, stealthy submarines and capable long-range, multi-purpose cruise missiles. The *Kilo* launching of *Kalibr* missiles seems to be only the prelude to what is coming: More capable submarines in terms of manoeuvrability, endurance, and geographic reach, coupled with immense fire power. As Owen Cote describes the problem, "it links a weapons platform whose inherent stealth makes it easy to defend and a weapon against which defence is extremely difficult."[63] In addition to the *Severodvinsk* class, the older submarines are also undergoing upgrades and retrofitting. In particular, the old *Akulas*, *Oscar IIs*, and *Delta IIIs* are being retrofitted to carry both anti-surface and land-attack missiles.[64] The *Oscar* class will replace the SS-N-19 *Shipwreck* anti-ship cruise missiles with *Kalibr* missiles, a mix of anti-ship and land-attack missiles – 72 missiles in total.[65] In comparison, the American *Ohio* class SSGN can carry 152 *Tomahawk* missiles, which is roughly considered to be a *Kalibr* equivalent. As part of the test and development regime, the operational *Severodvinsk* class has been firing *Kalibr* missiles at test ranges on the Kola Peninsula.[66] A *Severodvinsk*-class submarine also fired a *Kalibr* cruise missile from the Kola Peninsula in early 2019 while moored to a pier in Zapadnaya Litsa, some 60 km from the Norwegian border.[67] This means that these capable submarines do not have to leave port in order to pose a credible threat in a crisis or war.

In addition, the Russians are reportedly well underway in developing hypersonic missiles for launch from aircraft and missile silos on land and have also initiated work to place such missiles on submarines.[68] Hypersonic missiles are hard to defend against, not only because they manoeuvre at incredibly high speeds but also because they are harder to detect and track by existing missile defence sensors. The *Zircon* hypersonic anti-ship cruise missile will likely be the main armament in the next class of multipurpose submarines, the *Husky* class, which is planned to be launched in 2027.[69] The *Zircon* is expected to have slightly shorter range than the *Kalibr*. The *Severodvinsk* class will also reportedly carry the *Zircon* hypersonic cruise missile in the future.

The new missiles are eroding the distinction between attack submarines and multi-purpose submarines, further blurring the Allied understanding of roles and missions for each respective submarine class. When a scouting attack submarine, operating ahead of a naval force, is capable of engaging land targets with capable cruise missiles from distance, the threat scenario expands in scope. With a wider array of missiles on board, the respective submarine will pose a multidimensional threat, leading to a requirement to maintain situational awareness of more submarines than was earlier the case. In other words, the case for a robust ASW and underwater search force only strengthens. The new and extremely capable Russian missiles that can be fired from submarines are shaping the naval battle for the future. According to Pavel Podvig, the ambiguous combination of submarines capable of firing both nuclear and conventional cruise missiles will be a prominent feature of the Russian fleet in the near future.[70] The stand-off distance for a submarine attacking an adversary naval force is increasing, seemingly making it all the more important to be able to neutralize the submarine at a distance.[71] Airborne assets will play a crucial role in this effort, both with regard to ISR and targeting of submarines.

Russian strategy and posture

If Russian developments and deployments of submarine and missile technology prove anything, it is the fact that the bastion strategy remains alive and well. The concept is not only a topic for academic and military strategic discussion, but also an integral part of the political and media discourse on Norwegian security policy.[72] Russian naval operations and exercises have increasingly emphasized the concept as new and capable units are sent to sea.

Some naval experts doubt that the Russian fleet is sufficiently capable of providing adequate protection to their high-value units through the bastion defence strategy, due to the fact that the Russian fleet is significantly smaller compared to the fleet that comprised the original bastion defence of the late 1970s and 1980s.[73] But there is general consensus that the Russians are prioritizing quality over quantity, as is evident through the significantly improved modern Russian vessels when compared to those of the previous generations. Focus is now placed on smaller, multirole vessels with capable long-range weapons. The Russian Navy is operating what can be termed a "dual fleet," a combination of legacy vessels with new and modern units, as many units from the Cold War are still operational.[74] As we have seen throughout this thesis, this strategic ambivalence between the old school of thought, with a Mahanian focus on large surface combatants and aircraft carriers, versus the new school, with a focus on submarines as the primary fighting force of a lean, coastal defence navy, has actively influenced the operational capabilities of the Russian Navy.[75] The dual fleet seems to be a result of this.

The focus on the Arctic – and, as a continuation of this, the maintenance of the Northern Strategic Bastion as a baseline for naval operations in the North Atlantic – appears to be both foundational and enduring. Russia established the Arctic Command in 2015, co-located with the Northern Fleet in Murmansk.[76] This command aims to streamline the operations of all the branches of the Northern Fleet in its AOR in the Arctic under one hub for C2.

In addition to the new, capable, and quiet submarines of the Russian Navy, another prominent feature of its modernization and capabilities is the implementation of measures that deny access and freedom to operate in designated areas, so-called anti-access/area denial (A2/AD) measures. Elements of A2/AD have been operating for decades, one can say centuries, as they are essentially capabilities meant to deny one's adversary access to, or control over, a certain area of interest. What is new is the systematic approach to this type of denial, and the significantly improved capabilities of the weapons involved. Crucial elements of denial for air assets are missile threats, jamming of communications and jamming of positioning systems. One of the biggest concerns for NATO planners is the array of A2/AD elements in the Russian layered defence.[77] Below, we will discuss how this affects airborne surveillance in the High North.

It seems clear that in terms of naval combat, submarines provide the most bang-for-the-buck. Or as Kathleen Hicks explains: "the Russian Navy's use of submarines to signal presence, reach and power achieves an effect that is disproportionate to the resources committed."[78] Although submarines are unable to command, or control, the sea, they are able to clear it of hostile surface forces.[79] The capable missile technology now being fielded facilitates a continued responsibility on the part of Russian SSGNs to deny NATO surface and subsurface vessels entry into and freedom of manoeuvre in the North Atlantic, where they would be able to conduct cruise missile strikes against the Russian mainland. The primary mission of Russian attack and multirole submarines is therefore not only to protect the SSBNs on patrol behind them (pro-SSBN), but simultaneously to deny operations to adversary units such as aircraft carriers (anti-carrier).[80] The very essence of sea denial by submarines is the uncertainty projected on an approaching adversary based on the unpredictability of the submarines' whereabouts. This is further accentuated by the change from predictable submarine patrol patterns during the Cold War to an unpredictable submarine operation *modus operandi* in contemporary submarine operations.

The secondary mission of the submarine fleet is power projection over land. Very quiet, multirole submarines, capable of carrying several dozen *Kalibr* missiles, operating from virtually anywhere in the North Atlantic, will pose a credible threat to significant points of importance to NATO and its allies. The inherent potential in the offensive aspects of a layered defence, such as the Northern Strategic Bastion mentioned above, is also increasingly being discussed among strategists and scholars. The Russian bastions are seen as a key element of future offensive warfare, particularly in terms of long-range precision-guided cruise missiles.[81] Finally, the development of an anti-ship version of the *Kalibr* missile should be an indicator that the old SLOC mission of the Russian submarines has not been removed from their mission portfolio, contrary to Russian statements in the late 1990s.[82] Albeit not necessarily a SLOC mission in the traditional sense, where the threat to merchant shipping has been submarines roaming the Atlantic in search of Allied convoys to sink, the mission for modern Russian submarines likely entails neutralizing key Allied ports and airfields. In future conflicts, it seems likely that Russian submarines will pose a significant threat to the transatlantic logistics link without ever having to venture south of the GIUK gap. Some analysts suggest that if NATO is capable of controlling the GIUK gap, then the SLOCs will be secured.[83] Situated well above that line, a *Severodvinsk* class will still be able to hold all of Northern Europe's main ports at risk, as well as many of the important airfields used for logistics. As Magnus Nordenman points out, it seems likely that the modern approach to the SLOC mission is to hold NATO sea ports at risk, as opposed to patrolling the actual SLOCs at sea.[84]

Alliance steps to keep track of the threat

Although the political climate between Russia and the West, including Norway, for some time has been challenging at best, both Russia and Norway regularly point to the potential for cooperation on fishery regulation and administration, maritime trade, and technology.[85] There is also cooperation taking place between the two coastguards, as well as for search and rescue in the Arctic.

However, Norway has, since 2008, consistently been working to get NATO to re-focus more on the North Atlantic and the maritime domains.[86] The lack of warning times in a potential modern conflict in the High North necessitates presence and encompassing surveillance, an understanding that has settled at the highest political level in Norway.[87] And indeed, NATO has, in the past years, shifted its focus back to the maritime domain and ASW in particular.[88] Exercise *Trident Juncture* in autumn 2018 was the largest exercise held in Norway in several decades, and during the exercise a US aircraft carrier sailed north of the Arctic circle off Norway for the first time since 1991.[89] The recent rise in Western nuclear submarines calling on Norwegian ports is another sign of increased activity by Western submarines on highly classified missions in the High North.[90] Commentators also take note of NATO increasing the regularity and scope of the Alliance's ASW exercises since the Wales summit in 2014, in reaction to increased Russian submarine activities.[91] In mid-2017, the United States, the United Kingdom, and Norway signed a trilateral Statement of Intent (SoI) that focusses on cost-efficient logistics and enhanced cooperation for the three nations' P-8 operations.[92] The statement does not elaborate as to what this operational cooperation looks like in practice. However, it is likely that the three nations, flying the same aircraft, will work closely in peacetime as well as during crises to monitor the North Atlantic, both above and under the surface. As we have seen in the previous chapters, there is much experience on this type of collaboration to build on from the Cold War. The US Navy is again showing interest in a regular presence at Keflavik, after closing the base in 2006.[93] But the US Navy has also considered regularly utilizing Evenes in Northern Norway as a base for maritime airborne surveillance in the High North, indicating that Keflavik is still not close enough to the areas of interest.[94] However, Norway still enforces her basing policy, and it will be interesting to see if dynamics on allied aircraft operating from Norwegian soil will change in the near future. Supported by tanker aircraft in coordinated missions, both American and British SIGINT aircraft regularly cover the Barents Sea, particularly during exercises or other activities of particular interest.[95]

In addition to this, two Norwegian intelligence collection ships operate more or less continuously in the High North. The ageing *Eger* has the main responsibility of monitoring the northern parts of the Norwegian Sea and Western Barents Sea, while the much larger vessel *Marjata* monitors the Eastern parts of the Barents Sea.[96] The *Marjata* has been sailing in the Barents Sea since 2016, and is considered one of the most advanced vessels of its kind. The rationale for the more or less constant presence of these ships is to map all military activity and some civilian activity in the Barents Sea and areas close to Norway in order to establish what constitutes the normal situation or baseline activity.[97] The Norwegian Navy has also increased its presence in the High North, with regular frigate patrols, in particular.[98]

The re-establishment of the US Navy Second Fleet in Norfolk, Virginia, in May 2018, with a focus on high-end naval warfare in the Atlantic, was another reaction to the rise in Russian naval activities.[99] The Second Fleet will have the entire North Atlantic as its AOR, based on both the expansionist nature of Russian naval operations and the new long-range weapons that are being fielded.[100] The High North is, to an increasing degree, regarded as a potential battlespace of the future, as opposed to the protection of SLOC further south.[101]

Moreover, in the summer of 2018, NATO established the Joint Force Command (JFC) in Norfolk, with a focus on NATO command and control of naval operations in the North Atlantic. This, too, was established as a direct consequence of increased military activity in the North Atlantic and the Arctic.[102] NATO has also boosted the personnel of the Alliance's primary naval commander and advisor, MARCOM.[103]

In sum, both NATO and individual member states have taken steps to respond to the re-emergence of a stronger, more adversarial Russian naval presence in the High North. This is now recognized as a field of strategic priority.

Notes

1 Sveinung Berg Bentzrød, "Etterretningsstasjoner i Nord Anklages for Hemmelighold. Nå Slår E-Sjefen Tilbake," *Aftenposten*, October 8, 2016, www.aftenposten.no/norge/i/mwgog/Etterretningsstasjoner-i-nord-anklages-for-hemmelighold-Na-slar-E-sjefen-tilbake.

2 Bentzrød.

3 Trude Pettersen, "Black Sea Submarine Fires Cruise Missile from the Barents Sea," *The Barents Observer*, August 4, 2015, https://barentsobserver.com/en/security/2015/08/black-sea-submarine-fires-cruise-missile-barents-sea-04–08.

4 Per Anders Johansen, "Russisk Atomubåt Testet Atomraketter i Barentshavet i Morges," *Aftenposten*, December 12, 2015; Trude Pettersen, "Nuclear Missile World Record Attempt," *The Barents Observer*, March 2, 2016, https://thebarentsobserver.com/en/security/2016/03/nuclear-missile-world-record-attempt.
5 Alf Bjarne Johnsen, "E-Sjefen: Russland Øvde På Angrep Mot Nord-Norge," *VG*, March 5, 2019, www.vg.no/nyheter/innenriks/i/1kye4W/e-sjefen-russland-oevde-paa-angrep-mot-nord-norge.
6 Andrew Higgins, "Norway Reverts to Cold War Mode as Russian Air Patrols Spike," *The New York Times*, April 1, 2015, www.nytimes.com/2015/04/02/world/europe/a-newly-assertive-russia-jolts-norways-air-defenses-into-action.html?emc=eta1.
7 Norwegian Intelligence Service, "FOKUS 2024 – The Assessment of the Intelligence Service," Annual Unclassified Assessment (Oslo, 2024), 37, www.etterretningstjenesten.no/publikasjoner/fokus/Fokus24_innhold.
8 Phil Stewart, "Russia Jets Make 'Simulated Attack' Passes near U.S. Destroyer," *Reuters*, April 13, 2016, www.reuters.com/article/us-usa-russia-simulatedattack-idUSKCN0XA1UW; Ben Werner, "Russian Su-27 Fighter Buzzes U.S. Navy EP-3 Aries Over Black Sea," *USNI News*, January 29, 2018, https://news.usni.org/2018/01/29/30987; Tom Nilssen, "Nye Bildebevis Av Russisk Avskjæring," *NRK*, January 31, 2018, www.nrk.no/urix/nye-bildebevis-av-russisk-avskjaering-1.13894490; Barbara Starr and Brad Lendon, "Russian Fighter Came within 15 Feet of U.S. Air Force Jet," *CNN*, January 30, 2016, https://edition.cnn.com/2016/01/29/politics/us-russia-planes-intercept/index.html.
9 Marcus Jonathan, "US Confirms Drone Shot down by Iran," *BBC News Online*, June 20, 2019, www.bbc.com/news/world-middle-east-48700965.
10 Max Matza, "Everything We Know About the US Drone That Crashed in the Black Sea," *BBC*, March 16, 2023, www.bbc.com/news/world-us-canada-64972002.
11 Thomas Nilsen, "Alarm-Drill: 36 Russian Warships Sail out to Barents Sea," *The Barents Observer*, June 13, 2018, https://thebarentsobserver.com/en/security/2018/06/36-russian-warships-sails-out-barents-sea.
12 Karin Madshus and Rolf Lofstad, "Norsk Lasteskip Ble Anropt Av Russisk Slagskip," *Dagbladet*, April 15, 2019, www.dagbladet.no/nyheter/norsk-lasteskip-ble-anropt-av-russisk-slagskip/70985345.
13 Andreas Budalen and Henrik Ø. Heldahl, "Det Er Uvanlig Høy Russisk Aktivitet Utenfor Norskekysten," *NRK*, August 6, 2019, www.nrk.no/nordland/_-det-er-uvanlig-hoy-russisk-aktivitet-utenfor-norskekysten-1.14650593; Markus Thonhaugen and Kåre Riibe Ramskjell, "Russerne Med 'Uvanlig Stor' Militærøvelse: – Lenge Siden vi Har Sett Noe Slikt," *NRK*, August 13, 2019, www.nrk.no/nordland/russerne-med-_uvanlig-stor_-militaerovelse_-_-lenge-siden-vi-har-sett-noe-slikt-1.14658213.
14 Julie Vissgren, Tormod Strand, and Lokman Ghorbani, "Forsvarssjefen Om Russisk Militærøvelse: – En Nasjonal Utfordring," *NRK*, August 14, 2019, www.nrk.no/urix/forsvarssjefen-om-russisk-militaerovelse_-_-en-nasjonal-utfordring-1.14660898; Atle Staalesen, "30 Russian Naval Vessels Stafe Show of Force near Coast of Norway," *The Barents Observer*, August 15, 2019, https://thebarentsobserver.com/en/security/2019/08/30-russian-naval-vessels-stage-show-force-coast-norway; Thomas Nilsen, "Russian Navy Announces Comprehensive Naval Exercises off Northern Norway," *The Arctic Today*, August 6, 2019, www.arctictoday.com/russian-navy-announces-comprehensive-naval-exercises-off-northern-norway/.
15 Alf Bjarne Johnsen, "Russland i Gang Med Ubåtoperasjon i Nord: Største Siden Sovjet-Tiden," *VG*, October 29, 2019, www.vg.no/nyheter/innenriks/i/opxAP0/russland-i-gang-med-ubaat-operasjon-i-nord-stoerste-siden-sovjet-tiden?utm_source=vgfront&utm_content=row-2.
16 Tormod Strand, "Hemmelig Ubåt-Operasjon: 'Målet Er å Vise at Russland Kan Nå USA,'" *NRK*, October 29, 2019, www.nrk.no/norge/hemmelig-ubat-operasjon_-_malet-er-a-vise-at-russland-kan-na-usa_-1.14761298; Johnsen, "Russland i Gang Med

Ubåtoperasjon i Nord: Største Siden Sovjet-Tiden"; Tyler Rogoway, "Russia Sends Ten Subs into North Atlantic in Drill Unprecedented in Size since Cold War," *The Drive*, October 29, 2019, www.thedrive.com/the-war-zone/30728/russia-sends-ten-subs-into-north-atlantic-in-drill-unprecedented-in-size-since-cold-war.

17 Johnsen, "Russland i Gang Med Ubåtoperasjon i Nord: Største Siden Sovjet-Tiden"; "Hektisk Jakt På Ubåter," *Vesterålen Online*, October 31, 2019, www.vol.no/pluss/2019/10/31/Hektisk-jakt-p%C3%A5-ub%C3%A5ter-20289339.ece.

18 Norwegian Intelligence Service, "FOKUS 2019 – The Assessment of the Intelligence Service," Annual Unclassified Assessment| (Oslo, 2019), 26.

19 Norwegian Intelligence Service, "FOCUS 2023 – The Assessment of the Intelligence Service," Annual Unclassified Assessment (Oslo, 2023), 18.

20 Valery Gerasimov, "The Value of Science Is in the Foresight," *Military-Industrial Kurier (VPK)*, February 26, 2013, www.vpk-news.ru/articles/14632.

21 Patrick J. Cullen and Erik Reichborn-Kjennerud, "Understanding Hybrid Warfare," MCDC Countering Hybrid Warfare Project (Multinational Capability Development Campaign (MCDC), January 2017), 8. There is no agreed definition of hybrid warfare, not within NATO nor within academia. This MCDC report provides focussed description of the problem, and gives an excellent overview of the mechanisms in the application of power involved.

22 Molly M. McKew, "The Gerasimov Doctrine," *Politico Magazine*, October 2017; Michael Kofman, "Russian Hybrid Warfare and Other Dark Arts," *War On the Rocks*, March 11, 2016, https://warontherocks.com/2016/03/russian-hybrid-warfare-and-other-dark-arts/; Mark Galeotti, "I'm Sorry for Creating the 'Gerasimov Doctrine,'" *Foreign Policy*, March 5, 2018, https://foreignpolicy.com/2018/03/05/im-sorry-for-creating-the-gerasimov-doctrine/.

23 James Stavridis, "Maritime Hybrid Warfare Is Coming," *United States Naval Institute Proceedings* 142, no. 12 (December 2016).

24 Chris Kremidas-Courtney, "Countering Hybrid Threats in the Maritime Environment," *The Maritime Executive*, November 6, 2018, www.maritime-executive.com/editorials/countering-hybrid-threats-in-the-maritime-environment; Stavridis, "Maritime Hybrid Warfare Is Coming"; Jukka Savolainen et al., "Handbook on Maritime Hybrid Threats – 10 Scenarios and Legal Scans" (Helsinki, Finland: The European Centre of Excellence for Countering Hybrid Threats, November 2019).

25 Jan Harald Tomassen, "Skal Ha Bedt Russiske Sjøkapteiner Gjøre Seg Godt Kjent Langs Norskekysten," *NRK*, February 27, 2015, www.nrk.no/troms/skal-ha-bedt-russiske-sjokapteiner-gjore-seg-godt-kjent-langs-norskekysten-1.12234101.

26 Kathleen Hicks et al., "Undersea Warfare in Northern Europe," CSIS International Security Program (Center for Strategic and International Studies, July 2016), 4.

27 Thomas Nilsen, "Now, Russia Builds a Submarine Even Bigger than the Typhoon," *The Barents Observer*, May 3, 2017, https://thebarentsobserver.com/en/security/2017/05/russias-new-military-research-submarine-arctic-waters-will-be-worlds-largest#.WQnEKkjm64w.facebook.

28 AlfBjarne Johnsen, "NATO: Russerne Kan Sabotere Internett Med Ubåt," *VG*, February 17, 2018, www.vg.no/nyheter/utenriks/i/gPday5/nato-russerne-kan-sabotere-internett-med-ubaat; Thomas Nilsen, "Russia in the High North" (Matpakkeseminar, Norwegian MoD, November 14, 2019).

29 TeleGeography, "Submarine Cable Map" (TeleGeography), accessed October 23, 2019, www.submarinecablemap.com/.

30 Michael Kofman, "Fire Aboard AS-31 Losharik: Brief Overview," *Russian Military Analysis* (blog), July 3, 2019, https://russianmilitaryanalysis.wordpress.com/tag/gugi/; Bill Gertz, "Damaged Russian Sub Linked to Underwater Drone Program," *Soldier of Fortune* (blog), July 15, 2019, www.sofmag.com/damaged-russian-sub-linked-to-underwater-drone-program/; Heather A. Conley, Jeffrey Rathke, and Matthew Melino, "Enhanced Deterrence in the North – A 21st Century European Engagement Strategy" (Center for

Strategic and International Studies, February 2018), 13; Hicks et al., "Undersea Warfare in Northern Europe," 4.

31 Kofman, "Fire Aboard AS-31 Losharik: Brief Overview."

32 Sherry Sontag and Christopher Drew, *Blind Man's Bluff – The Untold Story of American Submarine Espionage* (New York: HarperPaperbacks, 1999), 376, 389.

33 Office of Naval Intelligence, "The Russian Navy – A Historic Transition" (Office of Naval Intelligence (ONI), December 2015), 17, www.oni.navy.mil/Portals/12/Intel%20agencies/russia/Russia%202015print.pdf?ver=2015-12-14-082038-923.

34 Hicks et al., "Undersea Warfare in Northern Europe," 9–11.

35 Office of Naval Intelligence, "The Russian Navy – A Historic Transition," 18.

36 Hicks et al., "Undersea Warfare in Northern Europe," 14.

37 Franz-Stefan Gady, "Russia's Pacific Fleet to Receive 2 Fast Attack Subs in 2020," *The Diplomat*, July 3, 2018, https://thediplomat.com/2018/07/russias-pacific-fleet-to-receive-2-fast-attack-subs-in-2020/; Conley, Rathke, and Melino, "Enhanced Deterrence in the North – A 21st Century European Engagement Strategy," 13.

38 Alexander Korolkov, "Russia's Top-Secret Nuclear Submarine Comes into Service," *Russia Beyond*, June 17, 2014, www.rbth.com/defence/2014/06/17/russias_top-secret_nuclear_submarine_comes_into_service_37483.html; Office of Naval Intelligence, "The Russian Navy – A Historic Transition," 18; Ministry of Defence of the Russian Federation, "Severodvinsk Nuclear Submarine of the Northern Fleet Carried out Missile Firing," August 18, 2017, http://eng.mil.ru/en/news_page/country/more.htm?id=12138913@egNews; Aleksandr Mozgovoy, "Russian Navy's Long Arm: Kalibr Missile Family," trans. J. Hawk, *Natsionalnaya Oborona/Southfront.Org*, August 10, 2015, https://southfront.org/russian-navys-long-arm-kalibr-missile-family/.

39 Hicks et al., "Undersea Warfare in Northern Europe," 15; Charlie Gao, "Russia's New Husky-Class Stealth Submarines: Armed with Hypersonic Missiles?" *The National Interest*, October 1, 2019, https://nationalinterest.org/blog/buzz/russias-new-husky-class-stealth-wsubmarines-armed-hypersonic-missiles-84626; H I Sutton, "Husky_SSN," *Covert Shores* (blog), December 30, 2019, www.hisutton.com/Husky_SSN.html; Dmitry Gorenburg, "Russian Naval Shipbuilding – Is It Possible to Fulfill the Kremlin's Grand Expectations?" (CNA Harvard University, October 2015), 1, www.ponarseurasia.org/sites/default/files/policy-memos-pdf/Pepm395_Gorenburg_Oct2015.pdf.

40 Gorenburg, "PONARS Eurasia Memo No. 395," 1–2.

41 Gao, "Russia's New Husky-Class Stealth Submarines: Armed with Hypersonic Missiles?"; Sutton, "Husky_SSN."

42 Gorenburg, "PONARS Eurasia Memo No. 395," 2.

43 Franz-Stefan Gady, "Putin's 'Red October': Russia's Deadliest New Submarine," *The Diplomat*, March 4, 2015, https://thediplomat.com/2015/03/putins-red-october-russias-deadliest-new-submarine/.

44 Office of Naval Intelligence, "The Russian Navy – A Historic Transition," 17; Gady, "Putin's 'Red October': Russia's Deadliest New Submarine"; Hans M. Kristensen and Robert S. Norris, "Russian Nuclear Forces, 2015," *Bulletin of the Atomic Scientists* 71, no. 3 (2015): 91; Pettersen, "Nuclear Missile World Record Attempt."

45 Mark Episkopos, "Meet the First Russian Submarine to Fire in Anger since World War II (and Its New Cruise Missiles)," *The National Interest*, January 12, 2019, https://nationalinterest.org/blog/buzz/meet-first-russian-submarine-fire-anger-world-war-ii-and-its-new-cruise-missiles-41292; Lee Willett, "Game Changer: Russian Sub-Launched Cruise Missiles Bring Strategic Effect," *Jane's By IHS Markit*, 2017, www.janes.com/images/assets/147/70147/Game_changer_Russian_sub-launched_cruise_missiles_bring_strategic_effect_edit.pdf.

46 Hicks et al., "Undersea Warfare in Northern Europe," 15–16.

47 Office of Naval Intelligence, "The Russian Navy – A Historic Transition," 19; Gorenburg, "PONARS Eurasia Memo No. 395," 2; Hicks et al., "Undersea Warfare in Northern Europe," 15–16.

48 Andresen and Bukkvoll, *Russian Weapons Development Towards 2020*, 47–48.
49 Willett, "Game Changer."
50 Office of Naval Intelligence, "The Russian Navy – A Historic Transition," 35.
51 Bård Wormdal, "Tidligere Forsvarssjef Om Nye Russiske Våpen: – En Trussel Som Må Tas Alvorlig," *NRK Online*, August 11, 2017, www.nrk.no/finnmark/tidligere-forsvarssjef-om-nye-russiske-vapen_-_-en-trussel-som-ma-tas-alvorlig-1.13767105.
52 Office of Naval Intelligence, "The Russian Navy – A Historic Transition," 35; Mozgovoy, "Russian Navy's Long Arm: Kalibr Missile Family."
53 According to PortTechnology.org, the top 5 ports in Northern Europe are Rotterdam, Antwerp, Hamburg, Bremerhaven and Felixstowe. See Matteo Natalucci, "Top 5 Ports in Northern Europe," *Port Technology*, August 30, 2019, www.porttechnology.org/news/top-5-ports-in-northern-europe/.
54 Robert Farley, "Is the Day of the Land Attack Cruise Missile upon Us?" *The Diplomat*, October9,2015,http://thediplomat.com/2015/10/is-the-day-of-the-land-attack-cruise-missile-upon-us/; Sebastian Roblin, "Why Russia's Enemies Fear the Kalibr Cruise Missile," *The National Interest*, January 22, 2017, https://nationalinterest.org/blog/the-buzz/why-russias-enemies-fear-the-kalibr-cruise-missile-19129.
55 Roblin, "Why Russia's Enemies Fear the Kalibr Cruise Missile"; Willett, "Game Changer."
56 Office of Naval Intelligence, "The Russian Navy – A Historic Transition," 34.
57 Mozgovoy, "Russian Navy's Long Arm: Kalibr Missile Family."
58 Dave Majumdar, "Cruise Missiles Strikes in Syria: Russia's Big Ad Campaign?" October 8, 2015, https://nationalinterest.org/blog/the-buzz/cruise-missile-strikes-syria-russias-big-ad-campaign-14032.
59 Roblin, "Why Russia's Enemies Fear the Kalibr Cruise Missile."
60 Marta Portocarrero, "Russia Fires First Submarine Missiles against Isis Targets in Syria," *Independent*, December 10, 2015, www.independent.co.uk/news/world/europe/russia-fires-first-submarine-missiles-against-isis-targets-in-syria-a6766426.html; BBC News Online, "Russia Hits Targets in Syria from Mediterranean Submarine," *BBC News Online*, December 8, 2015, www.bbc.com/news/world-middle-east-35041656.
61 Gady, "Russia's Pacific Fleet to Receive 2 Fast Attack Subs in 2020."
62 Mark Episkopos, "Meet the First Russian Submarines to Fire in Anger since World War II (and Its New Cruise Missiles)," *The National Interest*, January 12, 2019, https://nationalinterest.org/blog/buzz/meet-first-russian-submarine-fire-anger-world-war-ii-and-its-new-cruise-missiles-41292; Willett, "Game Changer"; Roblin, "Why Russia's Enemies Fear the Kalibr Cruise Missile."
63 Owen R. Jr. Cote, "The Third Battle – Innovation in the U.S.Navy's Silent War Struggle with Soviet Submarines," *Naval War College Newport Papers*, no. 16 (2003): 86.
64 Franz-Stefan Gady, "Russia to Upgrade 12 Nuclear-Powered Subs," *The Diplomat*, October 5, 2015, https://thediplomat.com/2015/10/russia-to-upgrade-12-nuclear-powered-subs/; Andrei Akulov, "Kalibr: Russia's Naval System Upping Cruise Missile Game," *Strategic Culture Foundation*, May 24, 2016, www.strategic-culture.org/news/2016/05/24/kalibr-russia-naval-system-upping-cruise-missile-game/.
65 Willett, "Game Changer."
66 Thomas Nilsen, "Watch Russian Submarine Test Fire Cruise Missile in Barents Sea," *The Barents Observer*, August 18, 2017, https://thebarentsobserver.com/en/security/2017/08/watch-russian-submarine-test-fire-cruise-missile-barents-sea.
67 Thomas Nilsen, "Russian Sub Launched Cruise Missile without Leaving Port," *The Barents Observer*, April 3, 2019, https://thebarentsobserver.com/en/security/2019/04/russian-sub-launched-cruise-missile-without-leaving-port#.XKQ_dO8JV7U.facebook.
68 R. Jeffrey Smith, "Hypersonic Missiles Are Unstoppable. And They're Starting a New Global Arms Race.," *New York Times Magazine*, June 19, 2019, www.nytimes.com/2019/06/19/magazine/hypersonic-missiles.html.

69 Matthew Bodner, "Russia Has Plans for Its Future Nuclear Subs, and It Involves Hypersonic Missiles," *Defense News*, June 1, 2018, www.defensenews.com/naval/2018/06/01/russia-has-plans-for-its-future-nuclear-subs-and-it-involves-hypersonic-missiles/?utm_source=Sailthru&utm_medium=email&utm_campaign=DFN%20DNR%206/1/18&utm_term=Editorial%20-%20Daily%20News%20Roundup.
70 Matthew Bodner, "Russia Adds 'Kazan' to Its Nuclear Attack Submarine Fleet," *Defense News*, accessed October 24, 2019, www.defensenews.com/naval/2017/03/31/russia-adds-kazan-to-its-nuclear-attack-submarine-fleet/.
71 See for example Willett, "Game Changer."
72 Alf Bjarne Johnsen, "Søreide: Ber NATO Trekke Nordover," *VG*, February 15, 2016, www.vg.no/nyheter/innenriks/i/wWazo/soereide-ber-nato-trekke-nordover.
73 Ståle Ulriksen, "Endring i Nord," *Bergens Tidende*, May 27, 2019.
74 Richard Connolly, "Towards a Dual Fleet? The Maritime Doctrine of the Russian Federation and the Modernisation of Russian Naval Capabilities," *Russian Studies, NATO Defense College*, no. 2/17 (June 2017): 2.
75 Liv Karin Parnemo, "Russia's Naval Development – Grand Ambitions and Tactical Pragmatism," *The Journal of Slavic Military Studies* 32, no. 1 (2019): 44.
76 The Kremlin, "Meeting with Defence Ministry Leadership and Military-Industrial Complex Representatives," May 12, 2015, http://en.kremlin.ru/events/president/news/49456.
77 Matthew Bodner, "NATO Deputy SecGen: Russia's Anti-Access/Area Denial Build-Up Is Biggest Worry," *Security and Policy*, February 14, 2016, http://securityandpolicy.com/nato-deputy-secgen-russias-anti-accessarea-denial-build-up-is-biggest-worry/.
78 Hicks et al., "Undersea Warfare in Northern Europe," 2.
79 James Holmes, "More Submarines and Less Aircraft Carriers: If the U.S. Navy Could Be Completely Rebuilt," *The National Interest*, December 3, 2017, https://nationalinterest.org/blog/the-buzz/more-submarines-less-aircraft-carriers-if-the-us-navy-could-23482?nopaging=1.
80 The discussion of anti-SSBN versus pro-SSBN can be said to have been laid to rest in the late 90s, where the anti-SSBN mission in the North Atlantic was effectively cancelled. See Kristian Atland, "The Introduction, Adaptation and Implementation of Russia's 'Northern Strategic Bastion' Concept, 1992–1999," *Journal of Slavic Military Studies* 20, no. 4 (2007): 519.
81 Andrew Krepinevich, Barry Watts, and Robert Work, "Meeting the Anti-Access and Area-Denial Challenge" (Center for Strategic and Budgetary Assessments (CSBA), 2003), http://csbaonline.org/uploads/documents/2003.05.20-Anti-Access-Area-Denial-A2-AD.pdf; John O. Birkeland, "The Unsinkable Aircraft Carrier – An American Response to the Chinese Anti-Access/Area Denial (A2/AD) Challenge" (Master of Military Operational Art and Science, Maxwell AFB, Montgomery, Alabama, Air Command & Staff College, USAF, 2013), file:///C:/Users/JohnO/Downloads/ADA603131.pdf; James Lacey, "Battle of the Bastions," *War On The Rocks* (blog), January 9, 2020, https://warontherocks.com/2020/01/battle-of-the-bastions/.
82 Atland, "The Introduction, Adaptation and Implementation of Russia's 'Northern Strategic Bastion' Concept, 1992–1999," 524.
83 Conley, Rathke, and Melino, "Enhanced Deterrence in the North – A 21st Century European Engagement Strategy," 36.
84 Magnus Nordenman, *The New Battle for the Atlantic – Emerging Naval Competition with Russia in the Far North* (Annapolis, MD: Naval Institute Press, 2019), 199–200.
85 Per Anders Johansen, "Putin Med Overraskende Norge-Flørt: På Tide å Utvide Horisonten Og Samarbeidet," *Aftenposten*, October 21, 2018, www.aftenposten.no/verden/i/1kJozl/putin-med-overraskende-norge-floert-paa-tide-aa-utvide-horisonten-og-samarbeidet.
86 Erik Lieungh, "Norge Trekker NATOs Øyne Mot Nordflanken," *NRK*, February 13, 2016, www.nrk.no/finnmark/norge-trekker-natos-oyne-mot-nordflanken-1.12801938; Alf Ole

Ask, "NATO Møter Utfordringer i Alle Himmelretninger," *Aftenposten*, February 15, 2016, www.aftenposten.no/verden/i/o60K/nato-moeter-utfordringer-i-alle-himmelretninger.

87 Øystein Antonsen and Arild Moe, "Alle Ser Mot Nord: -Arktis Gir Oss Flere Muligheter," *NRK*, January 22, 2018, www.nrk.no/troms/alle-ser-mot-nord_-_-arktis-gir-oss-flere-muligheter-1.13879346.

88 NATO, "Warsaw Summit Communique" (NATO, July 9, 2016), para 47, NATO, online archive.

89 Geoff Ziezulewicz and David B. Larter, "The Navy Sends a Carrier Back to Russia's Arcic Haunts," *Navy Times*, October 19, 2018, www.navytimes.com/news/your-navy/2018/10/19/the-navy-sends-a-carrier-back-to-russias-arctic-haunts/#.W820deP9FUo.email.

90 Kjell Persen, "Amerikansk Atomubåt Søkte Nødhavn i Norge," *TV2.No*, October 27, 2017, www.tv2.no/a/9433425.

91 Willett, "Game Changer."

92 George Allison, "US in Deal with UK and Norway to Form P-8 Poseidon Partnership," *UK Defence Journal*, July 4, 2017, https://ukdefencejournal.org.uk/us-deal-uk-norway-form-p-8-poseidon-partnership/.

93 Trude Pettersen, "U.S. Military Returns to Iceland," *The Barents Observer*, October 2, 2016, https://thebarentsobserver.com/en/security/2016/02/us-military-returns-iceland#.VsVfwxKHhBK.mailto.

94 NTB, "Evenes Flystasjon Kan Bli Overvåkingsbase for USA," *VG*, July 21, 2019, www.vg.no/nyheter/innenriks/i/bprjB/evenes-flystasjon-kan-bli-overvaakingsbase-for-usa#xtor=CS4-1.

95 Tyler Rogoway, "USAF's Nuke Sniffing Plane Is Flying on a Mission Near the Arctic Right Now," *The Drive*, February 22, 2017, www.thedrive.com/the-war-zone/7830/usafs-nuke-sniffing-plane-is-flying-on-a-mission-near-the-arctic-right-now?xid=emailshare; Sam LaGrone, "Russian Air Force Chief: U.S. Surveillance Flights Monitor Russia Daily," *USNI News*, December 17, 2014, https://news.usni.org/2014/12/17/russian-air-force-chief-u-s-surveillance-flights-monitor-russia-daily.

96 Sveinung Berg Bentzrød, "E-Sjefen: Derfor Trenger Norge to Spionskip i Nord," *Aftenposten*, September 26, 2016, www.aftenposten.no/norge/i/A9oqM/E-sjefen-Derfor-trenger-Norge-to-spionskip-i-nord.

97 Trude Pettersen, "U.S. Builds Up Arctic Intelligence Network," *The Barents Observer*, September 8, 2015, https://barentsobserver.com/en/security/2015/09/us-builds-arctic-intelligence-network-08-09.

98 Anders Brekke and Joakim Reigstad, "Dagens Forsvar Er Ikke Bærekraftig," *NRK*, April 11, 2016, www.nrk.no/norge/_-dagens-forsvar-er-ikke-baerekraftig-1.12894608.

99 David B. Larter and Mark D. Faram, "The US Navy's New Command Puts Russia in the Crosshairs," *Defense News*, May 4, 2018, www.defensenews.com/naval/2018/05/04/the-us-navys-new-command-puts-russia-in-the-crosshairs/#.WuzF9tk3w80.email.

100 Sam LaGrone, "CNO: New 2nd Fleet Boundary Will Extend North to the Edge of Russian Waters," *USNI News*, August 24, 2018, https://news.usni.org/2018/08/24/cno-new-2nd-fleet-boundary-will-extend-north-edge-russian-waters.

101 LaGrone.

102 Andrew ADM Lewis, "One Arctic: Maintaining Stability and Prosperity in the High North" (Henry Bacon Seminar, Washington, DC, May 8, 2019).

103 Deborah Haynes, "Putin's Submarines Spur NATO to Boost Its UK Nerve Centre," *The Times*, January 10, 2018, www.thetimes.co.uk/article/putin-s-submarines-spur-nato-to-boost-its-uk-nerve-centre-8h2bf95qp.

Bibliography

ADM Lewis, Andrew. "One Arctic: Maintaining Stability and Prosperity in the High North." Presented at the Henry Bacon seminar, Washington, DC, May 8, 2019.

Akulov, Andrei. "Kalibr: Russia's Naval System Upping Cruise Missile Game." *Strategic Culture Foundation*, May 24, 2016. www.strategic-culture.org/news/2016/05/24/kalibr-russia-naval-system-upping-cruise-missile-game/.

Allison, George. "US in Deal with UK and Norway to Form P-8 Poseidon Partnership." *UK Defence Journal*, July 4, 2017. https://ukdefencejournal.org.uk/us-deal-uk-norway-form-p-8-poseidon-partnership/.

Andresen, Rolf-Inge V., and Tor Bukkvoll. *Russian Weapons Development Towards 2020*. Forsvarets Forskningsinstitutt (FFI) FFI-rapport 2008/01957, January 12, 2009.

Antonsen, Øystein, and Arild Moe. "Alle Ser Mot Nord: -Arktis Gir Oss Flere Muligheter." *NRK*, January 22, 2018. www.nrk.no/troms/alle-ser-mot-nord_-_-arktis-gir-oss-flere-muligheter-1.13879346.

Ask, Alf Ole. "NATO Møter Utfordringer i Alle Himmelretninger." *Aftenposten*, February 15, 2016. www.aftenposten.no/verden/i/o60K/nato-moeter-utfordringer-i-alle-himmelretninger.

Atland, Kristian. "The Introduction, Adaptation and Implementation of Russia's 'Northern Strategic Bastion' Concept, 1992–1999." *Journal of Slavic Military Studies* 20, no. 4 (2007): 499–528.

BBC News Online. "Russia Hits Targets in Syria from Mediterranean Submarine." *BBC News Online*, December 8, 2015. www.bbc.com/news/world-middle-east-35041656.

Bentzrød, Sveinung Berg. "E-Sjefen: Derfor Trenger Norge to Spionskip i Nord." *Aftenposten*, September 26, 2016. www.aftenposten.no/norge/i/A9oqM/E-sjefen-Derfor-trenger-Norge-to-spionskip-i-nord.

Bentzrød, Sveinung Berg. "Etterretningsstasjoner i Nord Anklages for Hemmelighold. Nå Slår E-Sjefen Tilbake." *Aftenposten*, October 8, 2016. www.aftenposten.no/norge/i/mwgog/Etterretningsstasjoner-i-nord-anklages-for-hemmelighold-Na-slar-E-sjefen-tilbake.

Birkeland, John O. "The Unsinkable Aircraft Carrier – An American Response to the Chinese Anti-Access/Area Denial (A2/AD) Challenge." Master of Military Operational Art and Science, Air Command & Staff College, USAF, 2013. file:///C:/Users/JohnO/Downloads/ADA603131.pdf.

Bodner, Matthew. "NATO Deputy SecGen: Russia's Anti-Access/Area Denial Build-Up Is Biggest Worry." *Security and Policy*, February 14, 2016. http://securityandpolicy.com/nato-deputy-secgen-russias-anti-accessarea-denial-build-up-is-biggest-worry/.

Bodner, Matthew. "Russia Adds 'Kazan' to Its Nuclear Attack Submarine Fleet." *Defense News*. Accessed October 24, 2019. www.defensenews.com/naval/2017/03/31/russia-adds-kazan-to-its-nuclear-attack-submarine-fleet/.

Bodner, Matthew. "Russia Has Plans for Its Future Nuclear Subs, and It Involves Hypersonic Missiles." *Defense News*, June 1, 2018. www.defensenews.com/naval/2018/06/01/russia-has-plans-for-its-future-nuclear-subs-and-it-involves-hypersonic-missiles/?utm_source=Sailthru&utm_medium=email&utm_campaign=DFN%20DNR%206/1/18&utm_term=Editorial%20-%20Daily%20News%20Roundup.

Brekke, Anders, and Joakim Reigstad. "- Dagens Forsvar Er Ikke Bærekraftig." *NRK*, April 11, 2016. www.nrk.no/norge/_-dagens-forsvar-er-ikke-baerekraftig-1.12894608.

Budalen, Andreas, and Henrik Ø. Heldahl. "Det Er Uvanlig Høy Russisk Aktivitet Utenfor Norskekysten." *NRK*, August 6, 2019. www.nrk.no/nordland/_-det-er-uvanlig-hoy-russisk-aktivitet-utenfor-norskekysten-1.14650593.

Conley, Heather A., Jeffrey Rathke, and Matthew Melino. *Enhanced Deterrence in the North – A 21st Century European Engagement Strategy*. Center for Strategic and International Studies, February 2018.

Connolly, Richard. "Towards a Dual Fleet? The Maritime Doctrine of the Russian Federation and the Modernisation of Russian Naval Capabilities." *Russian Studies, NATO Defense College* 2, no. 17 (June 2017).

Cote, Owen R. Jr. "The Third Battle – Innovation in the U.S. Navy's Silent War Struggle with Soviet Submarines." *Naval War College Newport Papers* 16 (2003).

Cullen, Patrick J., and Erik Reichborn-Kjennerud. *Understanding Hybrid Warfare*. MCDC Countering Hybrid Warfare Project. Multinational Capability Development Campaign (MCDC), January 2017.

Episkopos, Mark. "Meet the First Russian Submarine to Fire in Anger since World War II (and Its New Cruise Missiles)." *The National Interest*, January 12, 2019. https://nationalinterest.org/blog/buzz/meet-first-russian-submarine-fire-anger-world-war-ii-and-its-new-cruise-missiles-41292.

Episkopos, Mark. "Meet the First Russian Submarines to Fire in Anger since World War II (and Its New Cruise Missiles)." *The National Interest*, January 12, 2019. https://nationalinterest.org/blog/buzz/meet-first-russian-submarine-fire-anger-world-war-ii-and-its-new-cruise-missiles-41292.

Farley, Robert. "Is the Day of the Land Attack Cruise Missile upon Us?" *The Diplomat*, October 9, 2015. http://thediplomat.com/2015/10/is-the-day-of-the-land-attack-cruise-missile-upon-us/.

Gady, Franz-Stefan. "Putin's 'Red October': Russia's Deadliest New Submarine." *The Diplomat*, March 4, 2015. https://thediplomat.com/2015/03/putins-red-october-russias-deadliest-new-submarine/.

Gady, Franz-Stefan. "Russia to Upgrade 12 Nuclear-Powered Subs." *The Diplomat*, October 5, 2015. https://thediplomat.com/2015/10/russia-to-upgrade-12-nuclear-powered-subs/.

Gady, Franz-Stefan. "Russia's Pacific Fleet to Receive 2 Fast Attack Subs in 2020." *The Diplomat*, July 3, 2018. https://thediplomat.com/2018/07/russias-pacific-fleet-to-receive-2-fast-attack-subs-in-2020/.

Galeotti, Mark. "I'm Sorry for Creating the 'Gerasimov Doctrine.'" *Foreign Policy*, March 5, 2018. https://foreignpolicy.com/2018/03/05/im-sorry-for-creating-the-gerasimov-doctrine/.

Gao, Charlie. "Russia's New Husky-Class Stealth Submarines: Armed with Hypersonic Missiles?" *The National Interest*, October 1, 2019. https://nationalinterest.org/blog/buzz/russias-new-husky-class-stealth-submarines-armed-hypersonic-missiles-84626.

Gerasimov, Valery. "The Value of Science Is in the Foresight." *Military-Industrial Kurier (VPK)*, February 26, 2013. www.vpk-news.ru/articles/14632.

Gertz, Bill. "Damaged Russian Sub Linked to Underwater Drone Program." *Soldier of Fortune* (blog), July 15, 2019. www.sofmag.com/damaged-russian-sub-linked-to-underwater-drone-program/.

Gorenburg, Dmitry. *Russian Naval Shipbuilding – Is It Possible to Fulfill the Kremlin's Grand Expectations?* CNA Harvard University, October 2015. www.ponarseurasia.org/sites/default/files/policy-memos-pdf/Pepm395_Gorenburg_Oct2015.pdf.

Haynes, Deborah. "Putin's Submarines Spur NATO to Boost Its UK Nerve Centre." *The Times*, January 10, 2018. www.thetimes.co.uk/article/putin-s-submarines-spur-nato-to-boost-its-uk-nerve-centre-8h2bf95qp.

Hicks, Kathleen, Andrew Metrick, Lisa Sawyer Samp, and Kathleen Weinberger. "Undersea Warfare in Northern Europe." CSIS International Security Program. Center for Strategic and International Studies, July 2016.

Higgins, Andrew. "Norway Reverts to Cold War Mode as Russian Air Patrols Spike." *The New York Times*, April 1, 2015. www.nytimes.com/2015/04/02/world/europe/a-newly-assertive-russia-jolts-norways-air-defenses-into-action.html?emc=eta1.

Holmes, James. "More Submarines and Less Aircraft Carriers: If the U.S. Navy Could Be Completely Rebuilt." *The National Interest*, December 3, 2017. https://nationalinterest.org/blog/the-buzz/more-submarines-less-aircraft-carriers-if-the-us-navy-could-23482?nopaging=1.

Johansen, Per Anders. "Putin Med Overraskende Norge-Flørt: På Tide å Utvide Horisonten Og Samarbeidet." *Aftenposten*, October 21, 2018. www.aftenposten.no/verden/i/1kJozl/putin-med-overraskende-norge-floert-paa-tide-aa-utvide-horisonten-og-samarbeidet.

Johansen, Per Anders. "Russisk Atomubåt Testet Atomraketter i Barentshavet i Morges." *Aftenposten*, December 12, 2015.

Johnsen, Alf Bjarne. "E-Sjefen: Russland Øvde På Angrep Mot Nord-Norge." *VG*, March 5, 2019. www.vg.no/nyheter/innenriks/i/1kye4W/e-sjefen-russland-oevde-paa-angrep-mot-nord-norge.

Johnsen, Alf Bjarne. "NATO: Russerne Kan Sabotere Internett Med Ubåt." *VG*, February 17, 2018. www.vg.no/nyheter/utenriks/i/gPday5/nato-russerne-kan-sabotere-internett-med-ubaat.

Johnsen, Alf Bjarne. "Russland i Gang Med Ubåtoperasjon i Nord: Største Siden Sovjet-Tiden." *VG*, October 29, 2019. www.vg.no/nyheter/innenriks/i/opxAP0/russland-i-gang-med-ubaat-operasjon-i-nord-stoerste-siden-sovjet-tiden?utm_source=vgfront&utm_content=row-2.

Johnsen, Alf Bjarne. "Søreide: Ber NATO Trekke Nordover." *VG*, February 15, 2016. www.vg.no/nyheter/innenriks/i/wWazo/soereide-ber-nato-trekke-nordover.

Jonathan, Marcus. "US Confirms Drone Shot down by Iran." *BBC News Online*, June 20, 2019. www.bbc.com/news/world-middle-east-48700965.

Kofman, Michael. "Fire Aboard AS-31 Losharik: Brief Overview." *Russian Military Analysis* (blog), July 3, 2019. https://russianmilitaryanalysis.wordpress.com/tag/gugi/.

Kofman, Michael. "Russian Hybrid Warfare and Other Dark Arts." *War On the Rocks*, March 11, 2016. https://warontherocks.com/2016/03/russian-hybrid-warfare-and-other-dark-arts/.

Korolkov, Alexander. "Russia's Top-Secret Nuclear Submarine Comes into Service." *Russia Beyond*, June 17, 2014. www.rbth.com/defence/2014/06/17/russias_top-secret_nuclear_submarine_comes_into_service_37483.html.

Kremidas-Courtney, Chris. "Countering Hybrid Threats in the Maritime Environment." *The Maritime Executive*, November 6, 2018. www.maritime-executive.com/editorials/countering-hybrid-threats-in-the-maritime-environment.

The Kremlin. "Meeting with Defence Ministry Leadership and Military-Industrial Complex Representatives." May 12, 2015. http://en.kremlin.ru/events/president/news/49456.

Krepinevich, Andrew, Barry Watts, and Robert Work. *Meeting the Anti-Access and Area-Denial Challenge*. Center for Strategic and Budgetary Assessments (CSBA), 2003. http://csbaonline.org/uploads/documents/2003.05.20-Anti-Access-Area-Denial-A2-AD.pdf.

Kristensen, Hans M., and Robert S. Norris. "Russian Nuclear Forces, 2015." *Bulletin of the Atomic Scientists* 71, no. 3 (2015).

Lacey, James. "Battle of the Bastions." *War on the Rocks* (blog), January 9, 2020. https://warontherocks.com/2020/01/battle-of-the-bastions/.

LaGrone, Sam. "CNO: New 2nd Fleet Boundary Will Extend North to the Edge of Russian Waters." *USNI News*, August 24, 2018. https://news.usni.org/2018/08/24/cno-new-2nd-fleet-boundary-will-extend-north-edge-russian-waters.

LaGrone, Sam. "Russian Air Force Chief: U.S. Surveillance Flights Monitor Russia Daily." *USNI News*, December 17, 2014. https://news.usni.org/2014/12/17/russian-air-force-chief-u-s-surveillance-flights-monitor-russia-daily.

Larter, David B., and Mark D. Faram. "The US Navy's New Command Puts Russia in the Crosshairs." *Defense News*, May 4, 2018. www.defensenews.com/naval/2018/05/04/the-us-navys-new-command-puts-russia-in-the-crosshairs/#.WuzF9tk3w80.email.

Lieungh, Erik. "Norge Trekker NATOs Øyne Mot Nordflanken." *NRK*, February 13, 2016. www.nrk.no/finnmark/norge-trekker-natos-oyne-mot-nordflanken-1.12801938.

Madshus, Karin, and Rolf Lofstad. "Norsk Lasteskip Ble Anropt Av Russisk Slagskip." *Dagbladet*, April 15, 2019. www.dagbladet.no/nyheter/norsk-lasteskip-ble-anropt-av-russisk-slagskip/70985345.

Majumdar, Dave. "Cruise Missiles Strikes in Syria: Russia's Big Ad Campaign?" October 8, 2015. https://nationalinterest.org/blog/the-buzz/cruise-missile-strikes-syria-russias-big-ad-campaign-14032.

Matza, Max. "Everything We Know About the US Drone That Crashed in the Black Sea." *BBC*, March 16, 2023. www.bbc.com/news/world-us-canada-64972002.

McKew, Molly M. "The Gerasimov Doctrine." *Politico Magazine*, October 2017.

Ministry of Defence of the Russian Federation. "Severodvinsk Nuclear Submarine of the Northern Fleet Carried out Missile Firing." August 18, 2017. http://eng.mil.ru/en/news_page/country/more.htm?id=2138913@egNews.

Mozgovoy, Aleksandr. "Russian Navy's Long Arm: Kalibr Missile Family." Translated by J. Hawk. *Natsionalnaya Oborona/Southfront.Org*, August 10, 2015. https://southfront.org/russian-navys-long-arm-kalibr-missile-family/.

Natalucci, Matteo. "Top 5 Ports in Northern Europe." *Port Technology*, August 30, 2019. www.porttechnology.org/news/top-5-ports-in-northern-europe/.

NATO. "Warsaw Summit Communique." NATO, July 9, 2016. NATO, Online Archive.

Nilsen, Thomas. "Alarm-Drill: 36 Russian Warships Sail Out to Barents Sea." *The Barents Observer*, June 13, 2018. https://thebarentsobserver.com/en/security/2018/06/36-russian-warships-sails-out-barents-sea.

Nilsen, Thomas. "Now, Russia Builds a Submarine Even Bigger Than the Typhoon." *The Barents Observer*, May 3, 2017. https://thebarentsobserver.com/en/security/2017/05/russias-new-military-research-submarine-arctic-waters-will-be-worlds-largest#.WQnEKkjm64w.facebook.

Nilsen, Thomas. "Russia in the High North." Presented at the Matpakkeseminar, Norwegian MoD, November 14, 2019.

Nilsen, Thomas. "Russian Navy Announces Comprehensive Naval Exercises Off Northern Norway." *The Arctic Today*, August 6, 2019. www.arctictoday.com/russian-navy-announces-comprehensive-naval-exercises-off-northern-norway/.

Nilsen, Thomas. "Russian Sub Launched Cruise Missile without Leaving Port." *The Barents Observer*, April 3, 2019. https://thebarentsobserver.com/en/security/2019/04/russian-sub-launched-cruise-missile-without-leaving-port#.XKQ_dO8JV7U.facebook.

Nilsen, Thomas. "Watch Russian Submarine Test Fire Cruise Missile in Barents Sea." *The Barents Observer*, August 18, 2017. https://thebarentsobserver.com/en/security/2017/08/watch-russian-submarine-test-fire-cruise-missile-barents-sea.

Nilssen, Tom. "Nye Bildebevis Av Russisk Avskjæring." *NRK*, January 31, 2018. www.nrk.no/urix/nye-bildebevis-av-russisk-avskjaering-1.13894490.

Nordenman, Magnus. *The New Battle for the Atlantic – Emerging Naval Competition with Russia in the Far North*. Annapolis, MD: Naval Institute Press, 2019.

Norwegian Intelligence Service. "FOCUS 2023 – The Assessment of the Intelligence Service." Annual Unclassified Assessment. Oslo, 2023.

Norwegian Intelligence Service. "FOKUS 2019 – The Assessment of the Intelligence Service." Annual Unclassified Assessment. Oslo, 2019.

Norwegian Intelligence Service. "FOKUS 2024 – The Assessment of the Intelligence Service." Annual Unclassified Assessment. Oslo, 2024. www.etterretningstjenesten.no/publikasjoner/fokus/Fokus24_innhold.

NTB. "Evenes Flystasjon Kan Bli Overvåkingsbase for USA." *VG*, July 21, 2019. www.vg.no/nyheter/innenriks/i/bprjB/evenes-flystasjon-kan-bli-overvaakingsbase-for-usa#xtor=CS4-1.

Office of Naval Intelligence. "The Russian Navy – A Historic Transition." Office of Naval Intelligence (ONI), December 2015. www.oni.navy.mil/Portals/12/Intel%20agencies/russia/Russia%202015print.pdf?ver=2015-12-14-082038-923.

Parnemo, Liv Karin. "Russia's Naval Development – Grand Ambitions and Tactical Pragmatism." *The Journal of Slavic Military Studies* 32, no. 1 (2019): 41–69.

Persen, Kjell. "Amerikansk Atomubåt Søkte Nødhavn i Norge." *TV2.No*, October 27, 2017. www.tv2.no/a/9433425.

Pettersen, Trude. "Black Sea Submarine Fires Cruise Missile from the Barents Sea." *The Barents Observer*, August 4, 2015. https://barentsobserver.com/en/security/2015/08/black-sea-submarine-fires-cruise-missile-barents-sea-04-08.

Pettersen, Trude. "Nuclear Missile World Record Attempt." *The Barents Observer*, March 2, 2016. https://thebarentsobserver.com/en/security/2016/03/nuclear-missile-world-record-attempt.

Pettersen, Trude. "U.S. Builds Up Arctic Intelligence Network." *The Barents Observer*, September 8, 2015. https://barentsobserver.com/en/security/2015/09/us-builds-arctic-intelligence-network-08-09.

Pettersen, Trude. "U.S. Military Returns to Iceland." *The Barents Observer*, October 2, 2016. https://thebarentsobserver.com/en/security/2016/02/us-military-returns-iceland#.VsVfwxKHhBK.mailto.

Portocarrero, Marta. "Russia Fires First Submarine Missiles Against Isis Targets in Syria." *Independent*, December 10, 2015. www.independent.co.uk/news/world/europe/russia-fires-first-submarine-missiles-against-isis-targets-in-syria-a6766426.html.

Roblin, Sebastian. "Why Russia's Enemies Fear the Kalibr Cruise Missile." *The National Interest*, January 22, 2017. https://nationalinterest.org/blog/the-buzz/why-russias-enemies-fear-the-kalibr-cruise-missile-19129.

Rogoway, Tyler. "Russia Sends Ten Subs into North Atlantic in Drill Unprecedented in Size Since Cold War." *The Drive*, October 29, 2019. www.thedrive.com/the-war-zone/30728/russia-sends-ten-subs-into-north-atlantic-in-drill-unprecedented-in-size-since-cold-war.

Rogoway, Tyler. "USAF's Nuke Sniffing Plane Is Flying on a Mission Near the Arctic Right Now." *The Drive*, February 22, 2017. www.thedrive.com/the-war-zone/7830/usafs-nuke-sniffing-plane-is-flying-on-a-mission-near-the-arctic-right-now?xid=emailshare.

Savolainen, Jukka, Terry Gill, Valentin Schatz, Lauri O Jala, Tadas Jakstas, and Pirjo Kleemola-Juntunen. *Handbook on Maritime Hybrid Threats – 10 Scenarios and Legal Scans*. Helsinki, Finland: The European Centre of Excellence for Countering Hybrid Threats, November 2019.

Smith, R. Jeffrey. "Hypersonic Missiles Are Unstoppable. And They're Starting a New Global Arms Race." *New York Times Magazine*, June 19, 2019. www.nytimes.com/2019/06/19/magazine/hypersonic-missiles.html.

Sontag, Sherry, and Christopher Drew. *Blind Man's Bluff – The Untold Story of American Submarine Espionage*. New York: Harper Paperbacks, 1999.

Staalesen, Atle. "30 Russian Naval Vessels Stage Show of Force near Coast of Norway." *The Barents Observer*, August 15, 2019. https://thebarentsobserver.com/en/security/2019/08/30-russian-naval-vessels-stage-show-force-coast-norway.

Starr, Barbara, and Brad Lendon. "Russian Fighter Came Within 15 Feet of U.S. Air Force Jet." *CNN*, January 30, 2016. https://edition.cnn.com/2016/01/29/politics/us-russia-planes-intercept/index.html.

Stavridis, James. "Maritime Hybrid Warfare Is Coming." *United States Naval Institute Proceedings* 142, no. 12 (December 2016).

Stewart, Phil. "Russia Jets Make 'Simulated Attack' Passes near U.S. Destroyer." *Reuters*, April 13, 2016. www.reuters.com/article/us-usa-russia-simulatedattack-idUSKCN0XA1UW.

Strand, Tormod. "Hemmelig Ubåt-Operasjon: 'Målet Er å Vise at Russland Kan Nå USA.'" *NRK*, October 29, 2019. www.nrk.no/norge/hemmelig-ubat-operasjon_-_malet-er-a-vise-at-russland-kan-na-usa_-1.14761298.

Sutton, H.I. "Husky_SSN." *Covert Shores* (blog), December 30, 2019. www.hisutton.com/Husky_SSN.html.

TeleGeography. "Submarine Cable Map." *TeleGeography*. Accessed October 23, 2019. www.submarinecablemap.com/.

Thonhaugen, Markus, and Kåre Riibe Ramskjell. "Russerne Med 'Uvanlig Stor' Militærøvelse: – Lenge Siden vi Har Sett Noe Slikt." *NRK*, August 13, 2019. www.nrk.no/nordland/russerne-med-_uvanlig-stor_-militaerovelse_-_-lenge-siden-vi-har-sett-noe-slikt-1.14658213.

Tomassen, Jan Harald. "Skal Ha Bedt Russiske Sjøkapteiner Gjøre Seg Godt Kjent Langs Norskekysten." *NRK*, February 27, 2015. www.nrk.no/troms/skal-ha-bedt-russiske-sjokapteiner-gjore-seg-godt-kjent-langs-norskekysten-1.12234101.

Ulriksen, Ståle. "Endring i Nord." *Bergens Tidende*, May 27, 2019.

Vesterålen Online. "Hektisk Jakt På Ubåter." October 31, 2019. www.vol.no/pluss/2019/10/31/Hektisk-jakt-p%C3%A5-ub%C3%A5ter-20289339.ece.

Vissgren, Julie, Tormod Strand, and Lokman Ghorbani. "Forsvarssjefen Om Russisk Militærøvelse: – En Nasjonal Utfordring." *NRK*, August 14, 2019. www.nrk.no/urix/forsvarssjefen-om-russisk-militaerovelse_-_-en-nasjonal-utfordring-1.14660898.
Werner, Ben. "Russian Su-27 Fighter Buzzes U.S. Navy EP-3 Aries Over Black Sea." *USNI News*, January 29, 2018. https://news.usni.org/2018/01/29/30987.
Willett, Lee. "Game Changer: Russian Sub-Launched Cruise Missiles Bring Strategic Effect." *Jane's By IHS Markit*, 2017. www.janes.com/images/assets/147/70147/Game_changer_Russian_sub-launched_cruise_missiles_bring_strategic_effect_edit.pdf.
Wormdal, Bård. "Tidligere Forsvarssjef Om Nye Russiske Våpen: – En Trussel Som Må Tas Alvorlig." *NRK (Online)*, August 11, 2017. www.nrk.no/finnmark/tidligere-forsvarssjef-om-nye-russiske-vapen_-_-en-trussel-som-ma-tas-alvorlig-1.13767105.
Ziezulewicz, Geoff, and David B. Larter. "The Navy Sends a Carrier Back to Russia's Arcic Haunts." *Navy Times*, October 19, 2018. www.navytimes.com/news/your-navy/2018/10/19/the-navy-sends-a-carrier-back-to-russias-arctic-haunts/#.W820deP9FUo.email.

10 The future of airborne ISR in the High North

The shift away from the predictable tracking patters of Soviet submarines during the Cold War to the more unpredictable operational movements of Russian submarines today has created new challenges for Norway and her allies in the High North. In particular, it has complicated the task of ASW.

In addition to reforming command structures, NATO and its member states must update their ASW setup in general, and their airborne ISR capabilities in particular. It is this author's contention that the framework for discussion presented earlier in this book, combined with historical precedence from the Cold War era and the decades following it, can point the way forward. This chapter thus provides suggestions for how new technology and new approaches to airborne surveillance might facilitate an improved situational awareness and basis for targeting in the High North.

Altitude

One benefit for airborne ISR is that when an MPA is flying along a sovereign national border, or some other line of restriction (e.g. stand-off distance to surface-to-air weapons), the aircraft is able to look beyond that border, for example, beyond the shoreline under surveillance.

The P-8 is capable of operating at a higher altitude than the P-3 (41,000 feet versus approximately 29,000 feet, respectively).[1] The theoretical radar range at 29,000 feet is approximately 210 NM.[2] At 41,000 feet, this range increases to approximately 250 NM. If an asset is capable of operating at 60,000 feet, the theoretical radar range increases to a little over 300 NM. Figure 10.2 emphasizes the increase in sensor range based on speed as a constant and altitude as a variable. With a baseline of 29,000 with a reference cruising speed of 330 knots, the area covered in 1 h by the radar sensor increases by 20% at 41,000 feet, and 40% at 60,000 feet. For area coverage and particularly for non-acoustic sensors, altitude thus matters significantly.

One historic example of assets at different altitudes working together was the British *Vulcan* bombers flying high and the *Shackleton* MPA flying low. In working together, they made the search and surveillance of contacts more efficient.

DOI: 10.4324/9781003626015-10

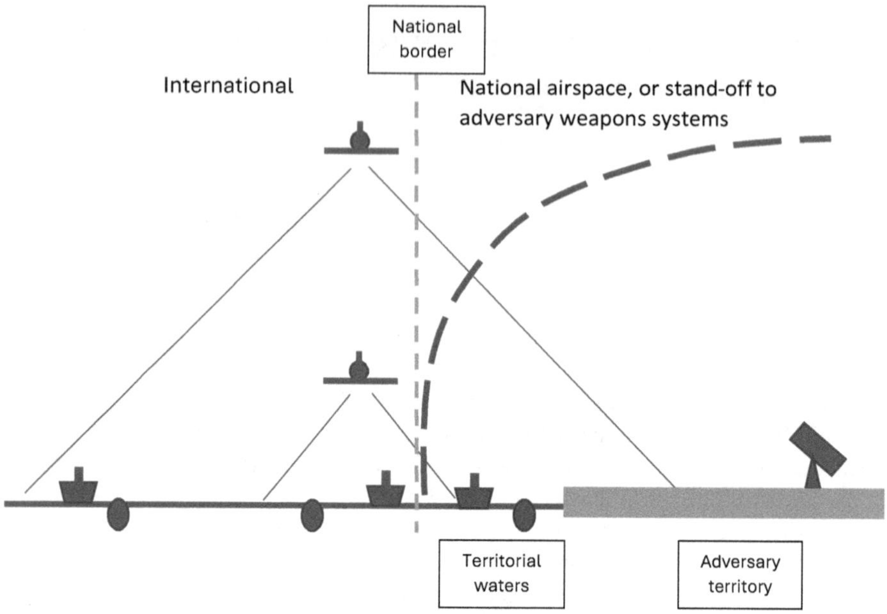

Figure 10.1 Increased sensor range with increased altitude

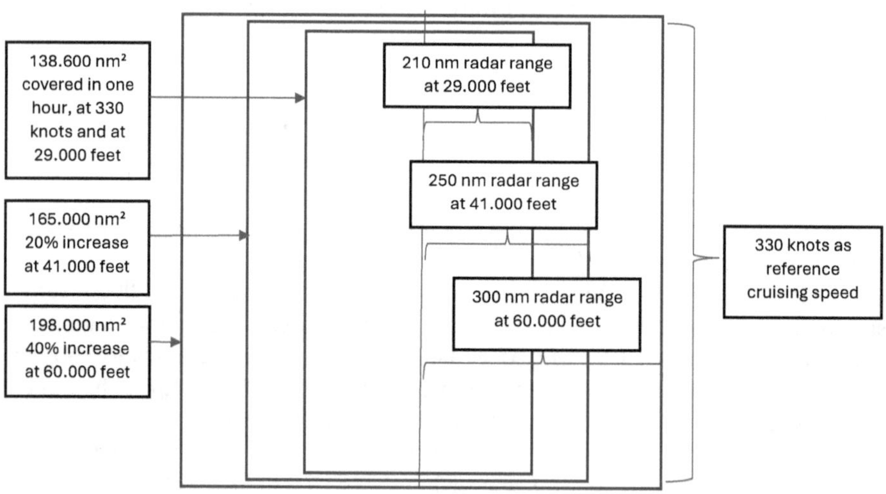

Figure 10.2 Increase in sensor coverage over time with increase in altitude

A modern example of such cooperation is a *high altitude/long endurance* (HALE) unmanned system flying at 60,000 feet, maintaining area overview in support of a manned MPA flying at lower altitudes, investigating and prosecuting contacts. NATO has dubbed such cooperation "*high boy/low boy*" tactics.[3]

Other systems with an obvious high ground with regard to surveillance are satellites. Extensively used for maritime surveillance already, the clear downside is the relatively short amount of time they can focus over a given area, and the lack of flexibility with regard to real-time tasking and changes to areas of focus and thus priority. In addition, satellites are limited to surface surveillance. But in combination with other assets, satellites provide extreme endurance, albeit only periodically, thus supporting the establishment of a pattern-of-life in an area. And perhaps most importantly, satellites can pass over adversary territory, without running the risk of being shot down.

Traditional airborne ASW, meanwhile, is characterized by aircraft flying at very low altitudes in order to react quickly to submarine movements, more precisely place sonobuoys into the water, and utilize sensors such as magnetic anomaly detection (MAD) and radar for fine-tuning a submarine's position. Newer generations of MPA will work at high altitudes in order to gain an overview of the area but will also increasingly be able to operate from medium to high altitudes during the search and tracking phase of the prosecution. Historically, MPA crews have been forced to undergo the cumbersome process of electronically relocating individual buoys at low altitudes at regular intervals and plotting them into the tactical overview due to the oceanic set and drift. Today, modern buoys with built-in Global Positioning System (GPS) will inform the aircraft tactical system of their position continuously throughout their lifespan, enabling tracking tactics to be executed from up high. This new technology also facilitates unmanned aircraft flying at high altitudes in executing ASW search and tracking.

Speed

For maritime airborne ISR, speed enables the aircraft to reach the operations area or a submarine's datum in a shorter amount of time. As a continuation of this, the sporadic nature of modern Russian submarine movements requires MPA to be able to move quickly from one place to another to investigate a contact. And, once positive contact has been achieved, there is presumably less time to acquire attack criteria and execute an attack due to the short acoustic ranges of the sounds emitted by the submarine, as will be discussed below. In this context, speed will be of critical importance. The combination of unpredictable operational patterns of the threat and short acoustic ranges indicates that the relevancy of a datum is shorter now than before. If the tracking aircraft (e.g. an unmanned aircraft carrying sensors but not weapons) cannot neutralize the target itself, then the platform employed to execute the attack (e.g. an MPA carrying torpedoes) must be able to transit quickly to the perishable datum.

A hypothetical datum, mid-way between the three core maritime air bases in the North Atlantic, lies approximately 950 NM from each base. A P-3 *Orion* will require approximately 2 h and 50 min to get to the scene, and a P-8 *Poseidon* will get there in 2 h and 10 min. Not only will the increased speed provide 1 h and 20 min longer on-station time for the P-8, but the submarine will have 40 min less time to evade the datum in the case of a P-8 hunting it. In a tactical example where

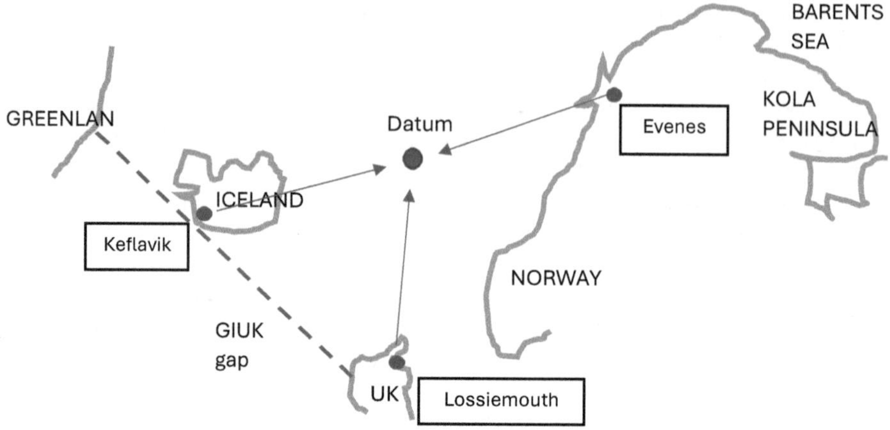

Figure 10.3 A hypothetical datum with equal distance to all three MPA bases in the North Atlantic

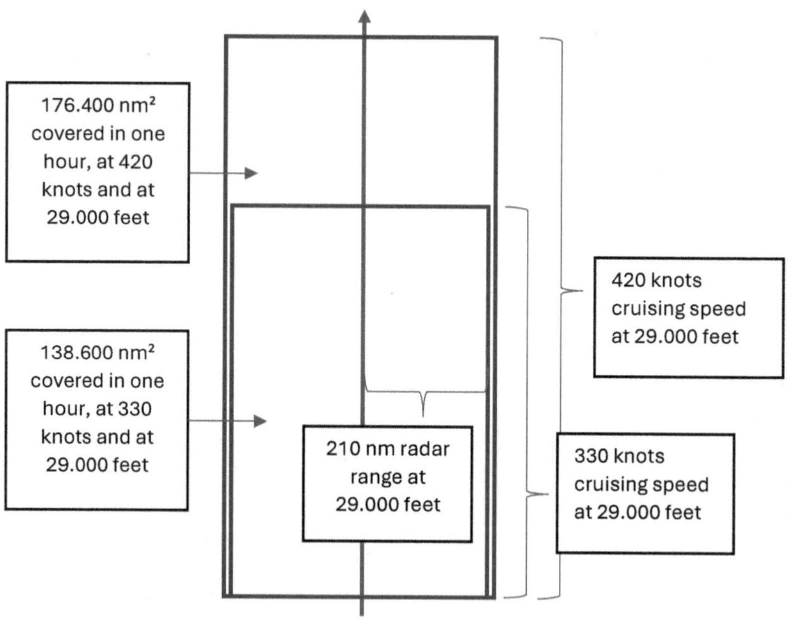

Figure 10.4 Increased sensor coverage with increased speed

the MPA is reacting to its own sensor hits, the difference is obviously less, though important. An MPA with a 210 NM radar horizon detecting a submarine at the edge of this range envelope will fly to investigate as soon as possible.

A distance of 210 NM is covered in 38 minutes when the aircraft is flying at 330 knots (P-3) and in 30 minutes when the aircraft is flying at 420 knots (P-8). In those

8 min of difference, the evading submarine will have transited anywhere from 1 to 3 NM (submarine speed between 5 and 15 knots), which with contemporary ranges (see below) can mean the difference between acquiring and losing contact.

Finally, speed allows a larger area to be covered in a shorter amount of time. For example, the P-8 flies faster than the P-3, with cruising speeds of 420 knots versus 330 knots, respectively. The following figure thus emphasizes the increase in sensor coverage based on altitude as a constant and speed as a variable.

Calculating with altitude, and thus sensor range, as a constant, an increase in speed will facilitate an increase in sensor coverage. In a direct comparison between the P-8 and the P-3, where both aircraft are flying at an altitude of 29,000 feet, an increase from 330 knots to 420 knots gives an increase in sensor coverage in 1 h of more than 27%.

Range

As we have seen time and again in this book, range in many ways epitomizes the very concept of maritime airborne ISR. The vast ocean areas that have been under surveillance by the aircraft discussed in previous chapters have shaped the MPA force, and range was the single most debated operational characteristic among the Group of Experts for the NATO *Atlantic* programme in the 1950s and 1960s. One simply must be capable of reaching the point of interest in order to be relevant.

In the case of the Barents Sea, arguably the most relevant area for intelligence collection in the North Atlantic for NATO MPA, the speed combined with range of the P-8 will be a factor for considering the relevance of the old 24°E guideline of the Norwegian Government. The objective of the restrictions was always to facilitate Allied activity and presence in combination with maintaining low tensions in the North. Svein Efjestad of the Norwegian MoD highlights the new surveillance technology as crucial in facilitating new activities as well as old activities in new ways in areas of Norwegian interest.[4] The new technology will necessitate a more flexible and dynamic approach to the guidelines.

Figure 10.5 shows the transit routes to the Barents Sea from the three main MPA bases in the North Atlantic, and the approximate position of the 24°E guideline that applies for foreign MPA operating from Norwegian soil. A change in transit speed from 330 knots for the P-3 to 420 knots for the P-8 means that the transit from Keflavik to the Barents Sea (approximately 1,000 NM) will be reduced from 3 h to less than 2.5 h. This results in a 40% increase in on-station time from 3 h to 4 h and 14 min.

Throughout the Cold War, the 24°E guideline gave the Norwegians a *de facto* monopoly in maritime airborne surveillance in the High North and the Barents Sea.[5] With new technology, new capabilities, and a new security environment, it seems likely that other actors will carry out intelligence operations in areas such as the High North if the intelligence requirements in the area outpace the capacity of local actors to deliver. The most prominent change in the framework of the Norwegian maritime air monopoly on intelligence collection in the Barents Sea will be experienced by the former P-3 operating nations, the United States and Norway.

As transit speeds of new aircraft increase, more on-station time is gained. Also, increasingly capable unmanned aircraft will be able to take off from outside Norwegian territory and operate in the Barents Sea over extended periods of time. New technology may therefore challenge and force changes to the Norwegian guidelines for airborne operations off the coast of Kola.

With a more forward-leaning Russian Navy, it seems natural that the notion that intelligence collection and presence in the Barents Sea shall be reserved strictly for Norwegian assets can once again be challenged in the near future by Norway's allies. The fact remains, however, that Norway will simply, due to geography, always boast the quickest response time and the longest on-station time in the Barents Sea.

The elements of hybrid warfare combined with the unpredictable *modus operandi* of Russian submarines are building the case for persistent surveillance in the High North. This persistency can only be facilitated by assets staying in the area for very long periods of time, and overlapping with a relieving asset. Satellites have regularity and predictability over a very long time period (several years), but they lack continuous persistency. They are continuously circling the globe, but over a given geographical area they are only intermittently present. MPA can stay aloft for a long time, but even with AAR, the onstation time in unlikely to pass much more than 20 h, if only due to human fatigue. Current and future technology therefore points to unmanned aircraft that can stay aloft for more than 30 h today, and much more in the future. Ground crews operating and/or monitoring the aircraft can function in shifts, thereby removing the element of human fatigue. The case for high-altitude, unmanned aircraft for persistent maritime surveillance in the High North is pertinent indeed. They do not, however, have the capability to conduct underwater ASW.

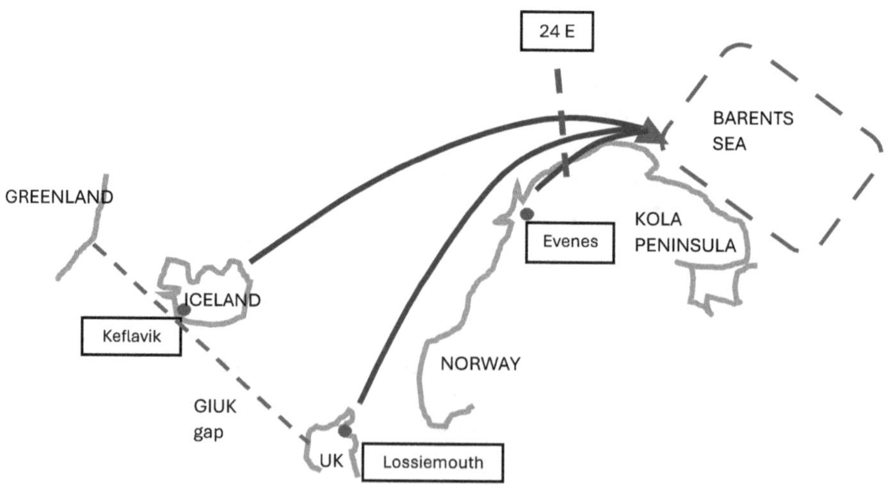

Figure 10.5 Transit and on-station times for MPA in the Barents Sea

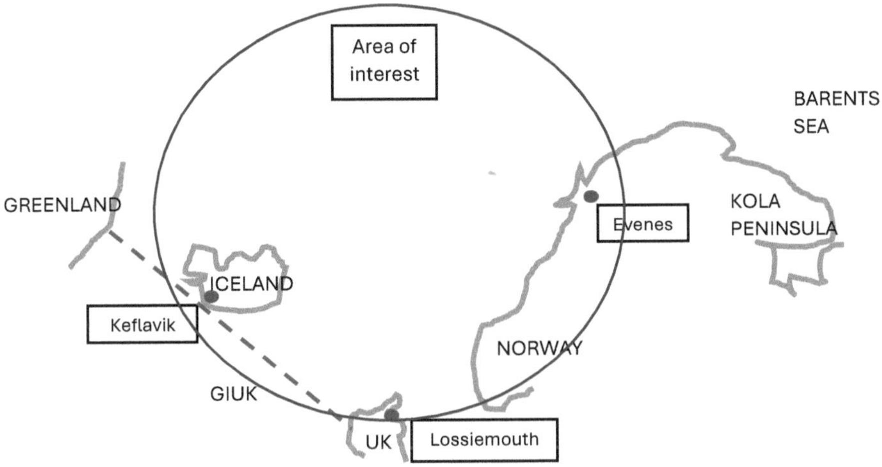

Figure 10.6 Short transit times to area of interest in the North Atlantic

Another point with the new *modus operandi* of Russian submarines is that all three MPA bases stand to be good launch sites for MPA missions in search of Russian submarines in the North Atlantic, as shown below.

In terms of air power, range facilitates two basic factors that play fundamental roles in the execution of a mission. First, range provides the ability to fly a certain distance, perform a mission, and then either return to home-base or fly to a different destination. Second, endurance provides the ability to stay in and above an area and thus gather information from that area *over time*. Tracking Soviet submarines during the Cold War demanded range in the context of the first factor. An MPA would fly to a point along the transit route, and be prepared to track the submarine for a certain distance along that route (situation A, Figure 10.7). Searching for and tracking modern Russian submarines is a different matter, however, as the operational area essentially encompasses the entire North Atlantic.

This, in turn, requires ISR assets to remain in the area to continue the search, and respond quickly to a sensor hit (situation B, Figure 10.7).

Given that the P-8 has approximately the same range and endurance as the P-3, the range factor in relation to ASW is reduced to the *modus operandi* of the target submarine.[6] As mentioned earlier, the predictability of the Cold War Soviet SSBN transit worked in favour of the Western MPA that were hunting it. The endurance available was utilized more efficiently than what is possible when the hunter effectively is searching large parts of the North Atlantic basin. This is further accentuated when taking into account that modern search tactics are even more dependent on cueing – other assets that loiter above, on, or below the surface of a given area in search of indications of targets of interest, which in turn cue more capable search, track and attack assets for the actual prosecution of the target.

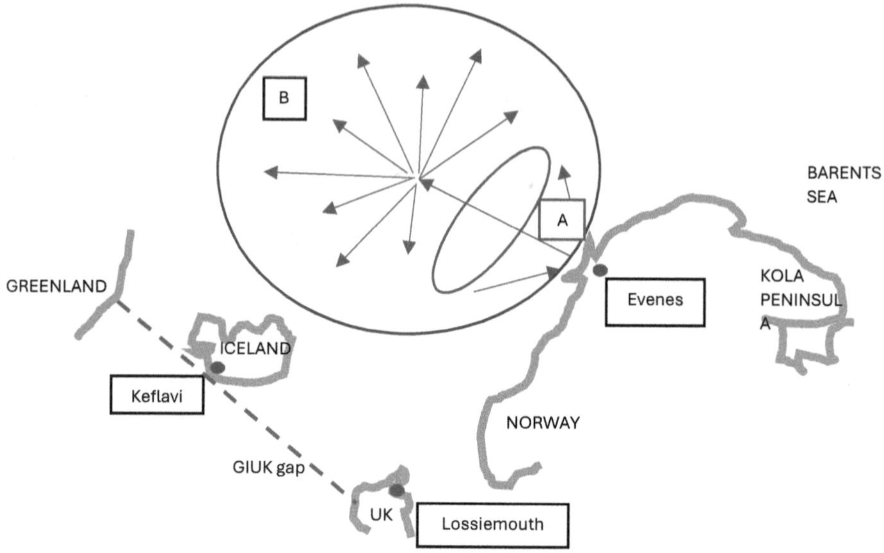

Figure 10.7 Efficient use of MPA range in a predictable submarine transit pattern (A). Demanding endurance requirements for unpredictable submarine modus operandi (B).

By flying faster and higher, however, the P-8 is able to cover a larger area in a shorter amount of time. By reaching the operations areas faster, the P-8 is able to loiter there longer than the P-3.

Communications

Traditional MPA operations did not require detailed analysis and reporting to be sent back to HQ at all times. MPA were mainly operating far away from other units, hunting submarines, only sporadically reporting home. Today, submarines are far more integrated into joint operations. In the case of Russia, we see integrated, cross-theatre submarine operations, which means that submarine operations, for example, in the Baltic Sea can and will most likely affect submarine operations in the North Atlantic. In order to better understand this complexity in adversary behaviour, it is paramount that headquarters is able to fuse movements, indications, and warnings nearly instantly, in order to understand what is happening and gain information superiority over one's adversary. Modern communication suites and support assets (such as satellites) are integral elements of this ability to build a tactical, operational, and strategic picture of the battle. For example, encrypted, broadband satellite communication facilitates exchange of acoustic analyses in order to engage subject matter expertise for solving it, in such cases where the aircraft sensor operator is struggling with in-flight analysis. Instant dissemination of ISR information also enables faster detection of anomalies, if many sensors together are capable of monitoring large parts of the area of interest. In a hybrid warfare environment at sea, where adversary actions are not necessarily the traditional and conventional naval movements, it will require more persistent and detailed information gathering in order to detect anomalies from the normal situation.

Given the nature of the operations that MPA traditionally execute over the open ocean, there was historically only a requirement for basic communication with other units and with headquarters and home base. The traditional setup consisted of HF long-range communications over voice and basic text messages (teletype). The NATO maritime *Link-11*, originally introduced in the mid-1960s, facilitated an exchange of tactical information, leading to a basic, common tactical picture. For radio communications with other military units, UHF was utilized, as remains common today. Essentially, transmission out of the aircraft consisted of short bursts of either voice or text transmissions, providing very short safety or tactical updates. In most cases, MPA operated by themselves and reported only after landing. An important element of the lack of communication has also been that radio silence removed the possibility of detection by enemy ELINT. Particularly, active transmissions from the aircraft would give away the presence of the MPA to a submarine at periscope depth checking whether it is safe or not to surface.

The communications suite of the P-8 *Poseidon* is a culmination of all the communications improvements that have been made to the P-3 *Orion* over the years. In addition to the traditional maritime NATO *Link-11*, the NATO *Link-16* has been integrated into the on-board systems. *Link-16* provides more detailed information, with a significantly higher rate of updates, compared to *Link-11*. This means that the MPA can, to a significantly higher degree, be integrated into the joint theatre of operations, and thus share a comprehensive tactical overview.

Internet connection is also integrated in the new MPA, enabling transfer of data to home base in addition to the military SATCOM connection, the latter being a necessity for integration into modern operations. A significant challenge for SATCOM communications in the High North is that the most commonly used communication satellites are *geostationary*: They orbit the Earth directly outside the equator, with the same speed as the rotation of the Earth, leading them to be perceived as static to an observer on the Earth's surface. This gives poor coverage in very high and very low latitudes, and thus poor coverage in the High North.

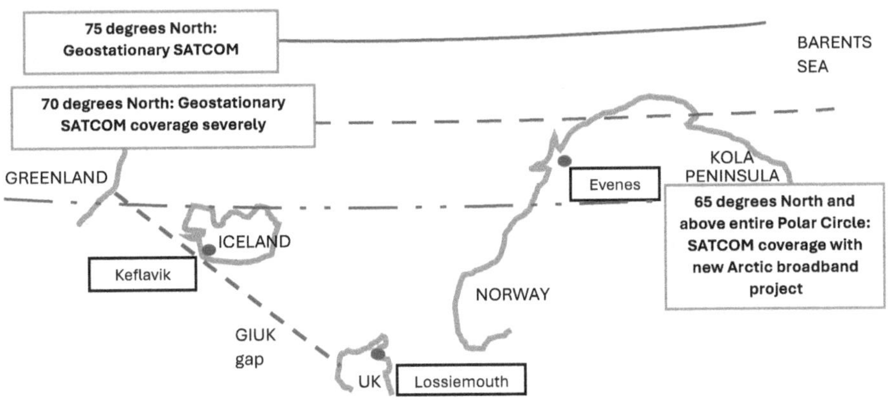

Figure 10.8 SATCOM coverage in the High North

There are commercial, global communication satellite networks available, but these are meant for civilian communications and usually provide a narrow bandwidth. In order to share intelligence products such as imagery, acoustic analyses, or video, much broader bandwidth is required. This is part of a military assessment. One can lease bandwidth from a commercial contractor, which evades the cost of launching a satellite. But then there is a requirement for end-to-end encryption, further decreasing the usually narrow bandwidth. One can also conduct civilian–military cooperation, where a military communications payload is placed on board a civilian satellite, and the overarching costs of launching and "flying" the satellite are shared. This is emerging as one of the most cost-efficient solutions to the problem. Finally, one can launch a satellite by oneself, leading to full control of the satellite but having to bear the entire cost of launch and daily operations. *Space Norway*, a company that is partly owned by the Norwegian government, is in the midst of a project that seeks to remedy this situation. In the fall of 2024, Space Norway launched two communication satellites into an orbit with sufficient inclination in comparison to the equator, providing good communications coverage in the High North. The satellites were operational in late 2024, ensuring NATO MPA monitoring the Arctic, among others, will have their means of communication vastly improved for the foreseeable future.

Survivability

The issue of survivability is frequently neglected in discussions of MPA operations. This fact seems to be a remnant from the time when MPA were operating by themselves, far out to sea, with limited or no real threat posed to them. In general, MPA feature in NATO military discourse as an asset that primarily can be utilized during peacetime and times of low tension, depending on the anti-air threat involved and the level of acceptable risk.

The main threat to aircraft flying over the ocean includes SAMs from submarines and surface ships, air-to-air missiles (AAMs) from carrier-based aircraft, and SAM and AAM threats from shore-based assets when operating within range of the adversary coastline. The SAM systems on board the *Kirov* battle cruiser include the SA-N-20 *Gargoyle*, which has a range of up to 83 NM/150 km.[7]

The *Udaloy*-class destroyer carries the SA-N-9 *Gauntlet*, which has a range of a little more than 8 NM/15 km.[9] The anti-aircraft armament of the relatively new *Admiral Gorshkov*-class frigate is the *Hurricane* SAM system, a derivative of the land-based S-400 system with a range of up to approximately 66 NM/120 km.[10]

AAMs inherently have long ranges due to their airborne launch platforms. In addition to the range of the aircraft, the Russian AA-13 *Arrow*, for example, has a range in excess of 150 NM.[11] This poses a significant threat to any adversary flying asset in the area.

Finally, when operating close to the Russian coastline, any air asset will have to negotiate a three-layered missile defence system, based on some of the world's most capable long-range, medium-range, and point defence systems. The S-200 and the S-400 create an outer-layer air defence bubble of more than 220 NM.[12] This is followed by a second, medium-range layer consisting of the S-300 or the *Buk* system. Then, the innermost layer consists of short-range systems such as 9K33 *Osa* and S-125 *Neva* for concentrated protection of key areas.

Figure 10.9 illustrates the generic SAM threats from surface vessels north of Norway and from shore-based systems on the Kola Peninsula. Not depicted in the figure are the shore-based interceptor and fighter aircraft that will carry AAMs and pose a significant threat to any aircraft venturing into the area in which the Russian ambition of control stands firm.

Considering all these threats, one would be hard-pressed to send a manned, large (with a significant radar cross-section, easily acquired by missile homing), relatively low-flying (below SAM ceiling of minimum 70,000 feet[13]), slow-moving (well below the speed of sound), and slow-manoeuvring aircraft, lacking proper self-protection measures, into the Barents Sea to conduct ISR operations, including ASW, during times of tension or war. The *Orion* or the *Nimrod* had modest self-protection measures, consisting of chaff and flare.[14] This means that as the tension rises the risk to the aircraft does as well, which leads to a retreat by MPA from the operational areas of responsibility. Continued MPA operations based out of Evenes and Andøya will, therefore, quickly become a part of an operational risk assessment.[15]

One natural reaction will be to move the MPA base further south, due to a combination of insufficient base protection and the lack of aircraft self-protection. The mere fact that the survivability of MPA can lead to a strategic intelligence asset effectively being removed from the theatre of operations in times of war should be a significant cause for concern. MPA will be moved back from the front lines when the need for accurate and actionable intelligence is greatest. This inherent weakness of the MPA is a factor that has not been sufficiently discussed, not in the MPA community, nor in military, nor civilian academia. There seems to be a sense of complacency towards MPA in that a future threat that will deny the utilization of

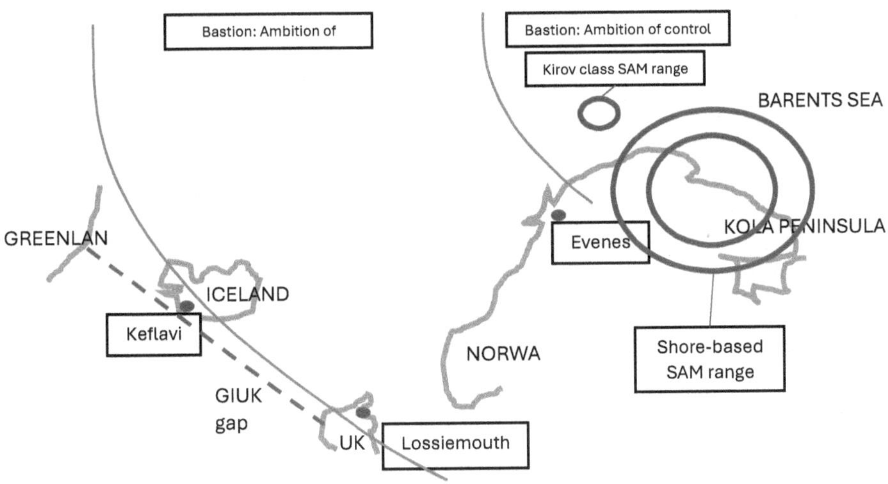

Figure 10.9 Russian SAM ranges in the Barents Sea[8]

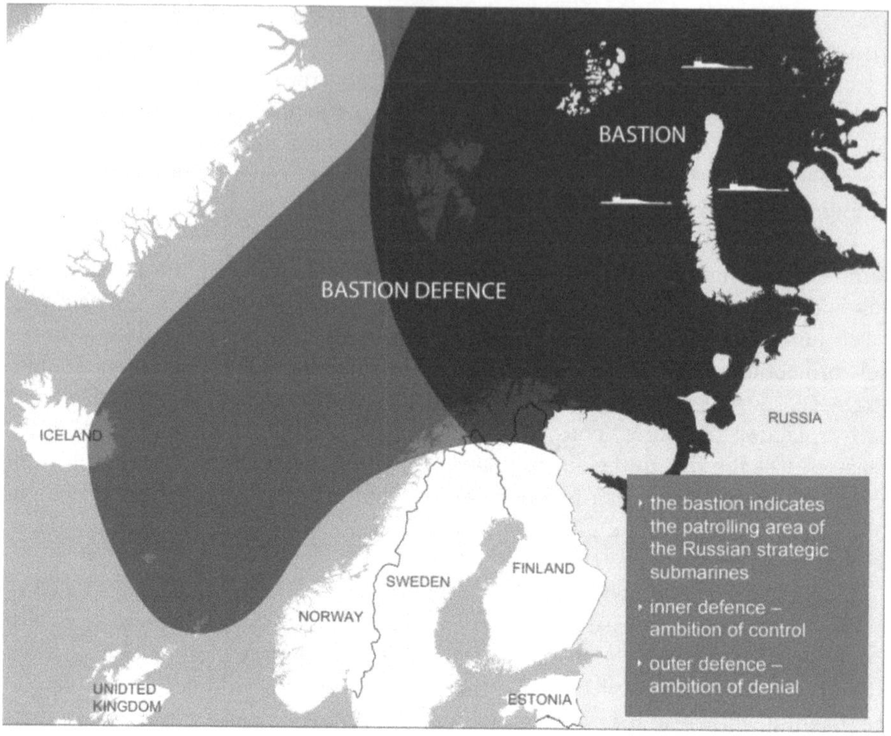

Figure 10.10 The Russian bastion and the reach of the bastion defence[16]

MPA in crises and war is not properly discussed in peacetime. The obvious danger being that when the MPA are pulled back from the front line, a gap in maritime situational awareness and understanding will emerge, as well as the ability to deliver long-range lethal force against a submarine threat.

This can be termed the *MPA mission portfolio paradox*: In times of war, the MPA fleet is geographically significantly restricted from executing its key mission: ACINT and ASW. There is significant risk related to executing ISR and ASW missions in the Barents Sea during war. Given the reach of modern surface-based SAMs and land-based aircraft with AAMs, the high-risk operational area in the High North will stretch beyond the Barents Sea, likely further down into the North Atlantic, as depicted in Figure 10.10. If the capability to perform long-range, airborne ASW is not supplemented, this key capability will diminish as soon as the risk level rises.

Supplements to long-range airborne ASW aircraft *on the grounds of survivability* vary, and can be facilitated by several factors: The sensor can operate beyond the reach of the A2/AD measures (e.g. satellites carrying non-acoustic ASW technology[17]); the sensor can operate without being seen by adversary defensive systems (e.g. submarines, manned and unmanned, and in the future unmanned aircraft that have an extremely low radar cross-section); the sensor can operate on a platform with extreme manoeuvrability (e.g. unmanned aircraft that can fly faster and that is more manoeuvrable than any known missile system); or the sensor can be carried by a platform with counter-measures that neutralize the respective A2/AD measures (e.g. a platform that carries credible and effective counter-measures to any known anti-air system, such as AI-based reaction to threats and powerful lasers to neutralize them[18]). The latter three will effectively mean that such assets can operate inside denied areas at a significantly reduced risk.

A2/AD threats are not denying ISR operations in peacetime. They are latent threats, waiting to be employed in times of crisis and war. However, the challenge to operating MPA in denied areas will not disappear. On the contrary, it escalates precisely when one requires capable ASW assets the most.

Sensors

The principles for hunting submarines are essentially the same today as they were during the Cold War (and even WWII). It is the significantly improved *capacity* of the respective sensors that take the search to a new level, particularly when combined with highly improved on-board computer processing capacity and the ability to share and triangulate sensor information nearly instantly with other units.

Most important to this discussion is the airborne platform's acoustic sensor. The most significant progress achieved for this sensor has, without a doubt, been the processing capacity on board the aircraft. The P-3 *Orion* is capable of monitoring 32 buoys simultaneously, while the P-8 *Poseidon* can monitor 64. New-generation MPA carry powerful processors. The operating environment for the sensor operators on board the aircraft has also been improved through reduced noise and fewer vibrations. This allows operators to work with more concentration and focus for a longer time.[19]

The acoustic sensitivity of the buoys themselves, however, has not improved significantly. Some of the reasoning for this lies in what can be termed the "acoustic challenge."

Figure 10.11 depicts how sound in the water is reduced over distance, and the distance in this case is that between a sound-emitting submarine and a sensor. The upper curve displays the sound from a relatively *noisy* submarine, and the curve below that from a relatively *quiet* submarine. Sound diminishes with the square of the distance, which means that by doubling the distance you cut the sound by a factor of four.[21] The upper dotted line illustrates the capacity of a *normal* sonar, and the lower dotted line that of an *improved* sonar. The normal sonar detects the *noisy* submarine at a range corresponding to that of point C, whilst an improved sonar detects the noisy submarine all the way out to point D. The sound curve of the *quiet* submarine starts lower on the graph, and quickly drops to the point of detection for both the normal sonar as well as the improved one. This means that improving the sonar will help significantly when tracking a noisy submarine, but only marginally when searching for and tracking a quiet submarine. As a corollary to this, more passive sonar equipment is required in order to cover the same area acoustically than was the case before.[22] Tactically, one therefore seems forced

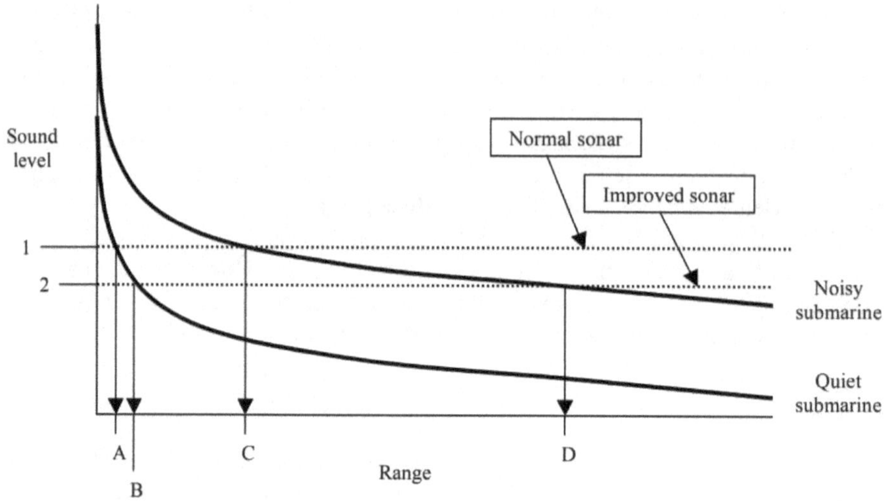

Figure 10.11 The acoustic challenge[20]

to switch to active buoys and sonar at an earlier stage in the search for the quiet submarine, or rely more on non-acoustic search methods.[23] In order to search for a quiet submarine using passive acoustic sensors from the air, one requires many and capable buoys, spread out over a large area, to be monitored over time. This, in turn, requires extended MPA endurance (to stay longer in the area), higher altitude (to pick up the signals from more buoys/several buoy fields at the same time), and more processing capacity (to process and analyse more buoys simultaneously).

In an example provided by a NATO report from 2016, passive sonobuoy ranges were, in the 1970s and 1980s, often around 3,000 yards (1.5 NM).[24] These ranges had, by the mid-2010s, been reduced to less than 500 yards (0.25 NM). When modern submarines reduce their sound emissions, the sensitivity of the sensor must increase significantly – alternatively, one can place more sensors in the water, reducing the distance between them.

Less sound emitted also leads to fewer instances (if any) of convergence zones at large distances. In addition, today's ocean is filled with more sounds, mainly from human activities, which leads to more ambient noise and clutter for the acoustic sensor operator to work through.[26] Several high-ranking American naval officials have explicitly stated in open sources that the *Severodvinsk* class will pose significant challenges to NATO's ability to maintain control of her whereabouts due to her capabilities and silent propulsion.[27] The succeeding ships of the *Severodvinsk* class are reportedly improved versions of the first vessel.[28]

Following on from this, the previously mentioned *modus operandi* of the adversary submarine also comes into play for acoustic sensors. There is a fundamentally lower chance of detection of sporadic movement pattern (B) than there is of predictable movement patterns (A).

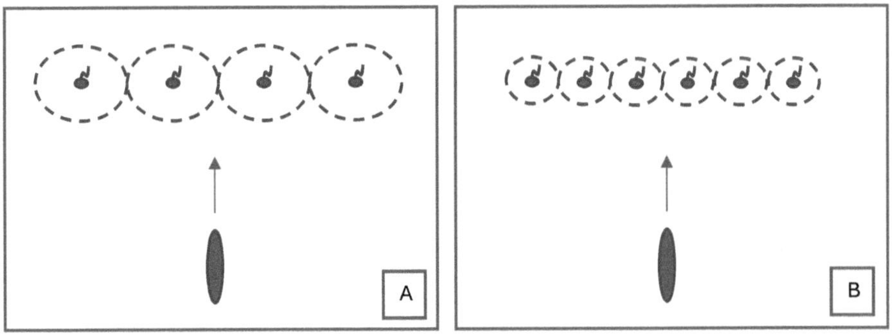

Figure 10.12 Sound range versus demand for buoys. The longer the range, the fewer buoys are demanded (A). The shorter the range, the more buoys are demanded (B).

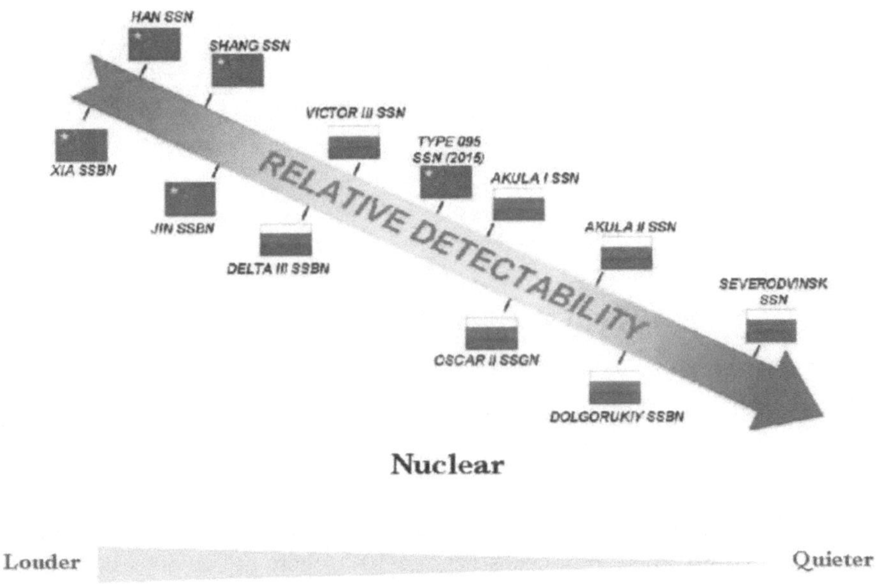

Figure 10.13 Development in detectability of Russian and Chinese submarines, as estimated by the US Navy Office for Naval Intelligence (ONI)[25]

If one can increase the sensitivity of the passive acoustic sensors, the probability of detection in a sporadic submarine movement pattern will increase accordingly (C).

Another challenge to the acoustic operator is the finite sonobuoy life. Sonobuoys used by airborne NATO units have a maximum life of approximately 8 h. After this, they are automatically scuttled and sunk. The limited buoy life is based on the limited timespan of an MPA sortie: When dropped into the water, the buoy

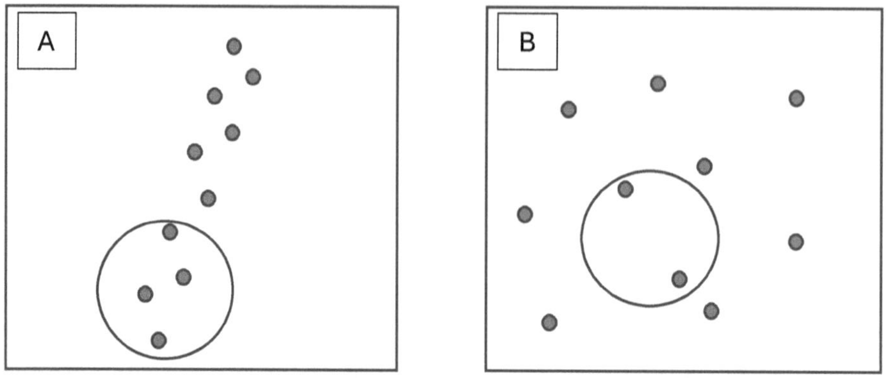

Figure 10.14 (A) Predictable submarine pattern: Higher chance of detection of systematic data. (B) Unpredictable submarine pattern: Lower chance of detection of sporadic data

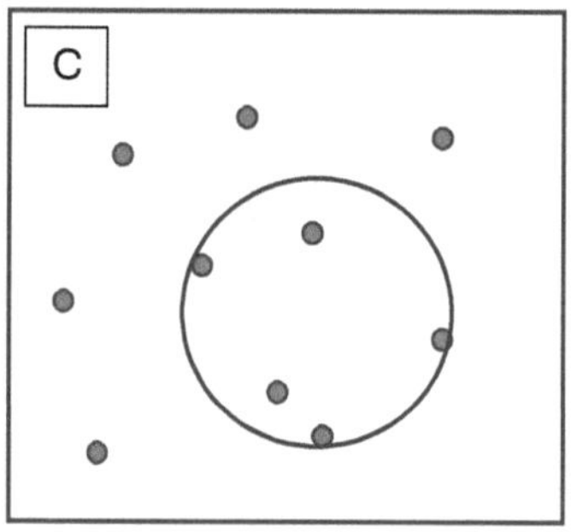

Figure 10.15 (C) Unpredictable submarine pattern: Higher chance of detection of sporadic data through increased sensor range

will rarely service the MPA with acoustic data for more than a few hours, and as a common MPA sortie rarely spans beyond 10 h, there are obvious cost-saving measures in reducing battery life. As discussed earlier, when a submarine contact was historically handed over from one MPA to another, the frequency channels of the most relevant buoys were given to the oncoming aircraft to continue the tracking process. But the oncoming aircraft would use the same frequency channels on its buoys as the previous aircraft, and if they were searching or tracking in similar areas over time, the buoys from the previous aircraft would interfere with the buoys of the oncoming aircraft. It would then be in the interest of the oncoming aircraft that as many buoys as possible that were not relevant for the continued tracking were scuttled ahead of time.

Therefore, one can increase the battery life of the buoys, which is more feasible than significantly increasing their sensitivity. When a modern submarine is evading detection, she now has the benefit of two factors against the airborne ASW platform: Her own reduced sound emissions, and the temporary nature of the airborne search. A submarine can sail and operate for 24 h, but a single MPA can only be onstation for approximately 10 h. And within the time available, the MPA has to choose where to place its finite number of sonobuoys. If, however, the buoys could be monitored by other assets that operated in the area after the MPA left the scene, there would be a more efficient utilization and exploitation of the buoys' capacity, in addition to the reduction in manoeuvrable areas for the adversary submarine, due to the large areas of buoy patterns being monitored over time. For example, an MPA can place several buoy patterns in areas of interest and monitor these until off-station time, after which the buoys are monitored by a high-altitude unmanned aircraft with an endurance of more than 30 h. Such tactics, however, demand an increased range of transmitting frequencies for the buoys than are available today, in order to deconflict all the frequencies in use.

The significantly improved processing capacity of modern MPA presents possibilities for the further development of multi-static technology and procedures. This system facilitates two important aspects in a submarine search. First, it improves the active search by expanding the area under influence of any given active buoy.

In a field of MAC buoys, one active transmission can be picked up and analysed by multiple buoys, either by detecting a return from the object (submarine), or by analysing the reduced signal and attenuation caused by the sonar shadow. One active buoy will, together with a series of passive MAC buoys, cover a larger area than is possible with a single active buoy. Second, a large area actively searched by MAC units will provide a stronger picture of where a submarine is *not* (a negative search). This approach to negative search results also supports area denial operations. Active transmissions are used to both search for adversary submerged units and deny those units the freedom to operate.

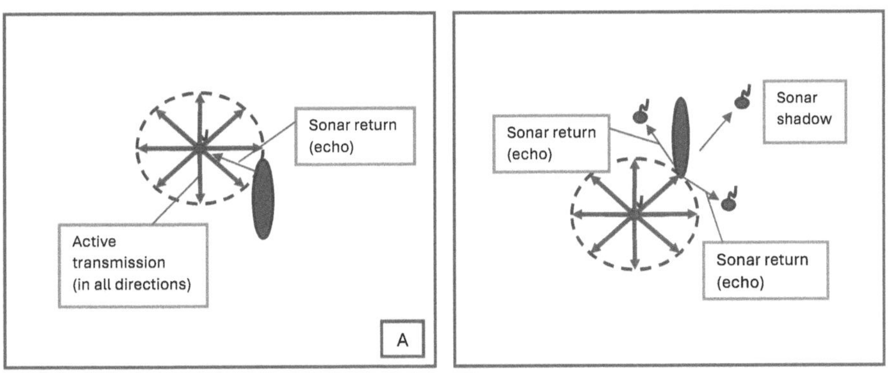

Figure 10.16 Increased search area for an active buoy in a MAC setup (B), compared to traditional use of active buoys (A)

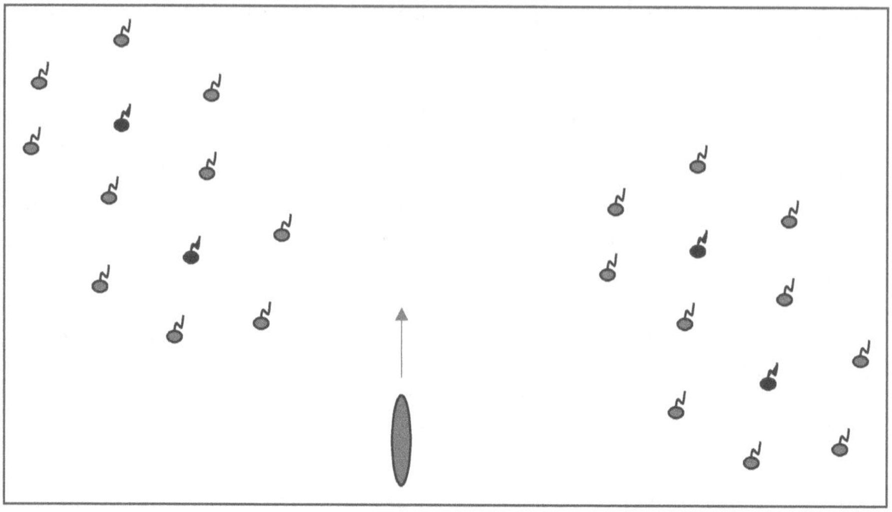

Figure 10.17 MAC fields supporting area denial operations

Maturity and time: Processing, Exploitation, and Dissemination

Finally, the first tier of our theoretical framework consists of the processes that transform raw sensor data into actionable intelligence at some level of maturity. For airborne platforms prosecuting a submarine, the relevance is in the ability to reach *tactical* sensor data maturity. The technology for acoustic analysis has improved over the years, in particular in regard to processing capacity, for example, for multi-static systems. However, the inherent problem with MAC is that it involves active procedures, which a lert the submarine of the MPA focus area. In terms of passive detectability, some progress has been made, but not what is required to meet the significant quieting progress in submarine development, as Figure 10.13 from ONI demonstrates. All in all, a modern MPA such as the P-8 will be able to reach tactical sensor maturity in less time than the P-3, when the sound-emitting submarine is used as a constant and the processing capability is used as a variable. However, when the search technology is used as a constant and the development of submarine silencing is used as a variable, it will take longer to reach tactical sensor maturity against a modern submarine than against the Soviet submarines that operated in the early 1980s. This gives us the following paradox:

In the 1980s, the louder noise levels of older submarines, combined with their outdated operational methods, allowed MPAs to reach attack criteria more quickly than they do today. Once again, two fundamentally negative factors, therefore, stand out. First, the adversary submarine has become significantly more quiet. Second, the adversary submarine is operating in an unpredictable pattern. As explained above, the quieting of submarines has more effect than the increased sensitivity of the sensor. The search for and subsequent tracking of a submarine by acoustic

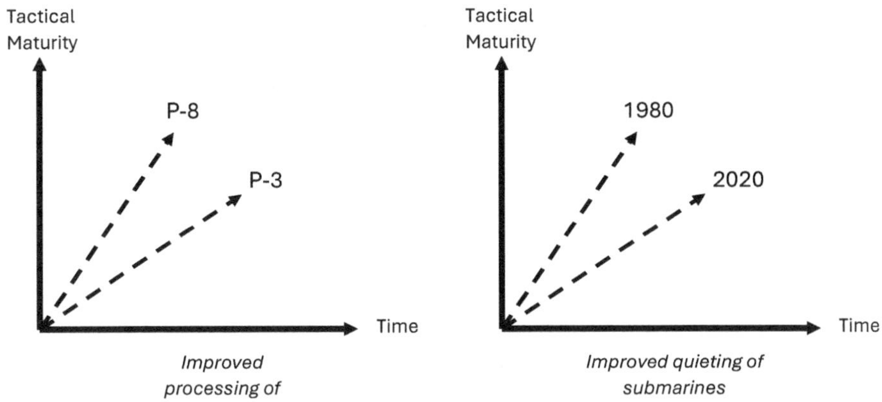

Figure 10.18 Acoustic range competition – sensor and processing capability vs. submarine quieting

means is what can be termed a wicked problem, perhaps best explained by a former leading figure in British maritime airborne surveillance:

> There are three types of problems. There is a simple problem, there is a complex problem, and there is a wicked problem. A simple problem is orders: Run over there, take that Nimrod to Kinloss. That is a simple thing. Takes a bit of effort, but, fire up the jet and off you go. A complex problem is a bit like writing an ATO.[29] It's lots of moving parts, it is difficult, but if you throw enough horsepower at it, it will get done. It is a process thing. ASuW[30] is like that. ASuW is a complex problem, that requires lots of correlation, just like an ATO, or an ACO,[31] but as long as you throw horsepower at it and do it diligently, it will happen. It is a bounded problem. ASW falls in the category of what I call wicked problems. When you start off on it, you don't know what the end result is going to look like. You haven't got a clue. You've got all the procedure you'll do, but it's a wicked problem because it is unbounded. And there will be lots of shades of grey that you have to interpret along the way.[32]

Processing passive acoustic sensor information is in itself demanding. In short, the operator requires whatever help that can be provided. Developing technology for processing so-called Big Data will likely be pivotal for sifting through immense amounts of oceanographic information and detecting what is relevant. Modern sensors are simply picking up too much information for any human to handle, and complex sensor analysis is, and will only become, more dependent on Big Data analysis and AI.[33] This will particularly be the case if drones are to take over ASW responsibilities from MPA. And with sonobuoys increasingly incapable of detecting and tracking modern submarines by themselves, airborne ASW assets are dependent on close cooperation and coordination with other assets, in space, in the air, on the surface, or below the surface. Finally, in today's fiscal environment, it seems impossible to mount a fully layered approach to ASW by one single nation alone. A multinational approach is required. The factors of silencing adversary submarines, their unpredictability in operating patterns, and the challenging sensor processing, therefore, demand an even more holistic approach to submarine search by air assets than was the case previously.

Three principled avenues for improving the chance of detection stand out. First, there is a requirement for *improved acoustic sensor processing*. It is telling that in 2020, when the world's most powerful Navy – backed by the most encompassing research and development environment in the world – is fielding the state-of-the-art MPA, which is the P-8 *Poseidon*, there is still a requirement for two *human* acoustic sensor operators on board. Acoustic tactical analysis is a wicked problem, and if one is to delegate that work to a computer, it will likely have to be one that runs on AI. ASW is not just a mathematical problem. Oceanographic factors that come into play are changes in salinity and temperature that combine with pressure increase with depth and cause multiple sound layers in the water column; ambient noises caused by human activities and unpredictable animals and life in the sea complicate the background for the frequency analysis; and of course the human factor of submarine tactics and general unpredictable behaviour lead to qualified guessing being an integral part of ASW.

Further building on the idea of assistance, the sensor analysis process will be vastly improved if two computers would be able to correlate their findings in geographical areas that are in proximity to each other. One example is two P-8 *Poseidons* working in tandem on a tracking challenge, with one being able to "complete the picture" of the other asset. Similarly, transmitting acoustic data to an analysis centre ashore offers potential benefits, with Big Data-capable computers and more subject matter expertise able to assist as a reach-back solution.

Second, building on experiences from the past, there is a requirement for a *layered, multi-asset approach to searching for and tracking modern submarines*. In order to meet the challenge holistically, there is a need to approach the search through six layers.

Together, these layers will cover as many search angles against the adversary submarine as possible. A layered approach like this will also provide redundancy in cases of elements, such as the MPA, being neutralized (or removed from the area of interest).

Finally, a comprehensive layered approach like the one described above requires *international cooperation*. Not even the mighty US Navy is able to muster all these layers in the High North at the same time. This involves intelligence gathering and acoustic searches for some of the most capable submarines in the world. An Allied approach is required to build this order of battle. With regard to C2, the fact remains, and the previous chapters have shown, that historical ASW operations in the North Atlantic have not been coordinated and executed by NATO *per se*, but by individual nations coordinating in a bi- and multilateral context.[34] They have been coordinated by national headquarters, controlling national assets on national missions. These

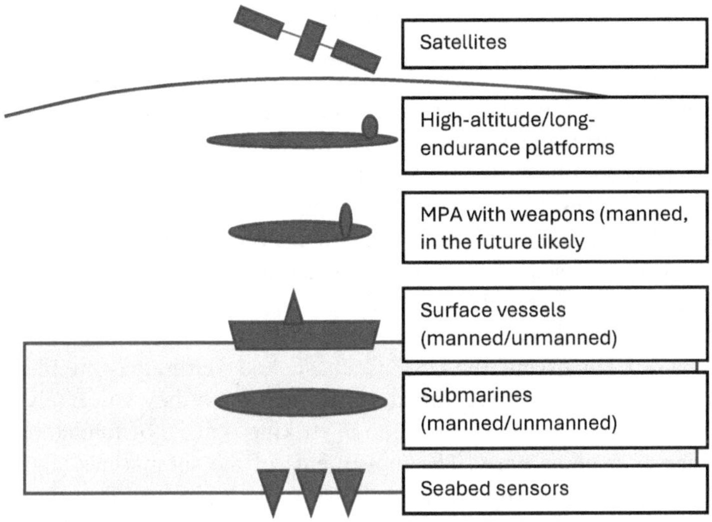

Figure 10.19 Six layers for a holistic approach to ASW

nations happen to be NATO members, and they adhere to procedures that have been established by operating together with Allies in exercises and daily operations. In order to be capable of seamless and efficient multinational coordination of complex ASW operations, NATO should integrate as many national headquarters as possible into the Alliance command structure in peacetime. This might lead to an Alliance command structure not sitting idly, waiting to have forces assigned to it, but an organization that is actively engaging in operations on a daily basis.

In addition, international cooperation will require close intelligence relationships, beyond those that exist within the framework of NATO. Proposals for deeper surveillance cooperation within the Alliance are commendable, but there are show-stoppers as to intelligence-sharing in these projects: This kind of intelligence cannot be shared with just anyone.[35] It is clear that the United States and the United Kingdom have a special relationship with regard to intelligence. Norway, too, has long-standing cooperation with the United States with regard to surveillance in the High North. As was the case for tracking Russian strategic submarines on their way to and returning from patrols during the Cold War, these three nations stand to gain significant benefit from close operational cooperation. All three nations operating P-8 *Poseidons* in the future will certainly assist in this respect.

The way forward

The main objective of this book has been to define the role of ASW within the airborne ISR mission set in the High North. The answer is that ASW is a fundamental and crucial element in airborne ISR in the High North, and without credible ASW search, tracking, and neutralizing capability, nations will be blind to the most significant naval threat in the North Atlantic. To further explain this, two factors stand out.

First, submarines are the capital ships of the Russian Navy, and have been so since the early Cold War. They remain the key factor for Russian naval operations, and will likely continue to be so in the future. Submarines should be considered the centre of gravity for the Northern Fleet, and most likely the Navy as a whole. They are likely to expand as the crucial factor for naval and joint operations for the Russian Armed Forces. The mission of the Northern Fleet's submarines is most likely first, to constitute an assured second-strike capability. Second, they are to comprise a forward and fundamental element of the bastion defence. Third, they are to pose a credible, dual-capable land-attack threat to both the European mainland (without crossing the GIUK gap) and the US East coast. And fourth, they are likely to pose a somewhat revised anti-SLOC threat, in the sense that they will likely not target the SLOCs directly, but indirectly through striking objects of fundamental strategic importance, such as ports. The movements of the submarines have changed, from a fairly predictable and repeated transit route to, in, and from known patrol areas, to operating over periods of time in large patrol areas posing a threat to all of Europe – without having to pass the GIUK gap. They are far more silent than before, and carry significantly more capable missiles.

Second, MPA are *by themselves* not a cost-efficient tool for conducting maritime airborne ISR in the High North in this contemporary context. It is fundamentally important to keep in mind that although the new MPA are more capable than their predecessors, they do not fully address the submarine threat: They must be a part of a system of systems. If politicians and military strategists are concerned that vast resources are being invested in new aircraft, they must also be made aware that this expensive investment is just one piece of the puzzle. First, the on-board processing systems and operators are not capable of handling the acoustic quietness of modern Russian submarines as they are configured today. The way to bridge that gap is through the utilization of AI, *Big Data* processing, and collaborative, real-time sensor processing in order to close in on the adversary submarine. In addition, the MPA are just one of six layers that are necessary to cover the search holistically. Satellites, unmanned systems in the air, on the surface, and below the surface, as well as seabed sensors, must also contribute to a comprehensive search picture. Finally, international cooperation is required. However, it is likely that established Alliance frameworks are not sufficient for the intelligence sharing agreements required. As before, a niche collection of partners must share the burden and work together.

In conclusion, the following recommendations for maritime airborne ISR are provided for Norway and her close allies in order to more effectively track Russian submarine movements in the High North.

High boy/low boy tactics

The US Navy is taking high boy/low boy tactics further with the BAMS concept, in which a MQ-4 *Triton* HALE unmanned aircraft operates at altitude, facilitating a significantly more efficient use of high-cost, shorter-endurance assets such as the P-8 *Poseidon*. For mitigating the complexities of modern maritime surveillance, this combination stands out as a requirement for credible airborne surveillance of large ocean areas, until long-endurance unmanned aircraft are able to perform the entire mission portfolio of contemporary MPAs, including ASW. With the current state of technology, one or the other will not suffice – we need both.

Satellites as part of maritime surveillance

Satellites are already extensively used for surveillance tasks, including in the maritime domain. Sensors from space do not penetrate water, but regular surveillance from space through the use of IMINT and ELINT sensors to cover more or less the entire area of interest in one pass will constitute the uppermost layer in a layered approach to covering vast ocean areas. The space dimension will support communications, surveillance, and targeting.

Speed must be extensively exploited

Increased speed means quicker reaction times to a datum and a larger area covered during a sortie. Decreased reaction time should be exploited and improved in the future. This can be achieved through sensors without the ability to act on their own findings (satellites, high-altitude unmanned aircraft) operate in an interconnected manner with weaponized platforms (today MPA, in the future weapons-laden unmanned aircraft) that can move quickly and deploy sensors and weapons into the water.

Norwegian security policy guidelines must be revised

The Norwegian Government should reflect on the guidelines for Allied activity in the High North and their relevance. New technology, particularly in terms of extremely long-endurance unmanned vessels and aircraft, coupled with new AAR capabilities, can shift the weight from Norwegian platforms to include Allied aircraft operating out of airfields far from Norwegian soil. This will lead to a more demanding dialogue in the future with Allies that want to operate in the High North.

International cooperation is key

Due to the vast areas in question, and the uncertainties associated with the Russian submarine modus operandi, close international coordination will be key to efficiently covering the North Atlantic with capable platforms. Norway should look to the past and geographical delineation arrangements that can inform future constructs for international cooperation.

Long range is key for pattern of life

Range is not only necessary in order to reach the respective points of interest in the vast North Atlantic, but it is also crucial to establishing the pattern of life in the area of interest. *The modus operandi* of the Russian Navy seems to include more elements of what can be termed hybrid warfare, which complicates the building of a comprehensive understanding of what is normal and what is not. If the goal is to notice anomalies, a solid understanding of normal day-to-day operations is required. Long range and more or less continuous presence are crucial to reach this understanding.

SATCOM for broadband communications and targeting

ASW ships and aircraft will benefit greatly from being able to instantly exchange acoustic data with other units nearby, or as an analysis reach-back function in order to be able to analyse the large amounts of acoustic data that become available in a modern ASW prosecution. This will require broadband, over-the-horizon communications, which can only be facilitated by SATCOM. Also, long-range maritime platforms will require encrypted and robust space-based communications for providing targeting to Allied units at distance.

Civilian–military cooperation for space-based assets

New Space has brought with it an expanding civilian market for space services. This must be exploited for the military. Commercial capabilities should be utilized where possible in order to save costs. Civilian–military cooperation is a cost-efficient framework for modern capabilities, be it in support of communication or surveillance.

MPA must be supplemented in A2/AD environments

The MPA mission portfolio paradox demands that in order to facilitate long-range ASW in areas threatened by A2/AD measures, MPA must be supplemented by platforms that are stealthy or have extreme manoeuvring capabilities. If MPA are to be utilized in offensive operations in crises and war, irrespective of the operating area, they must be protected (through either self-defence or support by other assets), capable of employing stand-off weapons, and/or capable of employing drones, either flying or sailing, that can operate within the range of A2/AD measures.

Long-life sonobuoys – multiplatform exploitation

Very long-life sonobuoys with GPS-tracking of their own position could constitute a long-life search field in an area of interest, such as a choke point. An improved communications suite that can handle more frequency channels will mitigate the challenge of interference between different buoy fields being monitored and analysed at the same time. This means that an MPA can fly out and deploy a field of sonobuoys, analyse the incoming acoustic data, and then hand off to a high-altitude, long-endurance asset that would remain in the area for much longer than the MPA. This will mitigate the finiteness of the MPA presence and complicate the options for movement available to the submarine.

Multi-statics

Fully operational multi-static technology and tactics will be a revolution for ASW acoustic sensor operators, and seems to be evolving into an absolute necessity for modern submarine prosecution, even though it is an active sensor. MAC systems will be key for executing denial operations against adversary submarines. In the near future, one might be able to link up several MAC prosecutions into a system of systems, in order to cover an even larger area holistically. This will require significant data processing capabilities, in addition to encrypted broadband communications between the search units.

AI and Big Data processing

AI and Big Data processing should be incorporated into airborne ASW platforms as soon as feasible, in order to handle the complexities of modern acoustic analysis. Allies should investigate the possibility for collaborative analysis of acoustic data, based on instant or near-real time dissemination of own acoustic data. Processing capacity and the ability to exchange large amounts of data more or less instantly, should be fully exploited in order to mitigate the acoustic challenge.

Layered persistent sensor framework

A layered approach to ASW must be adopted. MPA alone are insufficient to counter the emerging Russian submarine threat. Six layers for ISR and targeting of submarines should be sought: Space-based asset for sensors, communications and targeting; high-altitude long-endurance unmanned aircraft for near-constant presence; MPA for ISR and weapons employment; surface assets (manned and unmanned) for long-range, near-persistent presence with underwater sensors; underwater platforms (manned and unmanned) for near-persistent underwater presence, ISR, and carrying weapons; and finally seabed sensors for persistent ISR and cueing for other ASW platforms.

International cooperation and intelligence exchange

Historically, airborne ASW operations were coordinated by NATO alliance members, but all coordination was done under a national, bilateral, or multilateral framework. NATO procedures were used in addition to NATO-funded infrastructure. However, stating that the operations were conducted "by NATO" undermines an important element in describing the operations: There are key nations that were, and to a degree still are, able to handle ASW coordination of this sort. For the North Atlantic, this has been the United States, the United Kingdom, and Norway. Whether the infrastructure, procedures, and competency to coordinate and lead such operation are present in today's NATO Command Structure *per se* is quite a different matter. The close hold that is inherent in acoustic intelligence, as discussed in the historical chapters, is likely still a prominent factor, leading to a tension between intelligence collection and intelligence sharing. Close cooperation between selected partners thus stands out as crucial for releasing the potential inherent in international cooperation. As a continuation of this, NATO should integrate national headquarters closely with the NATO Command Structure writ large, in order to exploit the capability and situational awareness that is built on a daily basis outside the formal NATO networks.

Epilogue

***

The Norwegian Government published their new long-term plan in the Spring of 2024. The plan proposes to initiate an "international cooperation with close allies" in order to acquire "long-range drones with sensors and systems for persistent surveillance in maritime areas of interest in the North."[36] The plan is to phase the drones into the operational structure between 2029 and 2032.[37] With the addition of drones, the MoD thus proposes a structure for ASW that is similar to the layered approach presented earlier in this book, although satellites are presented as communication nodes and not as sensors, for unknown reasons.[38]

Following the publication of the white paper, the Chief of the Royal Norwegian Air Force Maj Gen Folland publicly questioned the decision of acquiring long-range drones. His main point was that high-altitude, long-range drones would be very vulnerable to enemy air defences in a high-intensity conflict, and the decision to acquire drones for long-range maritime surveillance seemed to be based on an outdated understanding of the security environment they would operate in. The general argued that satellites (orbiting in outer space) and stealth drones (better equipped to avoid enemy sensors and fire) would be much better suited for strategic surveillance. His main concern was, thus, the survivability of the sensors.[39] A month later, one of the chief's generals in the Air Staff, Brigadier General Lyssand, followed up with an article highlighting concerns from an organizational perspective. Lyssand's main concern was the significant asymmetry between ambition and tasks on the one hand, and actual available resources on the other. Phasing in long-range drones is demanding for the air force from both a manning and a competence perspective, and the Royal Norwegian Air Force (RNoAF) has no organizational experience in operating such systems.[40] BGEN Lyssand only mentioned in passing his boss' operational concerns of vulnerability. Together, the two articles stand out as a poorly coordinated contribution to the discussion, particularly considering that they represent a very clear dissent against their own ministry's proposal presented just a few months earlier. The poor coordination is particularly evident through the fact that the air chief does not seem to be particularly concerned about organizational issues. On the contrary, he instead proposes acquiring stealth drones, something that not even the US Air Force has operationalized at the time of this writing, let alone any smaller nations. But the general points to an important element presented in the discussion of maritime airborne surveillance in the High North, namely what this book presents as the "mission portfolio paradox" of many flying assets in peacetime. Big, relatively slow-moving aircraft with large radar cross-sections are outstanding, most analysts will say *indispensable*, contributors to peacetime surveillance. And it is similarly clear that these assets will be extremely vulnerable, almost unusable, in high-intensity conflicts, at least in the same geographical areas. From an acquisition standpoint, it stands out as a point of gross negligence that the general advocates for stealth drones, carrying a survivability characteristic that is required *only* in times of high-intensity conflict and war, and almost *irrelevant* during peacetime. Whereas high-altitude, long-range drones with little or no self-protection will be extremely relevant almost *all of the time* (peacetime) and be rendered less relevant *only* in times of high-intensity conflict and war. In addition, the ISR services that long-range drones provide will contribute significantly to the very foundation for any wartime intelligence and targeting that is required. This foundation will not be established in times of conflict, it will be a long-standing peacetime effort that will only be adjusted through the use of wartime intelligence. This important point was also underlined through articles that followed the air force leaders' interventions in the Summer of 2024.[41] One of the counter-arguments also underlines the fact that the air force confuses the requirement for long-range drones as a requirement in support for airpower, when it decidedly is in support of sea power.[42] This further underlines the point that any strategic surveillance at its core is in support of strategy and the joint effort, not the tactical efforts of assets mainly operating in their respective domain. Ultimately, and as just emphasized, both manned and unmanned aircraft performing maritime surveillance in the High North will mainly be operating in peacetime and crisis, in support of strategic intelligence.

Then, when we have discussed the operational requirement for airborne maritime surveillance and its merits in both peacetime and war, we can start discussing whether there is the organizational capacity to actually absorb the additions in manning, equipment, and operational ambitions.

Notes

1 Boeing, "P-8A Poseidon," Åpnet November 28, 2019, www.boeing.com/defense/maritime-surveillance/p-8-poseidon/index.page; Naval History and Heritage Command, "P-3 Orion," May 29, 2014. www.history.navy.mil/research/histories/naval-aviation-history/naval-aircraft/current-aircraft-inventory/p-3-orion.html.

2 The formula for theoretical radar range in nautical miles is $1.23\sqrt{\text{(height in feet)}}$.

3 AJP 3.3.3 *NATO Maritime Air Coordination* (NATO, December 2014), 2–5.

4 Efjestad, "Norwegian Self-Imposed Restrictions in the High North."

5 Klevberg, "Maritime Surveillance in the North," 375–77; Tamnes, *The United States and the Cold War in the High North*, 216–17.

6 The US Navy notes that the P-3 Orion is able to operate for 3 h at 1346 nautical miles, and the P-8 Poseidon for 4 h at 1200 nautical miles. See US Navy, "Fact file – P-3C Orion and EP-3 Aries"; US Navy, "Fact File – P-8A Posdeidon Multi-Mission Maritime Aircraft (MMA)," US Navy, December 3, 2018, www.navy.mil/navydata/fact_display.asp?cid=1100&tid=1300&ct=1.

7 Russian name: S-300FM. John Wood, *Russia, the Asymmetric Threat to the United States – A Potent Mixture of Energy and Missiles* (Santa Barbara, CA: Praeger Security International, 2009), 61.
8 These generic range circles have been adopted from the interactive map posted by Ian Williams, "The Russia – NATO A2AD Environment," Missile Threat Project (CSIS), January 3, 2017, https://missilethreat.csis.org/russia-nato-a2ad-environment/.
9 Russian name: 3K95 *Kinzhal*. Wood, *Russia, the Asymmetric Threat to the United States*, 62.
10 Russiantechnicaldesignation:9M96E/9M96E2.Military-Today.com,"AdmiralGorshkov Class–Multi-RoleFrigate,"*Military-Today*,ÅpnetDecember6,2019,www.military-today.com/navy/admiral_gorshkov_class.htm.
11 GlobalSecurity.org, "AA-13 ARROW / K-37/R-37 / RVV-BD."
12 Andresen og Bukkvoll, *Russian Weapons Development Towards 2020*, 56, 58; Missile Defense Project, *Russia Air and Missile Defense* (Center for Strategic and International Studies (CSIS), June 14, 2018), https://missilethreat.csis.org/system/russian-air-defense/.
13 Missile Defense Project, "Russia Air and Missile Defense."
14 Reade, *The Age of Orion – The Lockheed P-3 Story*, 55–56; Insinna, "The US military's chaff and flare industry is on fragile ground."
15 The Norwegian Government, "Presiseringer om utbyggingen av Evenes."
16 Tamnes mfl., "A Unified Effort to Strengthen the Defence of Norway," 21.
17 It is possible to shoot down a satellite with missiles fires from the surface of the earth. A kinetic anti-satellite kill is problematic, however, due to the thousands of fragments of the destroyed satellite then automatically entering their separate orbits, comprising space garbage and a constant threat to other elements in orbit. See, for example, David, "China's Anti-Satellite Test: Worrisome Debris Cloud Circles Earth."
18 Eshel, "Lifting the veil of Israel's classified laser weapon program"; CNN.com, "USAF: Fighter jet lasers 5 years away."
19 Fein og Jean, "Changing the Game."
20 U.S. House of Representatives, *Advanced Submarines Technology and Antisubmarine Warfare* (§ Committee on Armed Services, 1990), 60; Birkeland, "The Potential of LIDAR as an Anti-Submarine Warfare Sensor," 18.
21 Birkeland, "The Potential of LIDAR as an Anti-Submarine Warfare Sensor," 18.
22 Birkeland, 18–19; Perkins, *Alliance Airborne Anti-Submarine Warfare*, 39–40.
23 Birkeland, "The Potential of LIDAR as an Anti-Submarine Warfare Sensor," 18–19.
24 Perkins, *Alliance Airborne Anti-Submarine Warfare*, 40.
25 Office of Naval Intelligence, *The People's Liberation Army Navy – A Modern Navy with Chinese Characteristics* (Office of Naval Intelligence (ONI), August 2009), 22.
26 Perkins, *Alliance Airborne Anti-Submarine Warfare*, 38.
27 Dave Majumdar, "U.S. Navy Impressed with New Russian Attack Boat," *USNI News*, October 28, 2014, https://news.usni.org/2014/10/28/u-s-navy-impressed-new-russian-attack-boat; Franz-Stefan Gady, "Russian Sub Combat Patrols Nearly Doubled in 2015," *The Diplomat*, March 23, 2016, https://thediplomat.com/2016/03/russian-sub-combat-patrols-nearly-doubled-in-2015/.
28 Franz-Stefan Gady, "Russia's First Yasen-M Attack Sub to Begin State Trials in 2019," *The Diplomat*, December 8, 2018, https://thediplomat.com/2018/12/russias-first-yasen-m-attack-sub-to-begin-state-trials-in-2019/; Thomas Nilsen, "Sevmash Floats Out Second Yasen-M Class Nuclear Attack Submarine," *The Barents Observer*, December 25, 2019, https://thebarentsobserver.com/en/security/2019/12/sevmash-floats-out-second-yasen-m-class-nuclear-attack-submarine.
29 Air Tasking Order: The daily plan for utilizing air assets on any given day in a campaign or operation.
30 Anti-Surface Warfare: Warfare in the maritime domain aimed at neutralizing surface vessels.
31 Air Control Order: The plan and coordination of the utilization of parts of the air space for any given day in a campaign or operation.

32 Porter, *British Maritime Patrol Aircraft Operations During and After the Cold War*.
33 Shane P. Hamilton og Michael P. Kreuzer, "The Big Data Imperative – Air Force Intelligence for the Information Age," *Air & Space Power Journal* (Spring 2018), www.airuniversity.af.edu/Portals/10/ASPJ/journals/Volume-32_Issue-1/F-Hamilton_Kreuzer.pdf?mc_cid=a2eed12543&mc_eid=c0b972d50d.
34 For examples of somewhat misleading portrayals, see Hicks et al., "Undersea Warfare in Northern Europe," 20; Perkins, *Alliance Airborne Anti-Submarine Warfare*, 12; Nordenman, *The New Battle for the Atlantic*, 57–64.
35 Brooks Tigner, "NATO Considers P-8 as Temporary Gap Filler Until 2035," *Jane's Defense Weekly* 56, no. 15 (April 10, 2019); Magnus Nordenman, *NATO's Next Consortium: Maritime Patrol Aircraft*, Issue Brief, Washington, DC: Atlantic Council, Brent Scowcroft Center on International Security, May 2016.
36 Prop. 87 S (2023–2024), *Long-Term Plan for the Norwegian Armed Forces* (Norwegian Ministry of Defense, April 5, 2024), 58.
37 Prop. 87 S (2023–2024), 59.
38 Prop. 87 S (2023–2024), 69.
39 Rolf Folland, "Langrekkende droner – en riktig investering for fremtiden?" *Forsvarets Forum*, June 16, 2024, www.forsvaretsforum.no/drone-droner-luft/langrekkende-droner-en-riktig-investering-for-fremtiden/382824.
40 Kristian Lyssand, "Bemanning av ubemannede Droner," *Forsvarets Forum*, July 24, 2024, www.forsvaretsforum.no/drone-droner-luftforsvaret/bemanning-av-ubemannede-droner/388788.
41 Nils Haaheim-Saers, "Sjefen for Luftforsvaret bommer om langtrekkende droner," *Aftenposten*, July 1, 2024, https://www.aftenposten.no/meninger/kronikk/i/Xjbjjx/kronikk-av-soldat-sjefen-for-luftforsvaret-bommer-om-langrekkende-droner; Oliver Mendoza Kristensen og Christoffer Hveding, "Langtrekkende droner og ISR: Hvordan vi skaper effekt av anskaffelsen," *Stratagem* (blog), August 21, 2024, www.strategem.no/langtrekkende-droner-og-irs-hvordan-vi-skaper-effekt-av-anskaffelsen.
42 Haaheim-Saers, "Sjefen for Luftforsvaret bommer om langtrekkende droner."

Bibliography

AJP 3.3.3. *NATO Maritime Air Coordination*. NATO, December 2014.

Andresen, Rolf-Inge V., og Tor Bukkvoll. *Russian Weapons Development Towards 2020*. Forsvarets Forskningsinstitutt (FFI) FFI-rapport 2008/01957, January 12, 2009.

Birkeland, John O. "The Potential of LIDAR as an Anti-Submarine Warfare Sensor." M. Phil (R), War Studies, University of Glasgow, 2009. http://theses.gla.ac.uk/1252/1/2009birkelandmphil.pdf.

Boeing. "P-8A Poseidon." Åpnet November 28, 2019. www.boeing.com/defense/maritime-surveillance/p-8-poseidon/index.page.

CNN.com. "USAF: Fighter Jet Lasers 5 Years Away." *CNN*, December 15, 2015. https://edition.cnn.com/videos/us/2015/12/15/us-air-force-lasers-fighter-jets-orig-vstop.cnn.

David, Leonard. "China's Anti-Satellite Test: Worrisome Debris Cloud Circles Earth." *Space.com*, February 2, 2007. www.space.com/3415-china-anti-satellite-test-worrisome-debris-cloud-circles-earth.html.

Efjestad, Svein. "Norwegian Self-Imposed Restrictions in the High North." Interview, September 2, 2019. Norwegian Ministry of Defense.

Eshel, Tamir. "Lifting the Veil of Israel's Classified Laser Weapon Program." *Defense Update*, January 8, 2020. https://defense-update.com/20200108_hel_israel.html.

Fein, Geoff, og Grace Jean. "Changing the Game: US Navy Applies New Approaches to Submarines Threats." *IHS Jane's Navy International*, 2014. www.janes.com/images/assets/836/46836/US_Navy_applies_new_approaches_to_submarine_threats.pdf.

Folland, Rolf. "Langrekkende droner – en riktig investering for fremtiden?" *Forsvarets Forum*, June 16, 2024. www.forsvaretsforum.no/drone-droner-luft/langrekkende-droner-en-riktig-investering-for-fremtiden/382824.

Gady, Franz-Stefan. "Russian Sub Combat Patrols Nearly Doubled in 2015." *The Diplomat*, March 23, 2016. https://thediplomat.com/2016/03/russian-sub-combat-patrols-nearly-doubled-in-2015/.

Gady, Franz-Stefan. "Russia's First Yasen-M Attack Sub to Begin State Trials in 2019." *The Diplomat*, December 8, 2018. https://thediplomat.com/2018/12/russias-first-yasen-m-attack-sub-to-begin-state-trials-in-2019/.

GlobalSecurity.org. "AA-13 ARROW/K-37/R-37 / RVV-BD." February 6, 2018. www.globalsecurity.org/military/world/russia/aa-13.htm.

Haaheim-Saers, Nils. "Sjefen for Luftforsvaret bommer om langtrekkende droner." *Aftenposten*, July 1, 2024. https://www.aftenposten.no/meninger/kronikk/i/Xjbjjx/kronikk-av-soldat-sjefen-for-luftforsvaret-bommer-om-langrekkende-droner.

Hamilton, Shane P., og Michael P. Kreuzer. "The Big Data Imperative – Air Force Intelligence for the Information Age." *Air & Space Power Journal* (Spring 2018). www.airuniversity.af.edu/Portals/10/ASPJ/journals/Volume-32_Issue-1/F-Hamilton_Kreuzer.pdf?mc_cid=a2eed12543&mc_eid=c0b972d50d.

Hicks, Kathleen, Andrew Metrick, Lisa Sawyer Samp, og Kathleen Weinberger. "Undersea Warfare in Northern Europe." CSIS International Security Program. Center for Strategic and International Studies, July 2016.

Insinna, Valerie. "The US Military's Chaff and Flare Industry Is on Fragile Ground." *Defense News*, November 13, 2018. www.defensenews.com/industry/2018/11/13/the-militarys-chaff-and-flare-industry-is-on-fragile-ground/.

Klevberg, Håvard. "Maritime Surveillance in the North: 333 Squadron in Norwegian Security Politics." PhD, History, University of Oslo, 2011.

Kristensen, Oliver Mendoza, og Christoffer Hveding. "Langtrekkende droner og ISR: Hvordan vi skaper effekt av anskaffelsen." *Stratagem* (blog), August 21, 2024. www.strategem.no/langtrekkende-droner-og-irs-hvordan-vi-skaper-effekt-av-anskaffelsen.

Lyssand, Kristian. "Bemanning av ubemannede Droner." *Forsvarets Forum*, July 24, 2024. www.forsvaretsforum.no/drone-droner-luftforsvaret/bemanning-av-ubemannede-droner/388788.

Majumdar, Dave. "U.S. Navy Impressed with New Russian Attack Boat." *USNI News*, October 28, 2014. https://news.usni.org/2014/10/28/u-s-navy-impressed-new-russian-attack-boat.

Military-Today.com. "Admiral Gorshkov Class – Multi-Role Frigate." *Military-Today*, Åpnet December 6, 2019. www.military-today.com/navy/admiral_gorshkov_class.htm.

Missile Defense Project. *Russia Air and Missile Defense*. Center for Strategic and International Studies (CSIS), June 14, 2018. https://missilethreat.csis.org/system/russian-air-defense/.

Naval History and Heritage Command. "P-3 Orion." May 29, 2014. www.history.navy.mil/research/histories/naval-aviation-history/naval-aircraft/current-aircraft-inventory/p-3-orion.html.

Nilsen, Thomas. "Sevmash Floats Out Second Yasen-M Class Nuclear Attack Submarine." *The Barents Observer*, December 25, 2019. https://thebarentsobserver.com/en/security/2019/12/sevmash-floats-out-second-yasen-m-class-nuclear-attack-submarine.

Nordenman, Magnus. *NATO's Next Consortium: Maritime Patrol Aircraft*. Issue Brief. Washington, DC: Atlantic Council, Brent Scowcroft Center on International Security, May 2016.

Nordenman, Magnus. *The New Battle for the Atlantic – Emerging Naval Competition with Russia in the Far North*. Annapolis, MD: Naval Institute Press, 2019.

The Norwegian Government. "Presiseringer om utbyggingen av Evenes." The Norwegian Government, April 17, 2018. www.regjeringen.no/no/aktuelt/presiseringer-om-utbyggingen-av-evenes/id2598125/.

Office of Naval Intelligence. *The People's Liberation Army Navy – A Modern Navy with Chinese Characteristics*. Office of Naval Intelligence (ONI), August 2009.

Perkins, William. *Alliance Airborne Anti-Submarine Warfare – A Forecast for Maritime Air ASW in the Future Operational Environment*. Joint Air Power Competence Centre (JAPCC), June 2016.
Porter, Air Commodore (r) Garfield "Garry". *British Maritime Patrol Aircraft Operations During and After the Cold War*. London: Royal Air Force Club, May 31, 2017.
Prop. 87 S (2023–2024). *Long-Term Plan for the Norwegian Armed Forces*. Norwegian Ministry of Defense, April 5, 2024.
Reade, David. *The Age of Orion – The Lockheed P-3 Story*. Atglen, PA: Schiffer Publishing Ltd., 1998.
Tamnes, Rolf. *The United States and the Cold War in the High North*. Oslo: Ad Notam, 1991.
Tamnes, Rolf, Kate Hansen Bundt, Trond Grytting, Alf Håkon Hoel, Janne Haaland Matlary, Asle Toje, og Julie Wilhelmsen. *A Unified Effort to Strengthen the Defence of Norway*. Norwegian Ministry of Defense, April 28, 2015.
Tigner, Brooks. "NATO Considers P-8 as Temporary Gap Filler Until 2035." *Jane's Defense Weekly* 56, no. 15 (April 10, 2019).
U.S. House of Representatives. *Advanced Submarines Technology and Antisubmarine Warfare*. § Committee on Armed Services, 1990.
USNavy. "Fact File – P-3C Orion and EP-3 Aries." US Navy, December 3, 2018. www.navy.mil/navydata/fact_display.asp?cid=1100&tid=1400&ct=1.
USNavy. "Fact File – P-8A Posdeidon Multi-Mission Maritime Aircraft (MMA)." US Navy, December 3, 2018. www.navy.mil/navydata/fact_display.asp?cid=1100&tid=1300&ct=1.
Williams, Ian. "The Russia – NATO A2AD Environment." Missile Threat Project (CSIS), January 3, 2017. https://missilethreat.csis.org/russia-nato-a2ad-environment/.
Wood, John. *Russia, the Asymmetric Threat to the United States – A Potent Mixture of Energy and Missiles*. Santa Barbara, CA: Praeger Security International, 2009.

Index

FSC
www.fsc.org
MIX
Papier | Fördert
gute Waldnutzung
FSC® C083411

Kolibri 360 GmbH
Ferdinand-Jühlke-Straße 7
99095 Erfurt, Deutschland
produktsicherheit@kolibri360.de